T0356810

The Rise and Fall of King Coal

Hagley

HAGLEY LIBRARY STUDIES IN BUSINESS, TECHNOLOGY, AND POLITICS

Richard R. John, Series Editor

The Rise and Fall of King Coal

American Energy Transitions in

an Age of Markets, 1800–1940

MARK ALDRICH

Johns Hopkins University Press   Baltimore

© 2025 Johns Hopkins University Press
All rights reserved. Published 2025
Printed in the United States of America on acid-free paper
9 8 7 6 5 4 3 2 1

Johns Hopkins University Press
2715 North Charles Street
Baltimore, Maryland 21218
www.press.jhu.edu

Library of Congress Cataloging-in-Publication Data

Names: Aldrich, Mark, author.
Title: The rise and fall of King Coal : American energy
    transitions in an age of markets, 1800–1940 / Mark
    Aldrich.
Description: Baltimore : Johns Hopkins University Press,
    2025. | Includes bibliographical references and index. |
Identifiers: LCCN 2024025702 | ISBN 9781421451091
    (hardcover) | ISBN 9781421451107 (ebook)
Subjects: LCSH: Coal trade—United States—History—
    19th century. | Coal trade—United States—
    History—20th century. | Energy transition—United
    States—History.
Classification: LCC HD9545 .A43 2025 |
    DDC 333.8/22097309034—dc23/eng/20241125
LC record available at https://lccn.loc.gov/2024025702

A catalog record for this book is available from the British
Library.

*Special discounts are available for bulk purchases of this book.
For more information, please contact Special Sales at
specialsales@jh.edu.*

To Michele and Patricia

Old King Coal was a merry old soul,
     A merry old soul is he.
May he never fail in the land we love,
     Who hath made us great and free.

—*Charles Mackay*

Contents

Figures

Tables

Preface

This book tells the story of the rise and decline of Old King Coal as the dominant source of energy and power in the United States from roughly the late eighteenth century to World War II. More broadly put, it is a book about the various energy transitions during these years, but the emphasis is on coal as a source of fuel and power. By the advent of World War I, coal dominated. It was an enormous industry with more than 6,000 companies operating nearly 9,000 mines and employing nearly 785,000 men and women. At that time, coal was also a perennial source of public controversy. Progressives fretted endlessly over the anthracite cartel, and—incredible though it may seem to modern Americans—they also worried about running out of coal. Strikes, shortages, massive and deadly explosions, and Louis Hine's images of children picking slate in anthracite breakers symbolized coal's labor problems, while coal smoke blackened America's cities. Congress, federal agencies, and government commissions investigated nearly every aspect of the industry, including prices, wages, competition, waste, and resource availability. While the national debate over energy continues, all of these concerns have entirely disappeared from Americans' lives, and coal has retreated to the margins.

My interest in coal's rise and fall is both personal and professional. Ever since I realized that the lives of my parents were much easier than the lives of their parents had been and that, in turn, my life would be much easier than theirs, I have been curious: how and why have Americans' lives improved so dramatically in just a few generations? While that question led me to study economic history, it also caused me—as a longtime New Englander—to wonder about more practical matters, such as when we got a furnace and when we turned from burning coal to oil. My parents had heated with coal until sometime in the late 1930s when, like many Americans, they installed an oil burner. When my wife and I finally bought our own house in 1974, we found a gigantic old furnace that had once burned anthracite coal (the remains of which were all over the property) but had been converted to oil. We also discovered, to my wonder, that the house had a kerosene cookstove. I had no idea that such a thing existed.

As a young man, I had a glimpse of what the transition away from coal must have meant; my neighbor still burned coal, and I would tend his furnace when he was out of town. A screw drive fed fuel from a hopper to the furnace, and my job was to feed the hopper every few days. The coal, however, contained stone, which would jam the screw drive, break a shear pin, and stop the feed. I would arrive to a stone-cold house, descend to a cellar filthy with coal dust, and spend a half hour starting the fire with wood and charcoal. Such work deepened my appreciation of oil and gas heat.

As a professional, my early writing and teaching led me to coal's history in several ways. My work on railroad safety led me to investigate the carriers' innovations in wood preservation—a market response to a resource scarcity. As I began writing on work safety in mining, I discovered that it intersected with coal conservation in many ways. Historically the conservation and safety movements were branches of the same tree, while in mining, safety and conservation were sometimes complementary; strip mining, for example, improved both. Similarly, blasting, without undercutting, wasted coal by generating worthless dust, and it might lead to gas and dust explosions as well. As a teacher of environmental economics, I began to realize how much early writers worried over resource exhaustion—including coal, which seemed bizarre to me—as well as how little they understood resource markets.

As I began to read modern writers, I found much fine work on energy history, as well as a burgeoning literature on energy transitions, but the two do not usually come together. The work on transitions is often Olympian in its level of discussion, while the historical literature usually lacks a broader focus. There are histories of oil and oil companies; of natural gas, gas companies and pipelines; of the anthracite industry, coalmine workers, New Deal coal policy, and home heating. But there is not much on the rise and fall of coal as a fuel. Moreover, discussions of conservation nearly always emphasize market failures and public policies. Yet from my work on the way that market forces induced the railroads to improve safety and develop new methods of wood preservation, I began to suspect that markets must have been an important force for coal conservation as well, since the price system penalizes waste. Accordingly, it seemed that a bottom-up history emphasizing the role of markets in shaping past energy transitions, and how those choices have contributed to rising standards of living, would be an interesting and useful topic to explore.

It is probably impossible to write about energy in the early twenty-first century without being influenced by the current debates over climate change, but I have tried to avoid writing a presentist history; the energy concerns of people a generation ago were different from ours and deserve to be understood on their own terms. Yet our energy past does provide insights. One of them is that past transitions have been full of difficulties and often expensive dead ends: some energy experts in the 1920s thought the future would belong to low-temperature carbonization of coal, while governments flogged coal-based synfuels for years. A related point, no shock to economists, is that markets are enormously efficient, adjusting to surprises, killing bad ideas, and especially innovating to implement cheaper, better solutions to energy problems. Finally, while the transitions described in this book have made us richer, the current one—driven by policy, not individual choices—is likely to leave us poorer: as economists state the matter, putting a price on environmental damage will be a massive negative supply shock. Rich nations may accept this trade-off, but poor ones will not. Combating global warming worldwide will need cheaper, better sources of energy and power. America's energy history suggests that relying on markets to innovate cheaper, better forms of energy is our best bet.

The history of American energy choices is part of the broader pattern of economic development, for rising energy use was both a central cause and an outcome of America's Industrial Revolution, and I have tried to weave it into that story. It is also about the evolution of energy supply and demand and how changes in technology reshaped these patterns. The story is inevitably quantitative as well. How much coal? At what price? For such reasons, the following chapters contain tables and charts, calculations and energy conversions, although I have banished many of the details to appendices, which need be read only by those curious about details.

To some readers, numbers may imply precision, but, unfortunately, that cannot be the case in energy history. Even such comparatively modern data as coal production in the 1930s contain omissions. The Bureau of Mines never claimed to count unmarketed coal produced in small mines for local consumption, while bootleg (stolen) coal is almost by definition hard to measure. And, of course, the further back in time we go, the less accurate are the data. To the best of my knowledge, however, nothing here is misleading.

Supply and demand are merely aids to thinking about the forces shaping historical events, while the numbers are reflections of those forces. Thus, the history of American energy markets has been like a changing menu, with choices that became more or less available, expensive, and appetizing. Geology and geography set the table, but the offerings, their prices and quality, have all evolved as tinkers, engineers, scientists, entrepreneurs, and everyday men and women struggled to improve their lives. That is the story that interests me and that I hope will interest readers.

Initially, I planned to write only about the decline of coal, for that has been one of the central changes in modern energy use. I found, however, that I couldn't write about coal's decline without looking at what accounted for its rise because the same market forces drove both outcomes. My organization is therefore both chronological and by economic sector. I trace the domestic and industrial evolution of the kinds of fuels Americans used, and the ways they used them, and I emphasize that the benefits we have obtained from energy have grown much faster than fuel use itself.

My story begins with the National Governors' Conference, called by Teddy Roosevelt in 1908, to discuss conservation of a host of resources, energy prominent among them. The conference captured the Progressive vision that energy markets had been a source of great waste, which was endangering American prosperity. I call these views market skepticism and resource pessimism, and I term individuals who held these views, at that time and later, Progressives. Not all contemporaries shared these views, and I will argue that conservation has been part of Americans' energy history from the beginning, for much of what was termed waste was, in fact, an appropriate response to market signals.

I stop the story largely with the advent of World War II for two reasons. First, by this time, coal was everywhere in retreat, save for electricity generation (and as I write this preface, that interlude seems finally to be ending). Second, this book is mostly about the workings of the private economy until roughly 1940. After the war, political forces increasingly shape energy outcomes, and that story has been well told by others, but by 1940, the broad outlines of America's energy transitions were clear.

While the focus of the book is on the period from 1800 to 1940, I have limited my discussion to neither date when straying beyond them seemed important. Thus, the discussion of railroad fuel extends well into World War II and includes a glimpse of the postwar years because only then do the carriers' efficiency gains and fuel choices become clear.

I emphasize that this is not a general history of energy; it is a history of energy transitions with an emphasis on coal; accordingly, much is left out. There is little detailed policy history because politics played a minor role during these years. There is little on the labor wars that marked coal's history or the consequences of coal's decline for labor markets and local economies. Because my emphasis is on coal, there is not much on household electrification because that convenience only modestly affected coal's fortunes during these years and because others have treated the topic at length. For similar reasons, I ignore the explosive increase in automobiles and their fuel; while cars (and airplanes) eclipsed passenger rail, after about 1940, it was diesel fuel, not coal, that ultimately bore the brunt of such competition.

A number of individuals have helped me with this book. I am grateful to Jack Brown and Joel Tarr, each of whom read and improved parts of the manuscript. As usual, I am enormously indebted to the skill and helpfulness of Smith's librarians, especially Sika Berger, Susan Daily, and Chris Ryan. I also owe a debt to Constance Carter at the Library of Congress and Cammie Wyckoff and her staff in the Cornell University Library Annex.

The Hagley Museum provided access to the Pennsylvania Railroad Collection, which yielded insights into the carriers' early interest in burning oil.

Some of this material has appeared before, and I wish to thank the editors of the journals in which parts of these chapters first appeared. Part of chapter 5 was originally published as "An Energy Transition before the Age of Oil: The Decline of Anthracite, 1900–1930," in *Pennsylvania History: A Journal of Mid-Atlantic Studies* 85 (Winter 2018): 1–31, https://doi.org/10 .5325/pennhistory.85.1.0001. It's used here with permission from Penn State University Press. An earlier version of chapter 6 was originally published as "The Rise and Decline of the Kerosene Kitchen: A Neglected Energy Transition in Rural America, 1870–1950," in *Agricultural History* 94 (Winter 2020): 24–60. It's used with permission from the Agricultural History Society. Part of chapter 7 appeared as "Freeing the Furnace Slaves, or the Battle of the Basements: How Fuel Oil Displaced Anthracite Coal, 1925–1940," in *Oil-Industry History* 18 (2017): 115–140. It's used with permission from the Petroleum History Institute. Part of chapter 8 was originally published as "Conserving Resources, Saving Lives: Strip Mining Coal in America, 1880–1945," in *Mining History Journal* 30 (2023): 51–73. It's used with permission from the Mining History Association.

My wife, Patricia Sweetser, read much of the manuscript, for which I am deeply grateful. Her sharp eyes caught many blunders; any errors that remain are the sole property of the author.

The Rise and Fall of King Coal

Introduction

Greatest of All Is King Coal

Looking back from the vantage point of 1907, Henry Adams (1838–1918) characterized the preceding years. "Especially in the nineteenth century," he wrote, "society, by common accord, agreed in measuring its progress by the coal-output." Coal was indeed central to the lives of most Americans. Middle-class ladies, or perhaps their attendants, cooked on stoves fired with anthracite ("hard") coal—which was comparatively clean to handle and burn. It might heat a little water as well, and keep the kitchen warm—in the winter when you wanted it and in the summer when you didn't. Their less-well-to-do sisters also cooked with coal, but it was bituminous ("soft") coal, which was cheaper but dirtier. Coal stoves might heat the parlor, too; when Adams wrote, some affluent families even had coal-fired central heat. In large houses, they burned coal in prodigious quantities—perhaps fifteen tons in a season.[1]

Coal produced the gas that lit homes as well, and the electricity that was rapidly supplanting it. Coal fired the 20th Century Limited on its journey between New York and Chicago in the unheard-of time of only 18 hours station to station. Coal also powered the great Corliss engines that turned the wheels of industry, making America the workshop of the world. It fueled the blast furnaces and steelworks that built Teddy Roosevelt's Great White Fleet, which was then putting the world on notice that America had arrived, and coal powered that fleet as well. Berton Braley (1882–1966) caught the triumphalism of the age that linked American progress to coal:

> Steel has its empire of might, Coal is the maker of steel,
> Shaping it daytime or night, into the sword—or the keel,
> Rousing the magic of steam, driving the world to its goal,
> Making a fact of the dream, greatest of all is King Coal![2]

It was, however, a worried triumphalism, for even as Adams wrote, some were beginning to ruminate about waste and what would happen if America ran out of cheap coal.

Progressives Discover Coal

In 1908, President Theodore Roosevelt (1858–1919) called a conference of governors to address conservation issues, resulting in a published set of proceedings, as well as the three-volume National Conservation Commission's *Report* issued the next year. The idea of a conference had roots reaching back to 1903, as Roosevelt's ideas on conservation, which had initially focused on forestry, gradually extended to include a wide range of other resources—including, especially, coal. Indeed, two years before, he had caused the secretary of the interior to withdraw from entry and sale under the various land disposition acts some 64 million acres supposed to contain coal.[3]

There had been rumblings about conservation in the popular and technical press as well. Writing in 1900, *The Independent* had looked back on the previous 100 years as a "century of waste," while that same year, the *American Gas Light Journal* (AGLJ) warned of the "empty coal cellar." As the conference took shape, organizers decided to invite not only governors and other statesmen but also a virtual Who's Who of important Progressives, including scientists and representatives of scientific, technical, civic, and business organizations. The invitations included the presidents of scientific and engineering societies such as the American Chemical Society and American Society of Mechanical Engineers, as well as industry groups such as the American Mining Congress and even the National Hay Association. Of course, the press was there, including not only newspapers but representatives of popular and influential periodicals such as *Colliers*, *The Independent*, *Outlook*, and *Review of Reviews*. The technical press was also on hand, including the editors of *Iron Age*, *Engineering and Mining Journal*, and *Electrical World*. Yet if the conference included a broad swatch of individuals from business and industry, their job was to listen and learn: virtually every speaker at the conference was a federal, state, or university employee. The result was a certain imbalance in outlook.[4]

Roosevelt opened the conference, stressing that conservation was both moral and necessary to preserve the "wealth of this nation." He praised America's economic accomplishments: "nowhere has the [industrial] revolution been so great as in our country." Yet he went on to decry the lack of foresight that had led to "reckless and wasteful use" of natural resources. He spoke of dwindling forest resources and the loss of coal, iron, and oil reserves, emphasizing that "these resources are the final basis of national power" and calling for their "wise use." Roosevelt also briefly noted the need for "national efficiency," by which he meant policies that would improve health and safety. He concluded that conservation was the "right of the nation to guard its own future," and he cited with approval a Maine state court that upheld legislation restricting the right of an individual to cut trees on his own land where the result might be erosion elsewhere. The lesson was that "the property rights of the individual are subordinate to the rights of the community."[5]

These themes were not new. The year before, when he spoke to the National Editorial Association, Roosevelt had similarly lamented that

Americans' lack of foresight had led to "the reckless waste and destruction of much of our national wealth." Minerals were a peculiar problem, he thought, for "under private control there is much waste from short-sighted methods of working." Here he spelled out in more detail his hope that the future might bring wise use. He again emphasized the recent efforts of government "to get our people to look ahead, to exercise foresight and to substitute planned and orderly development of our resources in the place of a haphazard striving for immediate profit."[6]

For many, these were congenial themes, and the assembled luminaries listened to talks on the need for conservation of a host of natural resources. A major concern, as the president made clear, was the need to conserve fuel—especially coal—a point that Israel White (1848–1927), West Virginia's state geologist, drove home in a talk entitled "The Waste of Our Fuel Resources." White described in detail the "criminal wastes and wanton destruction" of coal, oil, and natural gas. Echoing the president, he asked, "How long can we hope to maintain this industrial supremacy in the iron and steel business of the world?" His answer was, "Just so long as the Appalachian coal-field shall continue to furnish cheap coal; and no longer," and he concluded that "unless this insane riot of destruction and waste of our fuel resources . . . [is] ended . . . our industrial power and supremacy" will end.[7]

Joseph A. Holmes (1858–1915) also addressed the problems of coal waste and the need for conservation. Historians remember Holmes as the first director of the United States Bureau of Mines (USBM). His training was in geology at Cornell, where he also studied physics and chemistry. Before he arrived at the bureau, he had directed fuel investigations at the Technologic Branch of the United States Geological Survey (USGS) that began in 1904 and focused on ways to save coal. His work there had made news, as had his warnings in 1907 of impending scarcity. In a news story headlined "National Waste," Holmes warned that cheap fuel was essential in "the struggle for industrial and commercial supremacy," and he concluded that "the nation . . . must safeguard the welfare of the citizen of tomorrow." In his address to the governors and in the National Conservation Commission's *Report*, Holmes reiterated his popular claims. Only 60 percent of potentially available anthracite was currently being recovered, and the wastes in burning coal were equally vast, as his work at the Geological Survey was demonstrating.[8]

Echoing Roosevelt's concern with "national efficiency," Holmes went on to point out a third form of waste: "even more serious than the question of waste of materials is the excessive loss of life in our mining and metallurgical operations." Holmes knew that he did not have to spell out the details to this audience. Two years earlier, in December 1907, three explosions had killed 635 men, and their story had made national headlines, while altogether, 2,534 men died mining bituminous coal that year. These failures of mining to conserve either coal or men provided yet another example to Progressives of the dangers of unfettered markets.[9]

Holmes explicitly defined conservation as the prevention of waste, not simply reduced resource use: "we cannot deny . . . the right of the present generation to use *efficiently* so much of these resources as it actually needs . . . [and] we cannot curtail present needs," he asserted. Moreover, "we cannot expect . . . this generation to mine or use these resources . . . unless there are profits." Holmes distinguished between "unnecessary" waste, which resulted from inefficiency, and "necessary" waste, which occurred because its curtailment was presently uneconomic. Like Roosevelt, Holmes also saw conservation as an ethical, as well as an engineering, problem. Nonrenewable resources like coal "should be regarded as property held in trust for the use of the [human] race rather than for a single generation." Thus, reducing waste where it was economic to do so was an ethical duty; however, much apparent waste was unavoidable because conservation was often unprofitable. Holmes urged legislation to reduce "unnecessary" waste, and he thought that research and education would lead the market to diminish currently necessary waste by making its reduction economic. Holmes became director of the Bureau of Mines in 1910; under his guidance and that of subsequent directors, the bureau would help fulfill this function, publishing much valuable research on coal mining and use and on mine safety. Yet Holmes was also a market skeptic. The National Conservation Commission's supporting assessment of coal described most of the waste of that mineral as necessary. But rather than accept the market's judgment, Holmes favored a coal cartel that could restrict "destructive" competition and allow higher prices that he thought would support less wasteful mining—an idea that would reappear in the 1930s.[10]

Other conference attendees included Gifford Pinchot (1865–1946) and Charles Van Hise (1857–1918). Pinchot had become the first chief of the United States Forest Service in 1905 and was probably the best-known conservationist of his day; Van Hise was a distinguished geologist and president of the University of Wisconsin. Like Holmes, both men worried about the impact of laissez faire policies on fuel reserves. At current rates of consumption, Pinchot had warned a popular audience in 1907, "our supplies of anthracite coal will last but fifty years." The problem, he later explained, was not simply consumption but waste.[11]

Van Hise's attendance at the conference was especially important, for he broadcast the commission's findings in his widely used text, *The Conservation of Natural Resources in the United States*. A later writer called it the leading authority on conservation for 25 years, claiming that it was through this text that the commission's findings had their greatest influence. Citing the commission's statistics, Van Hise asserted that "available and accessible" coal would be gone by 2027. "It has been said," he warned, "that the nations that have the coal and iron will rule the world." Accordingly, "to prolong . . . the life of coal, all waste should cease," and Van Hise favored education along with laws to "prohibit the uneconomic use of coal [and its] unnecessary waste . . . in mining." Van Hise also emphasized the need to conserve human life, claiming that coal mining killed a man for every 100,000 tons mined.[12]

This, then, was the problem of coal conservation as Progressive intellectuals saw it. They were pessimistic over resource availability and skeptical that markets could be trusted with conservation.[13] The issue was not simply resource exhaustion but rather the need to assure supplies of cheap energy, which was key to America's economic, military, and political power. Without cheap coal, they worried, America's industrial strength would shrivel, yet Americans were wasting it like drunken sailors.[14] Worse still, coal had no good substitutes, for natural gas and oil were—everyone imagined—even more limited in supply and would run out long before coal. Relying on USGS estimates contained in the commission's report, Pinchot concluded that oil "can not be expected to last beyond the middle of the present century." Holmes agreed, and he thought its chief waste was in uses for which coal should be substituted—a claim that would echo throughout the 1930s. His source on these matters, David Day (1859–1925) of the USGS, thought use of oil should be prohibited from everything but kerosene for lighting and as lubrication. Natural gas was equally limited, and Holmes claimed that supplies from existing fields might last no more than another 25 years. Fears that oil and gas reserves were extremely limited and likely to be "misused" continued to support coal conservation throughout the interwar years, even as these fuels gradually supplanted coal in an increasing number of uses.[15]

The conference received an immense amount of publicity, most of it largely uncritical. Indeed, contemplating American profligacy made some writers nearly hysterical: "the salvation of the American people depends on . . . the stopping of WASTE," *The Independent* shrieked. A more hopeful writer described the conference as the turning point toward a "more profitable and more moral national development." Perhaps, he gushed, "the great clock of time, whose hours are epochs, has struck."[16]

Engineers Object

Yet the story of fuel waste that emerged from the conference was seriously incomplete. None of the speakers and few writers in the popular press had any experience working in resource-based industries. Nor did they demonstrate any awareness that conservation was, in fact, a major concern of the engineers and others in the audience who did work in industry or that their efforts had a long history. Holmes had asserted, and Roosevelt would certainly have agreed, on the right of businesses to make a profit: did they believe that private businesses deliberately used resources in unprofitable ways? Why did the visionary entrepreneurs who built railroads, canals, and vast industrial enterprises suddenly lose that foresight when it came to the management of natural resources? And the same Progressives who decried the wastes of anthracite also denounced its producers as grasping monopolists. Why were such greedy businesses so wasteful?[17]

About a year after the convention met, all the major engineering societies held a special joint meeting on natural resource conservation. There, the mining engineer James Douglas (1837–1918) pointed out forcefully that while the commission had been talking about conservation, engineers had been practicing it. Douglas was a past president of the American Institute of Mining Engineers (AIME); he was also a businessman who had amassed a fortune from Arizona copper mining, been president of Copper Queen Mine, and made important innovations in ore processing. Douglas took issue with the vision of American history as a riot of unfettered waste. He summarized engineering for his audience: "by the very nature of our work . . . we are driven to employ as little material and as little energy as will serve our purpose . . . [for] both material and energy cost money." In short, while Holmes was worrying about "avoidable waste," engineers were busy avoiding it, because profit-making businesses were always on the prowl to save money. Nor was this a recent development: for "three quarters of a century . . . engineers . . . have combined in using their best skill and most competent efforts in . . . saving—not wasting." Ethical considerations reinforced professional responsibilities for Douglas, as well: the loss of coal byproducts was a "sin" and "every thinking man in a large institution . . . feels a sense of shame when he is conscious of waste." In short, the "gospel of efficiency" did not begin with the Progressive era, as historian Samuel Hays (1921–2017) suggests, but had, in fact, been around for quite some time. Nor did large corporations undertake conservation "because they could more readily afford it" but rather because it was profitable. Elsewhere, Douglas acknowledged that there had been much waste, but he did not blame this on lack of foresight; instead, he thought it a result of rapid development and inevitably imperfect knowledge. "It is better to make progress and thereby gain experience, even at the expense of such waste . . . rather than stand still," he asserted. The engineer Henry Petroski (1942–2023) argues that failure can be instructive, leading to improved designs, and Douglas's claim is that waste is the teacher of conservation. Douglas went on to rehearse the history of energy conservation in iron manufacture, concluding that "a glimpse back will show that the engineer has not been guilty of knowingly and willingly wasting."[18]

Other engineers also challenged the views expressed at the conference. In fact, the National Conservation Commission's own report on coal, assembled by Marius Campbell (1858–1940) and Edward Parker (1860–?) of the USGS and appearing a year after the conference, supported the contentions of Douglas and others: the riot of waste was a fantasy. The "principal loss or waste [of coal] is that *necessarily* left in the ground as pillars to support the roof," the geologists concluded. They also discussed the loss of poorer quality coal. Like Holmes, they thought such losses could only be prevented by higher prices and favored "legislative action" to raise prices.[19]

The governors' conference was a central moment in American energy history, not for what it accomplished, which was little, but for what it revealed about Progressives' understanding of their world. As Douglas pointed out, conference participants largely ignored the achievements of technical people working in industry, whose job it was to save money by conserving resources. With their focus on markets as a source of waste but not conservation, they were ill-prepared—one might almost say, they lacked the foresight—to see the energy transitions already going on around them or imagine that peak anthracite was then just over the horizon.

Historians of conservation, too, have mostly ignored coal during these years, perhaps because their usual interest has been on public policy, while at least until World War II, fuel saving was largely market-driven. The one topic relating to coal—pollution—that has been of most interest to modern writers was of secondary importance to contemporaries. The conference devoted two sentences to the "smoke nuisance," and the three volumes published by the National Conservation Commission barely mentioned the topic. Accordingly, it seems time for a history of the rise and decline of the age of coal that looks more closely at the role of market forces in these events and especially how they shaped conservation and waste.[20]

A Framework and Themes

My scaffold for this energy history comes from the economist Joseph Schumpeter (1883–1950), who understood capitalism better than anyone since Karl Marx. Capitalism, Schumpeter argued, "is by nature a form or method of economic change." It was also an engine of progress. Schumpeter quoted Marx approvingly, that "in scarce one hundred years . . . [capitalism] has created more massive and more colossal productive forces than have all previous generations together." Moreover, because capitalism is a process, its performance needs to be judged not at a moment in time but rather over time. Thus, in a snapshot, capitalism seems messy and unjust, but over time, "the capitalist achievement," Schumpeter famously observed, "does not typically consist in providing more silk stockings for queens but in bringing them within the reach of factory girls in return for steadily decreasing amounts of effort."[21]

In Schumpeter's vision, entrepreneurs drive this process. They were and are the men and women with a better idea—"foresight," to use a favorite Roosevelt word—and they bring inventions—discoveries—to market. These may be new products, old products in new places, new production processes, new selling methods, even new institutions. In a market economy, prices will provide signals of their values, and prices will also send signals to buyers and competitors and coordinate their activities. In Schumpeter's evocative phrase, successful innovation results in "creative destruction." Entrepreneurs are by definition creative, and some of them transform the world with their inventions—electric power, automobiles, the internet all come to mind. Such progress is messy, however, for innovations wreak havoc as well, destroying established ways of doing things. "Ford put America on wheels," an advertisement once claimed—but he also put the carriage industry out of business. Similarly, the innovations discussed in this book helped undermine coal as a way of life. Many innovations are simply flops, of course, and most are far less dramatic or important; they make incremental changes, but these, too, as they cumulate, are transformative.[22]

In the early history of American energy, many innovations derived from inventors and entrepreneurs with little technical training, but scientists and engineers soon began to play central roles—sometimes as innovators, sometimes as suppliers of the technical skill necessary to make the innovation work. For Schumpeter, entrepreneurs were mostly individuals, although his later writing stresses the role of large firms. More recently, however, William J. Baumol (1922–2017) argued that large corporations have institutionalized innovation. To do so, they hire these technically proficient men and women, often in research laboratories, but in energy history much innovation has also stemmed from "shop floor" modifications. Whether working as individuals or for corporations, however, as Douglas pointed out, the job of engineers is to save, not waste. It is part of their training—a professional duty and an ethic. The distinguished civil engineer Arthur Mellen Wellington (1847–1895) captured the essence of engineering. It is "the art of doing that well with one dollar, which any bungler can do with two after a fashion."[23]

Focusing devices induce innovation in certain areas. In such instances, markets can encourage a kind of crowdsourcing of innovation. Sometimes it is rising prices or shortages; there are also technological sequences: large-scale heat wastage leads to inquiry about how to capture it. Geographic imbalances between fuel supply and demand lead to innovation in transport. Nathan Rosenberg (1927–2015) used the phrase "technological convergence" to describe the fact that an improvement developed for a specific use might have broad applicability owing to similarities in procedures elsewhere. In the case of fuels used for heat and power, the inevitable commonalities among industries encouraged their potentially wide application.[24]

In applying this framework to America's energy transitions up to 1940, I develop the following major themes. First, and centrally, America's energy abundance, although shaped by geography and geology, has been socially constructed. The very term "energy abundance" presumes the ability to extract it economically. Coal, that is, did not simply hop out of the ground to nestle in stoves and furnaces, nor were gas and oil any more obliging; from the beginning, exploitation has been driven not only by entrepreneurs and inspired tinkers but also by scientists and engineers, who by the twentieth century constituted an "invisible college" that shaped energy markets. And marketing and advertising were central to the ways Americans marshaled their energy abundance.[25]

Second, private markets were the driving forces in America's energy history until World War II. Public policies did matter, and sometimes they mattered a lot; the price spikes and fuel shortages associated with World War I gave conservation a considerable nudge. Yet the great labor disputes that convulsed coal mining and the various federal and state commissions and inquiries that littered the interwar political landscape played only a minor role. Government policies during these decades were important largely because they augmented market activities, often providing information to improve outcomes.[26]

A third and related theme is that because these transitions were market-driven, they almost inevitably raised Americans' standards of living. Indeed, that was the motive, and while the result may seem obvious, it is worth emphasizing because so much writing about energy—including some by the present author—has emphasized energy's human costs in the form of pollution and injuries. Moreover, most of the gains from energy transitions in American households came in the form of comfort, cleanliness, and convenience. The national-product statistics do not capture such qualitative improvements, but they immensely lightened the burdens of everyday Americans.[27]

Fourth, while contemporaries and modern historians have often emphasized the profligacy of Americans' use of energy, conservation has been embedded in energy history from the beginning because the search for energy efficiency is built into market economies. Conservation, that is, did not simply rely on individuals' religious or ethical concerns with the environment—although these have mattered, too—and it did not have to wait until Progressives began to worry about marshaling the state in the service of the environment. Moreover, because of increases in energy efficiency, human welfare from energy use has improved much more than the use itself. The author can testify from personal experience that insulation of an old house dramatically increases comfort and reduces fuel consumption as well.

Waste is not a straightforward concept. As Holmes understood, some "waste" of energy may be currently impossible to reduce—perhaps because no one knows how to do it—and in such cases, criticism seems pointless. But not all technically feasible increases in energy efficiency (and so reductions in waste) are economically efficient because they cost more than they are worth. To reduce such waste, as some Progressives wished, is to save energy at the expense of Americans' standard of living. Yet innovation can also make it profitable to reduce currently uneconomic waste, for as both Douglas and Schumpeter understood, the very waste that arises from economic progress provides the signals and incentives for conservation. Uneconomic waste, that is, was a focusing device that powerfully encouraged innovation.[28]

Fifth, a point well known to energy experts is that sometimes energy conservation can *increase* fuel use, as the English economist Stanley Jevons (1835–1882) was probably the first to note. "*It is wholly a confusion of ideas to suppose that the economical use of fuel is equivalent to a diminished consumption. The very contrary is the truth,*" Jevons emphasized. His argument was essentially that more economical use reduced fuel costs, while cheaper fuel would expand demand for output and thus energy—the "rebound effect" in modern parlance. Yet as these chapters will demonstrate, sometimes conservation has indeed sharply reduced energy use.[29]

Finally, the thinking of energy experts evolved during these years. The importance of oil during World War I made its conservation a matter of national defense. Combined with routine predictions of shortages, this development shifted thinking about energy conservation away from coal toward oil and gas. In retrospect, it is clear these experts confused what was, in fact, a long-term transition away from solid fuels with what they imagined was a short-term market aberration that shortages of oil and gas would soon reverse. Accordingly, conservationists began to stress the need to save coal *markets* to reduce use of oil and gas. Some Progressives, in short, became reactionaries when it came to energy transitions; we should probably be happy that they were largely ignored.[30]

This story of market-driven energy transitions and their impact on conservation and living standards has not been a major focus of modern historians. Yet the tale seems well worth telling, in part because it is strikingly different from the transition occurring as this is written, which includes a highly selective reliance on markets and seems likely to reduce living standards.[31]

Schumpeter thought that entrepreneurial innovations were what drive capitalism, and energy transitions have been a central part of that story. The increasing use of high-quality energy in ever more efficient ways has, indeed, brought problems of pollution and climate change, but these are the hole in the donut, for they have also helped bring silk stockings within the reach of factory girls and unparalleled affluence to all of us who live in advanced countries.

Part I

THE RISE

1

The Dawning of the Coal Age, 1800–1860

> [Anthracite] coal is only partially used in dwelling houses, but would be in general demand . . . if it could be afforded.
>
> —*James Mease,* Picture of Philadelphia, *1811*

> As to coal, it abounds very much all over the western country and lies so near the surface of the ground that the wagon wheels often cut into it.
>
> —*Francis Baily,* Journal, *near Pittsburgh, 1796*

In 1911, a new trade journal appeared named *Coal Age.* It must have seemed an obvious title, for the first two decades of the twentieth century marked the apotheosis of coal's use for heat and power in the United States. Indeed, although no one saw it at the time, "peak coal" lay less than a decade in the future as conservation and substitute fuels would lay claim to its dominance. These energy transitions reflected America's evolution from a largely unmechanized, agricultural, and rural nation to an affluent, urban, industrial, and technologically sophisticated society, and they were a central part of that story. Coal's rise had been breathtakingly rapid: a century or so earlier—in 1800, say—coal played virtually no role in Americans' lives. Instead, heat came from millions of cords of wood, and power derived mostly from horses—supplemented in a few places by wind and waterwheels. This chapter briefly chronicles the beginnings of coal's rise to become America's dominant fuel. To understand how and why coal experienced its dramatic fall from power after World War I—the central focus of this book—we need to understand its rise, during which the same forces were at work. We begin with an overview of patterns of American energy consumption that yields some surprising conclusions. The following sections trace the slow spread of coal—especially anthracite—during these years as a domestic and industrial fuel, as well as the long retreat of wood. While coal was abundant, entrepreneurs, technologists, and marketers were central in its exploitation, and if markets drove the spread of coal, they also ensured that the pressure for conservation was ongoing. And conservation of coal undoubtedly spread its use.

Energy and Power: The Long View, 1800–1940

There is much anecdotal information about Americans' use of energy and power in the late eighteenth and early nineteenth centuries, but there is little in the way of hard data. The reader should take the data in table 1.1, derived from appendix 2, as approximations; especially before 1860, these data sometimes rest on weak foundations. They are, however, sufficiently precise for the purposes of this chapter. For additional details and an explanation of the sources and calculations, the reader should consult appendix 2. The usual units of energy measurement (cords of wood, barrels of oil) cannot be added up, but they all have a common denominator— calories or Btus. But because this is a book about coal, I typically express disparate energy sources as their bituminous coal equivalent (BCE). Thus, for example, the energy content of a cord of firewood about this time makes 1.25 cords roughly equivalent to one ton of bituminous coal.

Table 1.1 expresses the major sources of primary energy as percentages of each year's BCE.[1] Before 1850, the table omits wind and waterpower. Whale oil is too minor to include but is discussed later. Still, what table 1.1 depicts is surely true: in 1800, the overwhelming sources of energy were wood and animal power—along with Americans' strong backs. Probably 95 percent of the fuel wood at this time was burned just to cook and keep warm. Blacksmiths and iron forges turned a small amount into charcoal, but its use in steamboats and railroads lay in the future. Even on the eve of the Civil War, America's coal age was in its infancy, with renewable sources of energy (wood, horses and mules, wind and water) accounting for 87 percent of all energy. Thereafter, with the burgeoning of industrialization, coal drove all before it. Yet by the 1890s, alternative energy sources appeared, and by 1920, coal's share began to recede, while renewables over these years fell to just 31 percent of the total in 1900 and 10 percent in 1940.[2]

The shares of energy in table 1.1 tell us nothing about the amount of energy being consumed or how that consumption grew, so table 1.2 presents total energy use relative to GDP and to population. Although it may

Table 1.1. Coal and Its Competitors: Major Sources of American Energy, 1800–1940

Year	Horses & mules (%)	Wood (%)	Anthracite (%)	Bituminous (%)	All coal (%)	Oil (%)	Natural gas (%)	Wind (%)	Water (%)	Total (%)
1800	8.4	91.1	0.0	0.6	0.6	0.0	0.0	0.0	0.0	100
1840	7.9	88.7	1.6	1.8	3.4	0.0	0.0	0.0	0.0	100
1850	5.6	73.2	3.7	3.2	6.8	0.0	0.0	7.3	7.1	100
1860	6.3	67.1	7.1	6.1	13.3	0.1	0.0	7.0	6.3	100
1870	6.3	62.3	10.9	11.2	22.1	0.7	0.0	2.7	5.9	100
1880	7.5	50.8	13.0	18.9	31.9	2.7	0.1	2.0	4.9	100
1890	7.8	32.9	15.4	32.8	48.2	3.5	3.3	1.2	3.2	100
1900	6.7	19.5	14.1	49.0	63.1	3.6	2.4	0.7	4.1	100
1910	4.7	10.3	12.6	58.3	70.8	6.8	3.1	0.3	4.0	100
1920	3.5	6.7	9.5	62.1	71.6	10.8	3.6	0.2	3.7	100
1930	2.6	6.1	6.4	51.1	57.5	21.8	8.7	0.1	3.2	100
1940	1.8	5.1	4.9	44.9	49.8	29.3	10.7	0.0	3.3	100

Source: Table A2.1.

Table 1.2. American Energy Use per Person and per Million Dollars, Real GDP, 1800–1940

Year	All energy	Coal	All energy	Coal
	Per person		Per $ million real GDP	
1800	3.8	0.0	3,029	18
1840	4.1	0.1	2,469	83
1850	4.8	0.3	2,650	181
1860	4.7	0.6	2,174	288
1870	4.4	1.0	1,872	414
1880	4.3	1.6	1,241	396
1890	4.6	2.5	1,266	611
1900	5.2	3.5	1,234	778
1910	7.1	5.4	1,446	1,024
1920	8.6	6.2	1,592	1,140
1930	7.4	4.3	1,217	700
1940	7.8	3.9	1,046	521

Source: GDP in millions of 1996 dollars is from Carter et al., *Historical Statistics of the United States, Millennial Edition.* Coal and other energy are in tons of BCE. See appendix tables A2.1 and A2.2.

seem surprising that energy use per person grew slowly until the late nineteenth century, this was because of conservation and because much of the increasing use of coal was simply a substitution for wood. Also surprising, perhaps, is the decline in energy per dollar of GDP until 1880, even as coal use rose sharply; this also reflects the combination of substitution along with increasing energy efficiency. The burst of industrialization between 1880 and 1920 reverses this decline, but it picks up again sharply in the two decades after 1920. Coal's use relative to GDP, which explodes between 1880 and 1920, falls to half in the next two decades. In sum, perhaps Americans' energy history has been characterized by "prodigal waste," as the *New York Times* lamented in 1907. We have been "profligate energy consumers," a modern historian asserts. But the same acquisitiveness that yielded waste and profligacy also stimulated a constant quest for efficiency and energy conservation—a development that this and subsequent chapters will detail.[3]

The Setting

Coal is a complex mixture of almost infinite variability. For most purposes, though, it can be divided into three types: bituminous (soft), anthracite (hard), and lignite, which is subbituminous. These types have different energy (Btu) content, and when added together in a table, their shares reflect that fact. Anthracite is almost pure carbon, while bituminous coal contains volatile materials such as sulfur, making it likely to smoke. Both coals also contain incombustibles (termed "ash"). In 1800, coal consumption was mostly bituminous (see table 1.1); but by midcentury, it was 55 percent anthracite, and by 1900, it was 70 percent bituminous.[4]

Coal in Abundance

Americans were well aware that coal was the fuel of the English Industrial Revolution and were alert to its possibilities in the New World. Indeed, there were modest imports of English (bituminous) coal through the colonial period and into the early nineteenth century, although the federal government thoughtfully protected domestic production with modest tariffs after 1789. Many records refer to coal in tidewater Virginia during the eighteenth century. It was probably being mined—from open "cole pitts" with pick, shovel, and sledge—near Richmond, Virginia, and used locally in forges and fireplaces perhaps as early as 1704. With wood abundant and land-transport costs high, coal moved long distances only by water, however, and by midcentury, a small trade in Virginia coal flowed down the James River to Eastern Seaboard cities.[5]

American property law assigned mineral rights to owners of the surface property. By 1800, soft coal was being mined in what is now West Virginia. There were advertisements for coal lands there of "inexhaustible" quantity as early as 1790, and one of the streams was termed "Stone Cole Lick." In Kentucky, a newspaper referred to "coal in abundance" as early as 1798. A survey of 1785 reported coal in Ohio, and entrepreneurs were mining and rafting it down the Ohio River by 1803. Maryland coal was being mined and used to make nails as early as 1789. In eastern Pennsylvania, anthracite proved hard to burn in local forges as early as 1769, and experiments employing it in early blast furnaces were also unsuccessful. In western Pennsylvania, near Pittsburgh, surveyors described "great quantities" of "Stone-Coal." By about 1790, bituminous coal was being shipped from the Pittsburgh region down the Monongahela and Ohio Rivers on barges, but most of it was consumed locally as domestic fuel and in salt production and, especially, iron manufacture. In Pittsburgh, as elsewhere, the cost of fuel wood and coal determined the extent of their markets. In 1802, one writer explained the shift to coal in that region as due to the scarcity of labor, not wood, because cutting wood was more labor intensive than digging coal. A decade later, coal was being used as a "common fuel" in that region, as wood was selling at $2.00 a cord while coal could be purchased at $0.06 a bushel or $1.50 a ton—yielding more energy for less money.[6]

One result of cheap soft coal around Pittsburgh was that complaints about smoke began to appear as early as 1800; indeed, Cliff Davidson has calculated that by then, smoke levels generated around 2 mg./m^3 of suspended particulate matter—similar to levels found in smoky Chicago in 1913 (and probably contributing modestly to that city's mortality rate). Yet in this respect, Pittsburgh was the exception during the antebellum decades because nationwide, bituminous coal had not yet come to dominate (see table 1.1). Accordingly, writing in 1854, one traveler referred to Pittsburgh as the "blackest, smokiest and most dingy city in America."[7]

This early interest in coal suggests four important points. First, the colonies, and later the United States, were not only coal rich; the coal itself was easy to find. On a trip that took him south from Pittsburgh into Kentucky in 1797, one traveler noted, "This whole country abounds in coal, which lies almost upon the surface." Similarly, many others described coal outcroppings, and most early mines began when local people tried digging into exposed beds. Thus, while supporters of state geological surveys invariably stressed their economic value, in most states, mining preceded them. Second, there was much experimentation with the new fuels, especially in the case of anthracite, not all of it successful. Third, if the technology could be made to work, fuel choice was overwhelmingly a matter of economics. Fourth, geology, geography, and technological change held the keys to coal's development.[8]

If the whole country abounded in coal, nineteenth-century America was also developing the entrepreneurs, technologists, marketers, and institutions necessary to exploit it should the incentive arise. Few, it seems, would have agreed with James Fenimore Cooper's sheriff when he cried, "Damn your coal; who wants to find coal in these forests?" Coal had at least one major advantage over wood as a fuel: it was a far more dense energy source, taking up about a quarter of the space of its energy equivalent in wood.[9] Thus, coal could travel farther, and it required less handling. Its rise to become the dominant American fuel in the century after 1800 was the outcome of this comparative ease of transport combined with its wide geographical dispersion, the rising cost of wood, increases in wages that favored coal over wood, and the spread of the transportation network. Each of these was shaped in turn by the work of an emerging community of what Silvio Bedini has termed "thinkers and tinkers."[10]

An Emerging Technological Community

Eighteenth- and nineteenth-century America was part of a transatlantic scientific and technical community. Although the United States was a technological follower until the late nineteenth century, Americans could import European ideas, and a striking number of innovations occurred almost simultaneously in this country and abroad. Very early on, a rising commercial and technical press routinely reported European as well as American work, as did organizations such as the Franklin Institute and the American Philosophical Society (APS).

Sometimes European ideas arrived with Europeans, for, as we will see, many important tinkers were immigrants. An important perquisite to this diffusion of learning was, of course, literacy. In 1840, the German engineer Franz Von Gerstner (1796–1840) toured the United States and wrote of Americans' inventive activities. A key, he thought, was "the schools, which diffuse general knowledge and enable every man correctly to judge his enterprise and calculate the results." Indeed, by 1850, the primary school enrollment rate in the United States was the highest in the world.[11]

As Von Gerstner appreciated, much early work reflected the activities of a host of brilliant inventors who were literate but largely innocent of scientific training—"tinkers." But men and a few women with formal technical training became increasingly important. Scientists such as Benjamin Silliman Sr. (1779–1864) and J. Peter Lesley (1819–1903) emerged from the University of Pennsylvania, and Yale graduated Benjamin Silliman Jr. (1816–1885). In 1818, Silliman Sr. founded the *American Journal of Science* (AJS), the first scientific journal in the United States that provided a forum for much research into energy use. The Sillimans and Lesley also exemplified the increasing importance of scientific consultants to emerging businesses. Both Sillimans made important contributions to the early petroleum industry. Lesley had received valuable training on Pennsylvania's geological survey and wrote widely on coal. Other state surveys were important training grounds as well, exemplifying the ways state governments augmented market activities. Geologists founded the American Association for the Advancement of Science in 1848. West Point was America's first engineering school, founded in 1802, and it was followed by Rensselaer Polytechnic Institute in 1824. Harvard and Yale established the Lawrence and Sheffield schools in 1847. The American Society of Civil Engineers emerged in 1852. Graduates from these surveys and schools increasingly joined the many entrepreneurs and tinkers with little formal education to midwife the age of coal. If the whole country abounded in coal, as early settlers had discovered, America was also innovating the technological network needed to bring it to market.[12]

A Wooden Age Gives Way

The early nineteenth century was, as Brooke Hindle put it, a "wooden age," and table 1.1 makes clear just how important wood was as an energy source. In 1800, virtually all of it went into domestic heating, but by 1860, railroads and steamboats were also consuming modest quantities, as were iron furnaces and other industrial users. To modern Americans, trees are wonderful and beautiful; indeed, we have coined the term "tree huggers" to describe modern environmentalists. One rarely found tree huggers among men and women of the early nineteenth century, however. Forests were the enemy, for trees blocked progress, which required land clearing.[13]

Farmers hated trees: describing a farmer clearing land in upstate New York about 1816, one French traveler observed, "He seems to have declared war upon the whole species [of trees]." Unsurprisingly, therefore, "the bulk of the fuel" resulted from this source. Heating with wood was therefore an early form of recycling of what was in effect an agricultural waste product.

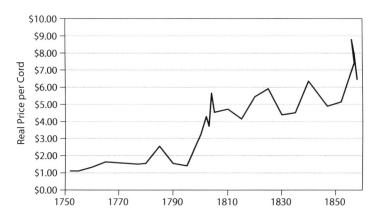

Figure 1.1. Wood prices in Massachusetts, 1752–1858. Increasingly expensive firewood stimulated efforts to bring anthracite to market. Massachusetts Bureau of Statistics of Labor, *Sixteenth Annual Report*, 1885 (Boston, 1889), 201–301. Data for 1777 on deflated prices are using David-Solar price index (published in Carter et al., *Historical Statistics*, series Cc2; 1777 = 100).

This also implied that the cost of burning wood was only the *extra* cost associated with cutting it into four-foot lengths, splitting it, and hauling it to the fireplace or market—which was one reason wood was cheap. Such activities, done in fall or winter, when farm labor was available, made it an important cash crop. In 1834, one journal noted that "the steamboat, with something like magical influence, had converted . . . [trees from useless obstacles] into objects of rapidly increasing value" to farmers.[14]

The American colonies, and then the United States, comprised a series of loosely (initially, very loosely) interconnected markets for firewood. Accordingly, rising prices in urban areas began to reflect local scarcities as early as the eighteenth century (fig. 1.1). The price increases drew supplies from longer distances inland and from the coastal trade, ameliorating scarcities in cities and elevating prices elsewhere. Moreover, while domestic consumption of firewood was the dominant cause of its scarcity, railroads, steamboats, and manufacturers employing stationary steam engines felt the same pinch. Transportation developments, in turn, shaped both energy supply and demand. Steamboats and railroads that had initially burned wood contributed to its scarcity, yet they transported it and increasingly shipped coal.[15]

Mining and Moving Coal

Although rising wood prices helped spur the shift to coal, there was little use of the latter before about 1820 because high transportation costs sharply constricted markets. During these years, production of bituminous coal exceeded that of anthracite simply because it had been found in 12 states by then, whereas anthracite was mined only in eastern Pennsylvania.

Hoping to discover more coal and other valuable resources, several states instituted geological surveys, beginning with the Carolinas in 1823 and 1824. Before the Civil War, major coal-mining states such as Pennsylvania, Ohio, Illinois, and Virginia all undertook surveys, but there is little evidence of their economic importance. Most historians of science have focused on the surveys' contributions to geology rather than geology's

Figure 1.2. Early mining often took advantage of outcrops of coal; it was typically small-scale and unmechanized, employing only human or animal power. (Harry Fenn, "A Family Coal Mine,—A Pittsburgh Sketch," published in *Every Saturday*, March 11, 1871, 233. Public domain)

A FAMILY COAL-MINE. — A PITTSBURGH SKETCH.

contribution to coal. The surveys probably helped entrepreneurs avoid some unprofitable ventures, and they trained a host of scientists who provided consulting expertise to subsequent mining ventures.[16]

The expansion of bituminous production in western Pennsylvania was typical of that in other states in the antebellum period. As I have noted, coal was often not difficult to find, sometimes appearing in outcrops, and by the 1780s, western Pennsylvania settlers found it amenable to an early form of strip mining or drift mining (a horizontal entry into a coal seam).

Once a profitable seam had been discovered, mining technology was simple: picks, shovels, black powder, animal power, carts, and wheelbarrows (fig. 1.2). In 1844, steam power appeared to drive the first anthracite breaker. By the 1850s, as shaft mines became deeper, steam engines arrived to run elevators for lifting men and coal, but there was otherwise little

Figure 1.3. Anthracite coal price and production, 1826–1860. The enormous expansion of output held down prices even as firewood was becoming more expensive. Prices are for June of each year, adjusted for inflation. (Arthur H. Cole, *Wholesale Commodity Prices in the United States, 1700–1861,* supplement [Cambridge, MA: Harvard University Press, 1938]; Carter et al., *Historical Statistics,* consumer price index Cc2; 1860 = 100)

mechanization before the Civil War. With coal easy to find and cheap, and underground workings comparatively unmechanized, neither geology nor engineering seemed very important to many small companies. One writer describes such operators as "ill equipped and inexperienced," concluding that this led to "dangerous and wasteful practices." These were the sorts of operations that James Douglas probably had in mind when he later argued that "it is better to make progress and thereby gain experience, even at the expense of such waste . . . rather than stand still."[17]

In all states, much of the consumption was local for domestic heat, but by the early 1800s, coal was being employed as fuel for making glass, brick, and iron as well, and coal that was close enough to rivers was being shipped downstream. The overlapping canal and railroad booms from the 1830s through the 1850s, along with local industrialization, sped up this slow growth as it drove down delivery prices and enormously expanded markets in Pennsylvania and elsewhere. In the two decades after 1820, bituminous coal production in the United States quadrupled (see table A1.4).

Yet the growth in hard coal far outstripped that of bituminous production during these years: in 1840, anthracite production was 250 times greater than it had been 20 years earlier, and by 1860, it was 2,500 times as great (table A1.4). As with bituminous coal, improved supply conditions were central to this explosive growth, for even as output surged, prices sagged (fig. 1.3). In anthracite, as in bituminous coal production, there are indications of modest technological progress in mining in the antebellum years. Here, too, steam engines allowed improved drainage, especially in deep mines, thereby offsetting some of the penalties of geology. Far more important than geology or mine engineering in increasing mine production was low-cost transport, for while geography placed anthracite close to eastern markets, its location in the rugged eastern Pennsylvania hills initially shut it out of populous, industrializing seaboard cities.

Local use of anthracite dated from at least the 1720s. As a domestic fuel, it was superior to wood: it was cleaner and generated fewer dangerous sparks, took less space, and lasted longer, thereby reducing labor costs. But hard coal was hard to burn: it did not light well, and with inadequate draft, the fire might simply go out. Innovations and marketing solved the problem.

About 1808, Jesse Fell (1751–1830), a Pennsylvania nail maker, invented a new grate that improved anthracite combustion. That same year, two entrepreneurial brothers, John (dates unknown) and Abijah Smith (1764–1826), rafted some anthracite down the Susquehanna River and, employing Fell's grate, gave demonstrations on how to burn anthracite. The market, however, remained tiny because transport costs were high. The War of 1812 cut off coal supplies from Virginia and elsewhere, however, so Philadelphia iron makers experimented with the use of anthracite. After the war, two of them—Josiah White (1781–1850) and Erskine Hazard (1790–1865)—bought coal lands and began improving navigation on the Susquehanna River with an eye to eastern markets. Yet when White and Hazard sent their first rafts of coal down the Susquehanna to Philadelphia, they could hardly give it away. Thus, more than was true for soft coal, the expansion of anthracite sales depended on marketing by producers and on technological improvements that facilitated burning the bulky fuel. Accordingly, White and Hazard joined with other producers and began a publicity campaign to sell the new product. Jacob Cist (1782–1825) was a mine owner and merchant; he went door-to-door promoting stoves and, along with Hazard and White, published a pamphlet that included testimonials by local businessmen to the wonders—and the economy—of anthracite. The object, as White had noted earlier, was "to make a noise in Philadelphia." Marketing, it seems, shaped American energy transitions from the beginning.[18]

Finally, in the mid-1820s, a canal-building boom provided tidewater access for anthracite. So lucrative did the anthracite market appear that these canals were largely privately financed; White and Hazard's Lehigh Coal and Navigation Co. opened the first one circa 1821. The Schuylkill Canal followed in about 1825, and the Delaware & Hudson Canal, with assistance from state loans, opened in 1829. Several other canals soon followed, bringing coal to Baltimore, Philadelphia, New York, and, more broadly, to tidewater, while railroad shipments began in the 1840s. At a time when wood was becoming increasingly scarce, these new supplies of anthracite brought down its price (see fig. 1.3), even in the face of explosive urban growth. Philadelphia, for example, expanded from about 63,000 residents in 1820 to more than 565,000 in 1860.[19]

Marketing remained central to the new fuel's spread: about 1831, Pottsville's mine companies gave $100 to the Franklin Institute to encourage cookstove innovation, and Philadelphia's Fuel Savings Society promised a reward for the development of a cheap cooking stove. In response, stoves and fireplace grates proliferated, and buyers learned how to use the new fuel. "Dealers in coal and wood sprang up" as markets up and down the

East Coast exploded. One likely benefit of this substitution of anthracite for wood was a reduction in smoke both indoors and out, for although hard coal burned largely smoke free, wood—especially if it was green—might emit clouds of smoke.[20]

Yet if canals had midwifed the age of coal, railroads brought it to maturity. It is hard to imagine the eventual widespread use of coal as a source of heat and power without the railroad, for—like wood—it was hard to transport. Without railroads, its use would likely have been confined to regions close to mines and to river and canal towns and coastal areas. The railroad made its debut in 1829, when the Stourbridge Lion—an import from Great Britain—became the first locomotive to run in America. Appropriately enough, its owner was the Delaware & Hudson Canal Co., and its job was to haul coal. By 1850, what came to be called the anthracite railroads were bringing more coal to market than were the canals. One of those lines—the Philadelphia and Reading—which had moved no coal in 1840, carried nearly 1.5 million tons a decade later. Transport of soft coal followed a similar path; initially, river shipments dominated, while the Allegheny Mountains barred it from eastern markets. But with the completion of the Pennsylvania Railroad from Philadelphia to Pittsburgh in 1854, soft coal moved east to compete with anthracite.[21]

Domestic Fuel

Much of the early demand for coal was to replace wood as a source of domestic heat. A writer in 1831, probably referring to Philadelphia, described "the rapidly increasing consumption [of hard coal] in private families and public offices." In 1834, another householder claimed that wood heat had required 24 cords at $5.43 each while he needed only 14 tons of coal at $5.00 a ton. Looking back from a 50-year vantage point, Carroll Wright (1885–1905) of the Massachusetts Bureau of Labor claimed that by 1835, anthracite was in common use in the homes of Boston artisans. Indeed, from 1835 through 1838, Massachusetts imported each year about 87,000 tons of anthracite and 64,000 tons of bituminous, or approximately a fifth of a ton per person. Wood, accordingly, began its long, slow retreat as a domestic fuel. It remained important in rural areas, and its economics improved the greater the economic distance a household was from the mines, but a large-scale survey of urban households about the time of World War I showed that wood typically accounted for less than 1.5 percent of urban family fuel bills.[22]

Scarcity raised the payoff to learning about fuel efficiency; this was reflected in a scientific investigation of the relative value of various kinds of wood and coal published by the American Philosophical Society in 1823. The work, by Marcus Bull (1787–1851), was an early example of the application of experimental analysis to energy problems. Bull was associated with a coal company, and one writer describes him as having a "scientific disposition," but his education is unknown. Benjamin Silliman Sr. of Yale College, then on his way to becoming one of America's premier scientists, praised Bull's findings, and they were widely and respectfully reprinted. The market seems to have anticipated at least some of his findings, however, as both sellers and buyers did their own cost/benefit calculations to price fuel by energy content. The data from which figure 1.3 derives also yield occasional information on prices by species of wood, and pine invariably sold at a discount—as Bull and others since have found that it should—while hardwoods were always at a premium.[23]

Although Bull had evaluated a few coal samples, and Europeans had been interested in the evaporative power of coal for some time, there seems to have been little immediate follow-up on Bull's efforts. Finally, in 1838 and again in 1841, Walter Johnson (1794–1852), who taught chemistry and physics at Pennsylvania College, published some of the first American work. In 1841, Henry Darwin Rogers (1808–1866) of the Pennsylvania Geological Survey also published a chemical analysis of numbers of coal and iron samples. Johnson's work must also have caught the eye of the United States Navy, which commissioned a large-scale analysis of the heating power of coal that Johnson published in 1844. *Hunt's Merchant's Magazine* promptly reprinted Johnson's findings and those of Rogers as well, observing that the need for such chemical analysis would rise as the use of coal spread. Indeed, later in the century such analyses would become more valuable because they included such commercially important considerations as the water content of the fuel.[24]

Scarcity also encouraged innovation on combustion efficiency, for households might go through 15 to 20 cords of wood a year. Estimates are that Americans used about 23 million cords of firewood a year in 1800 (table A1.1), largely as domestic fuel, or about 4.34 cords (3.5 BCE) per person. With 80 to 90 percent of the heat simply going up the chimney, it represented a colossal squandering of fuel and labor. To use the terminology of Joseph Holmes, it represented a "necessary waste," which innovation would soon reduce. One result was Count Rumford's inventions that increased fireplace efficiency in the 1790s. Rumford (Benjamin Thompson, 1753–1814) was a brilliant self-taught British American physicist; his fireplace design was shallower and sported a narrower throat, a smoke chamber, and beveled sides and back that threw more heat into the room.[25]

Even more important and earlier, about 1741, was Benjamin Franklin's stove. Franklin claimed that his stove used a fourth of the wood to make the room twice as warm, and in 1823 Marcus Bull's experiments found stoves about ten times as efficient as fireplaces. Although he grumbled that it concealed the fire and took up space, Henry David Thoreau explained that he had bought a cookstove because of its fuel economy "since I did not own the forest." Stoves were the superior innovation, and as one writer observed in 1847, "The fire place has very generally gone out of use wherever fuel has become costly and stoves . . . have taken its place."[26]

Many inventors followed in Franklin's wake, with more than 800 patents issued between 1790 and 1845. Cookstoves quickly followed those for heating, for by 1860, the census listed 250 stove makers, and such competition greatly improved quality, drove down prices, and generated a near infinite variety of stoves for heating and cooking. By this time, a few churches, commercial establishments, and houses of the well-to-do also began to install a furnace in the basement, employing convection to supply central heat. According to the National Association of Stove Manufacturers, there were 25,000 stoves made in 1830; 100,000 in 1840; 375,000 in 1850; and a million in 1860. Fitting an equation to these data and solving it for production for each year indicates that there were about 9.4 million stoves produced over the period. Nor did the stoves sell themselves; producers differentiated their products and integrated forward into wholesaling and retailing. By the 1840s, pictorial advertisements began to appear in popular magazines. Assuming stoves lasted 10 years, there would have been about 6.5 million in use in 1860, about 1.3 per white family. While such calculations are at best illustrative, they suggest that stoves must have been ubiquitous on the eve of the Civil War.[27]

Writing years later, in 1882, Nathaniel Hawthorne lamented "this almost universal exchange of the open fireplace for the cheerless box stove." He was saddened that "we now make our fire in an air-tight stove, and supply it with some half a dozen sticks of wood between dawn and nightfall," concluding that "I shall never be reconciled to this enormity." One may infer that Hawthorne did not cut his own wood: elsewhere, he noted that the previous owner of his manse, who used fireplaces, had been supplied with sixty cords of wood a year. Despite these various complaints, it is hard to overstate just how much the invention and widespread use of the stove for heating and cooking increased Americans' welfare. It did not, of course, bring heaven on earth: William Dean Howells recalled that their box stove had burned wood with "inappeasable voracity," and because the building was so porous, it "then did not heat the room." Firing the stove with coal, as Howells probably discovered, although it did nothing for the porosity of the room, at least helped appease the stove's appetite.[28]

Modern figures suggest that Franklin's and Bull's estimates of a stove's comparative efficiency were a bit enthusiastic; stoves are roughly 60 percent efficient (versus 20 percent for a traditional fireplace). Accordingly, stoves required a third as much fuel as did fireplaces to produce a given amount of usable heat. Even as stoves immeasurably improved Americans' lives, they probably reduced domestic energy use per person. Table 1.3 presents estimates of domestic fuel consumption for 1800 and 1860, and although firewood still predominated in rural society, anthracite and bituminous

Table 1.3. Domestic Energy Use, 1800 and 1860

	1800		1860	
Fuels	Domestic	BCE tons	Domestic	BCE tons
Wood (cords)	22,000,000	18,000,000	119,700,000	95,760,000
Anthracite (tons)	—	—	2,730,182	2,648,277
Bituminous (tons)	59,000*	59,000	3,632,745	3,632,745
Whale Oil (bbl.)	2,671	297	11,257	1,251
Mfg. Gas (Mft.^{3})	—	—	(2,191,668)	(273,959)
Kerosene (bbl.)	—	—	(171,429)	(189,993)
Total	—	18,059,297	—	102,042,273
Per capita (tons)	—	3.4	—	3.24

Source: See text and appendix table A2.9.
Note: Figures in parentheses for gas and kerosene are excluded from total and per capita BCE because both are included in bituminous.
*Assumed to be one-half of total production that year.

coal had increasingly entered the picture. But whatever the fuel, much of it must have been burned in stoves, for per capita fuel consumption (BCE) apparently declined from its 1800 levels.[29]

Thus, Americans took the efficiency gains from stoves in two ways: they burned less fuel per person and, following Franklin, made their rooms a lot warmer, for each cord or ton generated perhaps three times the comfort it would have produced two generations earlier. The gains from energy conservation, that is, are not simply energy saved but improvements in the lives of ordinary persons as well. A modern writer argues that this improvement of stoves over fireplaces also eased demands on forests. Moreover, while stoves saved energy, they *encouraged* the shift to coal, thereby improving comfort and reducing demand on forests in another way. The reason was that while stoves burned less wood than did fireplaces, the wood required perhaps twice as much labor to cut it into two-foot instead of four-foot logs, partly offsetting the labor saved from burning less wood. In the 1850s four-foot firewood sold for $1.00 a cord while cutting it to stove length doubled the price. As noted, coal was also a denser fuel than wood, so hauling it required less labor per Btu—an advantage discovered by industrial users as well. Finally, a coal fire in the kitchen stove would last the night, so early risers would no longer have to "shiver and shake with cold," as one writer enthused in 1844. Domestic coal use was thus a labor-saving innovation; coal and stoves were complementary innovations, with each increasing the other's payoff and both improving Americans' welfare.[30]

One important result of the shift from fireplaces to stoves for cooking and heat, largely neglected by historians, must have been an improvement in families' health. Modern research on indoor air pollution in developing countries, where families often use wood or soft coal for fuel and cook on open stoves, finds high levels of particulate pollutants, which are associated with elevated levels of diseases and mortality. For example, studies find levels of particulate matter (PM_{10}) ranging from 300 to 3,000 mg/m^3, whereas the US standard is 150. One study of children in India exposed to such levels of pollution found the level of acute respiratory infection to be 50 percent higher than those not exposed. When Americans cooked and heated with wood or soft coal in open fireplaces, pollution levels must have been similar. Moreover, modern exposures are in warm countries, are often for only three to seven hours a day, and are largely confined to women (while cooking) and to children. In colder areas of the United States, exposures would have been much longer. The shift to stoves burning anthracite must have sharply reduced exposures and, accordingly, diseases—because stoves would yield fewer pollutants for the fuel burned, because they used less fuel, and because anthracite was less smoky than wood. It is impossible to quantify these health gains, but they were surely real and significant.[31]

Fuel for Industry

Anthracite coal played a central role in propelling the early Industrial Revolution in the United States. Fuels perform three functions in industrial processes: they supply heat to process materials; they are a source of power; and they provide the raw materials for other products (coke, gases, dyes, plastics, pesticides, etc.). Process fuel in the United States, until the 1820s, usually meant charcoal, which is nearly pure carbon. It was used in salt making, brick kilns, and the making and working of iron.

Heat for Industrial Processes

As canals and railroads began to open the anthracite fields, iron manufacturers experimented with hard coal as fuel in the forges, rolling mills, and blast furnaces that dotted New York and eastern Pennsylvania. The lure of coal was, as usual, its lower cost compared to charcoal. As I noted above, by 1808 the nail manufacturer Jesse Fell had invented a grate that would burn the recalcitrant fuel, and by the 1820s, as coal became cheaper, forges were rapidly substituting anthracite for charcoal. The process was more difficult at rolling mills, where puddling furnaces melted pig iron (the result of smelting iron ore in a blast furnace) to transform it into wrought iron for rails and other shapes. Puddling used lots of fuel—in 1850, roughly four tons of anthracite per ton of product—so the payoff to substituting hard coal for charcoal was substantial. While early producers encountered difficulties in designing a furnace that would burn anthracite, they seem to have succeeded by about 1840.[32]

The most difficult puzzle was how to substitute coal for charcoal in the blast furnaces that smelted pig iron from ore. Companies experimented with raw bituminous coal and coke—which is soft coal heated with little oxygen to drive off volatile elements, and it, too, is nearly pure carbon. But because of impurities in the fuel, soft coal and coke made poor iron, and by 1841, Walter Johnson claimed no one in Pennsylvania was using coke as fuel any more. One of the pioneering efforts to use anthracite was by the Lehigh Coal Company—suggesting that suppliers were trying to expand their market. It also failed but for a different reason. The traditional cold-blast techniques that blew unheated air through the mix of fuel, limestone, and iron ore in the furnace would not smelt the ore with anthracite as the fuel. By about 1830, however, English experiments using heated air with coke for fuel proved dramatically successful. As reported in the American press, use of the hot blast doubled furnace output because it sped up the process, and the improved combustion cut coke use from about eight tons per ton of pig iron to fewer than three! It thus boosted productivity relative to all inputs. Smelting with anthracite came to the United States about 1836, when Frederick Geissenhainer (1771–1838), a German immigrant and Lutheran clergyman, successfully made pig iron with anthracite using the hot-blast technique. In the late 1830s, the process took hold, apparently with results similar to those in Britain. Thus, the story of the hot blast is another example of the way markets encouraged discovery and diffusion of new knowledge. By 1863, *Scientific American* was reporting that raising blast pressures from four to eight psi more than tripled output, while fuel economy rose modestly during the next several decades as well. Here again the rise of coal reflected not only its abundance but also the ability of entrepreneurs and tinkers to find ways to exploit it. It also shows how conservation can expand fuel use.[33]

It is impossible to know with any precision how much coal went into iron making during the antebellum decades, for little information has survived. But some rough estimates are possible. In 1860, it required about 2.12 tons of coal—roughly 81 percent of which was anthracite—to produce each of the 641,439 tons of pig iron produced with coal. Coal was also employed at other stages in iron making, most importantly in the production of wrought iron via the puddling process. Pennsylvania data for 1849 demonstrate about half the pig iron produced using coal was puddled and that it took about four tons of coal to produce a ton of wrought iron. Hence, a rough estimate of coal used to produce pig and wrought iron in 1860 would be about 2.7 million tons of coal. This figure, which excludes the probably considerable amount of coal used in forges and casting operations, amounted to roughly 19 percent of all hard coal produced that year.[34]

Gradually, the expansion of iron making created a demand for scientific knowledge of the materials and processes involved, resulting in the studies of iron ore and coal noted above. By 1850, textbooks began to appear that covered not only the properties of fuels, ores, and fluxes but the chemistry of iron making as well. The fifth annual report of Pennsylvania's geologic survey, published in 1841, contained a chapter on the chemistry of coal and ore samples. In 1850, the self-styled "mining engineer"

Frederick Overman (1803–1852) published *The Manufacture of Iron*, which contained information on ore composition along with an extensive discussion of various fuels, much of it taken from the work of Walter Johnson. In 1855, producers formed the American Iron Institute; its secretary, J. Peter Lesley, promptly began to gather statistics, publishing *The Iron Manufacturer's Guide* in 1859.[35]

The changeover from charcoal to anthracite provides a glimpse of how coal would shape the future Industrial Revolution, not only for iron but for other fuel-using industries as well. An acre of wood might yield 20 cords if clear-cut—and then nothing for perhaps 50 years. An acre of coal yielded far more fuel and might last for decades. Hence, charcoal furnaces were small, mobile, and rural. As iron making began to employ anthracite, producers could establish larger, more permanent facilities. The industry's geographic center of gravity also shifted toward cheaper transportation along rivers and canals, which provided low-cost access to coal and markets. All of this suggests that anthracite played a central role in the evolution of large, low-cost iron manufacture that would in turn produce the machine tools, steam engines, and locomotives that powered the early Industrial Revolution.[36]

Yet anthracite's domination of iron production would be brief. Transport costs precluded its use west of the Alleghenies. There, despite the availability of cheap bituminous coal, charcoal iron dominated until the Civil War because available coal contained impurities that yielded poor-quality iron. In 1859, however, the Connellsville Pennsylvania coalfields opened, making cheap, high-grade coke from soft coal available for the first time. Iron making, accordingly, migrated west. By the twentieth century, bituminous coal would drive anthracite almost entirely from iron making and, indeed, from most other manufacturing industries as well. But while anthracite was largely confined to Pennsylvania, soft coal was spread widely, being mined in 28 states by 1900 and making for a broader geographic diffusion of industry than reliance on hard coal would have allowed.[37]

Power for Factories

As anthracite became increasingly cheap relative to wood, entrepreneurs also began to use it as a source of power for the new steam engines, and parallel experiments occurred with bituminous coal west of the Alleghenies. Experiments with steam date from the eighteenth century, and Philadelphia was using steam engines fired with Virginia coal to pump water by about 1800. Yet the spread of the steam engine was a slow process, so the growth of coal as a source of power proceeded slowly as well.

In an 1833 United States Treasury report on manufactures (the McLane Report), the Rhode Island textile manufacturer Samuel Slater (1768–1835) observed that "steam [power] has so far been very little used." Although

Slater was referring to his own state, he might just as well have been speaking of the entire country. The later United States Treasury survey of 1838 put the national total of stationary steam engines at 1,860, of which about half had been built after 1835.[38]

Steam spread slowly because it had to compete with waterpower in the East. Not surprisingly, the geographic distribution of steam engines in 1838 indicates that they were disproportionately used in the West, where waterpower was scarce and both wood and coal were cheaper. Yet if the availability of waterpower slowed the growth of steam in the East, using water for power had a number of disadvantages. Before the age of electricity, its power could not be shipped long distances; most sites were small, and all were subject to disruption from droughts and floods. Thus, steam loosened the link between power and factory location that water had forged. Factories driven by steam could now locate in cities close to markets and labor supplies. Water might still play a role in location, but now it was to provide cheap transport of coal to river or tidal cities. About 1834, Cincinnati manufacturers "obtained [coal] cheaply from the Pittsburgh district by great boats." In his 1829 essay on mechanics, Zachariah Allen (1795–1882) explained the economics of fuel choice. Allen was a Brown College–educated tinkerer who was also a lawyer, a mill operator, an insurance entrepreneur, a mechanical engineer, and an inventor who developed a home furnace and made original contributions to steam engines. In Pittsburgh, Allen informed readers, the cost of coal per horsepower-day was a tenth the cost of wood. One unintended byproduct of this was Pittsburgh's famous air pollution.[39]

In the East, the availability of waterpower slowed the spread of steam, but where steam was employed, "coal seems [to] have been used . . . [as fuel] almost from the beginning," one modern writer concluded—no doubt because transport to Eastern Seaboard cities made it cheap, whereas wood was expensive. The English civil engineer David Stevenson (1815–1886) concurred. Writing in 1838 of his American travels, he noted that coal fired the steam engines located in cities because of the scarcity of wood. Thus, anthracite not only smelted the iron that built the steam engines; in eastern factories, it powered them as well. In 1831, two thirds of Philadelphia's steam engines burned coal. The 1833 report of Secretary Louis McLane (1786–1857) on manufactures demonstrated widespread use of bituminous and anthracite coal in forges, blacksmith shops, and mills (probably for space heating); occasionally, it was also used to fire steam engines. As steam and coal spread among eastern manufacturers, they reshaped industrial location there as they did in the West. In New England, cotton mills shifted southeast, toward Fall River, where the first coal-fired steam-powered mill arrived in 1843. By the 1850s, new cotton mills were employing coal-fired steam.[40]

One advantage of steam power was that the exhaust might be employed to heat the mill. In the 1840s, Joseph Nason (1815–1872) and James Walworth (1828–1912) developed a system based on the work of Anglo-American inventor Angier Perkins (1799–1881). The system circulated high-pressure waste steam in small pipes throughout the building and began to be employed not only by textile companies but in public buildings as well. Its high pressures made the Nason & Walworth system unsuitable for domestic purposes, although it seems to have been the model for later, successful domestic central steam heat.[41]

It is impossible to know precisely how much coal these stationary steam engines consumed, but some plausible estimates are possible for manufacturing and mining (see tables A2.6 and A2.7). The best estimates are that there were 700,000 horsepower of steam engines installed in manufacturing in 1860 and 120,000 in mining. These burned about 8.4 pounds of coal per horsepower-hour. If they all burned coal and operated standard work hours and days, they would have generated roughly 2.6 billion horsepower-hours or roughly half of what farm horses generated that year (table A1.2). Doing so would have consumed roughly 11 million tons of fuel in 1860, amounting to half of all coal use in the latter year. Because of breakdowns, bad weather, labor problems, and materials shortages, this is surely an overstatement, and a better estimate might be eight to nine million tons. Still, by the Civil War, coal and factory steam power were becoming inseparable—a marriage that would last about fifty years.[42]

Power for Water Transport

Appropriately enough, in a thinly settled country with vast distances and miserable roads, steam power in America was applied most rapidly to transportation by water. Robert Fulton's (1765–1815) North River Steamboat (*Clermont*) was the first successful such vessel in the United States. Fulton could read and write but otherwise had little schooling. One of his biographers and the leading mechanical engineer of his day was Robert Thurston (1839–1903), who noted that many others at the time were also working on steam navigation. By these others, he meant John Fitch (1743–1798) and Robert Stevens (1787–1856)—two more of the largely uneducated tinkers who made major contributions not only to steamboat technology but (in the case of Stevens) to railroads as well. Hence, Thurston asserted that Fulton "was not in any true sense 'the inventor of the steamboat,'" but he goes on to say that "his services in the work of introducing that miracle of our modern time cannot be overestimated."[43]

With the support and backing of the wealthy Robert Livingston (1746–1813), Fulton began experimenting with steam navigation about 1801. In its maiden voyage in 1807, the *Clermont* accomplished a round trip between New York City and Albany (about 300 miles) in 62 hours at a time when a sailing ship would have taken 48 hours one way. Fulton's voyage inaugurated a steamboat craze. Louis Hunter put the number of boats on western rivers alone at 17 in 1817, while steamboat tonnage in the United States increased from 25,000 in 1823 to 771,000 in 1860.

It did not take long for innovators to apply steam to ocean transport as well. Moses Roberts seems to have been the first; a charter member of the Savannah Steamship Company, he installed a one-cylinder, 90 horsepower engine in the sailing ship *Savannah*—the company's one vessel. The ship carried 75 tons of coal and 25 cords of wood, and the engine sat on deck, driving two side-paddle wheels. In 1819, the *Savannah* became the first transatlantic vessel to employ steam; the steam was just a supplement to the sails, and the voyage marked a very modest beginning. It took 29 days and 11 hours, of which only 90 hours were under steam. The *Savannah*'s fuel consumption amounted to a staggering 24 pounds of BCE per indicated horsepower-hour.

Steam power improved speed, raising the productivity of both capital and labor, and had other advantages, as a poet indicated:

> When our port before us lies,
> Though winds and tides oppose—
> Take off the half-stroke, give her steam,
> And then right in she goes.

But coal took up valuable cargo space, which generated strong incentives for economy. Improvements in ship design, steam engines, and boilers, as well as the gradual substitution of iron for wooden construction and screw propellers for paddle wheels, raised fuel efficiency. Yet until the Civil War, progress was modest: in 1860, sailing ships still constituted four-fifths of America's merchant marine.[44]

A low-pressure steam engine based on the English design of Matthew Boulton and James Watt powered Fulton's boat. These were large, heavy engines and produced little power; high-pressure engines modeled on the design of Oliver Evans would soon push them aside. Like Fulton, Evans was yet another tinkerer with little education, but he had a startlingly original mind. Evans invented a continuous-process flour mill, contributed to vapor-compression refrigeration, and fathered the high-pressure steam engine. These engines were smaller, lighter, and cheaper to build than the Boulton & Watt models. With a greater ratio of power to weight, they were a better fit for steamboats and locomotives. But Evans's engines had no condenser and exhausted steam at high pressure, so they burned prodigious quantities of fuel. The adoption of high-pressure engines instead of the Boulton & Watt design illustrates how cheap coal shaped Americans' technological choices.[45]

Yet Evans also experimented with and advocated the use of steam expansively rather than exhausting it at high pressure. This process cut off steam admission to the cylinder at one-fourth or even one-eighth of the stroke and let it drive the piston by expansion, thereby saving fuel. In 1824, a French engineer conducted a careful engineering survey of American steamboats. Most burned wood and probably did not cut off steam, for by later standards, they were wildly inefficient. The *Maryland*, for example, with a 60-horsepower engine, burned 7.3 cords of wood (equivalent to 5.84 tons of coal) in its nine-hour trip between Baltimore and Easton, Maryland, or about 16 pounds per horsepower-hour. In 1829, one traveler reported that western boats consumed 35 to 45 cords of wood per day, and loading had become so laborious that deck passengers were drafted to help load as part of their fare. In 1848, *Niles' National Register* reported that the *Empire*, plying the Great Lakes between Chicago and Buffalo, devoured 600 cords a trip. There were 16 such steamboats averaging 13 trips a season and consuming about 125,000 cords of wood a year.[46]

Despite claims that western boats were especially profligate in their use of fuel, efficiency was improving. In the 1830s, treatises by engineers and scientists such as Columbia professor James Renwick (1790–1863), Dionysius Lardner (1793–1859), and Frederick Mone (dates unknown) began to appear, explaining the value of using steam expansively. According to Louis Hunter, the leading modern expert, although expansive use of steam spread slowly, most eastern steamboats employed it by 1840, and it seems to have spread in the West as well. Insulation also proved its value; Mone claimed that in one experiment, it reduced fuel consumption by 40 percent. Boat design also mattered. Writing in 1830, Renwick noted that the most recent boats built for the Long Island Sound were longer and thus had improved fuel economy. Moreover, to assess economic efficiency, fuel use must be compared with some measure of output. Modern work by James Mak and Gary Walton calculated that fuel use on the round trip between Louisville and New Orleans *rose* between 1815 and 1860, but because steamboat payloads rose even more, wood use fell from 1.53 to 0.58 cords per ton carried over the period.[47]

As I noted above, initially, much wood fuel came from land clearing and burning and was essentially a form of waste recycling. As a later writer perceptively observed, "The inducement [to farmers] . . . consisted chiefly in the price received for chopping and hauling; the timber itself was not considered of much value." Yet once such wastes near navigable waters were depleted, wood prices had to rise to cover the full costs of clearing and more distant transport. As early as 1828, *Niles' Weekly Register* estimated that New York steamboats burned about 180,000 cords of wood a season and opined that "this vast destruction must . . . exhaust the lands within reach of navigable waters in the course of a few years"; he concluded that "the day is not far distant when they would turn to coal." Such foresight was what inspired the myriad experiments in burning coal even when it was not initially economic. A rough estimate of western steamboat wood consumption toward the end of the antebellum decades is possible. Hunter provides evidence on the number of steamboats on western rivers for 1843, 1857, and 1859 by size classes, allowing estimation of average size. Mak and Walton provide data on fuel consumption per boat-ton per day and the number of days the boats worked. If all boats had burned only wood, the resulting calculation indicates about 1.1 million cords would have been burned in 1843 and 2.7 million in 1857 and 1860 (table A2.3). Thus, steamboat fuel, although it might be locally important, accounted for, at most, 2 percent of the 126 million cords burned in 1860.[48]

There are scattered reports of western boats burning bituminous coal as early as 1812, but with wood cheap and coal not always available, its use spread slowly. By the 1830s, as western coal beds opened, coal use increased on the upper Mississippi and Ohio Rivers, and many boats featured fireboxes that could burn either wood or coal, allowing companies to hedge against price or availability surprises. Writing in 1838, Secretary of the Treasury Levi Woodbury submitted his report on steam engines to Congress. Most steamboats had been built to burn wood, he observed, "but in later years bituminous coal has in many instances been substituted and in several, anthracite coal." In 1838, W. W. Mather (1804–1859) of Ohio's Geological Survey informed the legislature that "many of our boats [on the Ohio River] now use coal." The reasons were not only cost, he claimed, but also because coal took one-ninth the space, one-third the weight, and one-fourth the labor required of wood. As a steamboat captain explained, with wood at $2.50 a cord and coal $1.75 a ton, he could save 30 percent by switching fuels. Anticipating Stanley Jevons's argument that conservation would increase coal use, one writer predicted in 1840 that coal "will lead to the introduction of steam navigation on routes where wood has been too dear to justify the experiment." Hunter claims that "by 1860 it [coal] was well established on western rivers."[49]

While buyers seem to have initiated the shift to coal in the West, anthracite producers took the lead in the East. Lehigh Coal & Navigation Company had been experimenting with coal burning in its canal tow boats since 1826, and in 1831, it purchased the steam vessel *Pennsylvania* and modified its firebox to accept anthracite. Hard coal did not work well in fireboxes designed to burn wood, however. It needed more draft and might be slow to raise steam. Anthracite cut the *Pennsylvania*'s fuel bill in half but took an hour to raise steam. Once again, marketing and innovation came to the rescue, for at about the same time, the Delaware & Hudson Coal Company began to pay some companies to modify boilers, and it supplied free coal to others but with disappointing results. The company finally contracted with Eliphalet Nott (1773–1866), a clergyman, inventor, and stove manufacturer who had designed a new vertical tube boiler with blowers to burn the ornery fuel. A trial run had proved successful in 1835, and in 1836, the company financed the *Novelty*, which was powered by hard coal burned in Nott's boilers and completed a trip from New York to Albany in 12 hours. Philip Hone (1780–1851), former president of the Delaware & Hudson and the mayor of New York City, was on board. The trip consumed about 20 tons of coal, costing $100; he thought burning wood would have required 40 cords at $6.00 each, more than twice the expense. Once again, coal's ability to save labor was a special drawing point; it cut in half the required number of firemen. With technical problems solved, and as wood supplies receded, coal use on eastern rivers and in the coastal trade predictably spread. In 1844, one writer estimated that eastern steamboats burned 150,000 tons of anthracite a year—about 7 percent of that year's hard-coal production—and by the end of the decade, "scarcely anything but anthracite was used" on Hudson River steamboats and those on Long Island Sound.[50]

These early experiments with fuel use owed comparatively little to science and much more to cut-and-try. An exception was the investigation into an unpleasant consequence of fuel use: steam boiler explosions, which began to occur on both steamboats and railroads with the advent of the high-pressure engine. In 1830, when the steamboat *Helen McGregor* blew up at the dock in Memphis, Tennessee, killing more than forty people, it galvanized the Franklin Institute to investigate boiler explosions, and the United States Treasury soon subsidized these efforts. The study showed that the fundamental cause was simply the development of too much pressure for the boiler, and it demonstrated that overheating iron would weaken it, which revealed the dangers of boiler encrustation or low water. The study also refuted such popular theories as the claim that water under pressure decomposed into hydrogen and oxygen. These findings did not end the problem of boiler explosions on either steamboats or railroads, but they did focus efforts on the need to wash boilers and maintain water levels. They also provided a glimpse into the important role that science might play in the quest for safety and efficiency.[51]

Power for Land Transport

Before the railroad, land transport was horse-drawn or you walked ("shanks' mare"), and local movement of people and goods remained that way into the twentieth century. Although there were more than seven million horses and mules on farms in 1869 (table A1.2), fragmentary data suggest that there were also more than one million nonfarm-work animals by that date (table A1.3). Some horses and mules supplied factory power or drove cordwood saws or hauled mine wagons; probably most were dray horses, moving goods such as coal in urban markets. For heavy work, they might supply four hours a day, with more for lighter duties. If they averaged four hours a day, six days a week, they provided 1.5 billion horsepower-hours in 1860. The food energy to provide such work, I calculate (table A1.3) to be roughly equivalent to 1.5 million tons of coal, which was

7.5 percent of that year's production and amounted to about two pounds per horsepower-hour, which was far more efficient than the steam engines of that day.[52]

Yet however efficient a horse might be, it could not deliver the sustained horsepower-hours of the iron horse. As I have noted, the Delaware & Hudson Canal Co. introduced the railroad to America, and naturally enough, the fuel was anthracite. The next year, the Baltimore & Ohio's (B&O) Tom Thumb also burned hard coal on its maiden run. As with steamboats, however, coal use by railroads spread slowly and for the same reasons: wood was cheap, and anthracite was difficult to light, burned out fireboxes, and might not provide steam in a hurry. Soft coal was easier to burn, but it, too, shortened the life of fireboxes and had other drawbacks as well. Although perhaps no worse than a wood burner, which according to *Scientific American* "scatters fire and smoke broad-cast on all behind," soft-coal firing also generated clouds of smoke and cinders that might burn passengers' clothes and farmers' barns.[53]

The carriers' gradual discovery of the best way to employ the new fuels reveals a kind of collective learning with much experimentation by "mechanics" with little or no formal training in engineering or science. Yet in the case of anthracite, while many railroads contributed to an understanding of the new fuel, the pioneering role of the Philadelphia & Reading stands out, perhaps because it had access to hard coal and was a major carrier of anthracite as well. In 1847, that company estimated that if it could shift from wood to anthracite in locomotives, it would save $125,000 a year. The company emphasized that hard coal also saved space compared to wood and allowed longer runs with fewer stops. There were seemingly endless experiments: companies built locomotives with vertical boilers, blowers to add draft, and improved grates. By 1849, as a report to the Reading by the engineer George Whistler Jr. (1822–1869) made clear, while problems remained, anthracite was economic. Whistler estimated that the Baltimore, an anthracite-fired locomotive, could pull a train of 800 tons at a cost of about $64 less per round trip than if it had been fired by wood. Extra repairs from burning hard coal amounted to $456 a year, which would be made up in seven trips. Innovations continued, however, the most successful being that of James Millholland (1812–1875), a master mechanic of the Reading in the 1850s who produced locomotives with long, wide fireboxes that proved more satisfactory. Experimentation continued well after the Civil War.[54]

Bituminous coal burned more readily than anthracite; its main drawback was cost. Accordingly, innovations in woodburning probably retarded its spread. Alexander Holley (1832–1882) was a Brown University graduate, one of the most distinguished railroad engineers of his day and a founder of the American Society of Mechanical Engineers (ASME). He reported in 1861 that improved firebox grates on New York Central wood burners had increased efficiency from 35 to 47 miles per cord. Still, Holley thought coal was likely to save money. He recounted wood's drawbacks: more frequent stops, decay, greater fire dangers. And there was theft: with numerous, inevitably unguarded, woodsheds, one company estimated that a quarter of the population within half a mile of the shed burned company wood. Moreover, "wood is constantly becoming more scarce and expensive."[55]

Yet use of soft coal also spread slowly. As occurred with use of anthracite, inventors generated a profusion of modifications in grates, boilers, and fireboxes that improved combustion but were expensive. Writing in 1861, Holley estimated that it would cost $2,000 to $3,000 each to convert wood burners to soft coal. Paradoxically, the uncertainty that this profusion of options must have generated may have slowed conversion to coal, as companies adopted a "wait and see" approach. The *American Railway Times* caught this problem. In 1858 and 1859, the journal estimated that of the 6,000 locomotives in the country, "full one half of these would find it more economical to burn coal if the best, simplest and most economical plan could be determined." Yet the *Times* estimated that no more than 400 did so (Albert Fishlow puts the figure at 500). Experiments continued, however, and gradually, companies shifted. The Baltimore & Ohio began experiments with soft coal in 1837 and by 1845 claimed to be "burning the Cumberland [bituminous] coal in the most satisfactory manner." When it found a source of low-sulfur coal in 1856, the Illinois Central made the shift to avoid the "onerous outlay . . . for fuel" and cut such costs by three-fourths. Railroads' choices, of course, were not simply wood versus coal or anthracite versus bituminous, for all fuels were themselves highly variable. Accordingly, companies learned from the work of others. George Vose (1831–1910) was a distinguished civil engineer, and his *Handbook of Railroad Construction* (1857) published information on fuel qualities obtained from various railroads, as well as from the works of Walter Johnson and Frederick Overman. In 1859, the Chicago Burlington & Quincy reported that 25 of its 62 locomotives burned coal with more to follow as they cut fuel bills 47 percent.[56]

Holley noted in 1861 that the previous year had witnessed a "general rush" in New England and the Old Northwest (Michigan, Illinois, and contiguous states) to substitute coal for wood. He attributed the rush to rising wood prices and falling coal prices. In addition, George Griggs (1805–1870), mechanical superintendent of the Boston & Providence Railroad, had invented a brick arch that could be retrofitted into woodburning locomotive boilers at very little cost. The arch greatly improved combustion, thereby reducing smoke (nothing would eliminate it altogether, as engineers reluctantly concluded). On Griggs's railroad, it cut fuel costs from $0.35 per mile while burning wood to $0.08 when burning coal, which was surely part of the explanation for the New England rush. Experiments by the Pennsylvania Railroad in the 1850s had established that coal was a more economical fuel than wood and tested various coals to see which was best. By 1864, the Pennsylvania Railroad explained to stockholders that the brick arch "converts the old wood burners into our best coal burners" at a cost of about $50 per locomotive, and it was implementing them as quickly as possible. That year, it burned 151,000 tons of coal in locomotives compared to 53,000 cords of wood.[57]

As the above discussion suggests, American railroads were not large consumers of coal during antebellum times. Albert Fishlow has calculated that the 500 woodburning locomotives in 1859 consumed about 2.16 million cords (1.7 million tons BCE) of the roughly 126 million cords of wood burned that year. In contrast, the estimated 500 coal burners of 1859 probably averaged 60 pounds of coal to the mile and worked 20,000 miles a year. If so, they burned about 300,000 tons, or roughly 2 percent of the more than 14 million tons of hard and soft coal mined that year. The energy consumed by urban horses nearly equaled that of railroads, while the manufacture of gas and kerosene consumed far more coal in 1860. Thus, coal's importance as a railway fuel lay in the future, for during antebellum times, the carriers were far less important as coal consumers than they were in increasing demand from other users by lowering its delivered cost.[58]

Switching away from wood when the technology and economics of coal became favorable was one way antebellum railroads worked to reduce energy costs, but the American railroad press reveals considerable interest in raising fuel economy as well. The railroad journals printed many reports of English efforts to increase fuel economy and provided much discussion of locomotives' technical efficiency, along with the need for better roadbeds and more careful work by the firemen, but there is little evidence of any scientific investigation into locomotive efficiency before the Civil War.

Economic efficiency did improve, however. Locomotives became bigger, and companies increased freight-car size and train length, both of which decreased fuel use per ton or passenger mile. Although he had no formal education, Zerah Colburn (1832–1870) became a distinguished engineer, locomotive designer, and technical journalist, earning a Telford Medal from the British Institution of Civil Engineers. In 1854, he urged New England railroads to consider the fuel and other economies resulting from larger engines. By then, however, Colburn was preaching to the choir. *Mining Magazine* claimed that, adjusted for inflation, the cost of carrying coal had fallen to half from 1843 to 1853, mostly because of "the doubling of the capacity of the engines." Albert Fishlow's figures indicate that railroad output (passenger and ton-miles) per cord of wood rose about 58 percent between 1839 and 1859.[59]

Fuel for Light

After about 1800, coal began to provide light, as well as heat and power.[60] While lighting via gas never became a major use for coal, its importance is twofold: it yielded an enormous advance in human well-being, and it provided an early glimpse of the value of chemical engineering and the range of products that derived from carbon-based fuels.[61]

Previously, Americans had burned pine knots or fish oil, as well as lard oil and whale oil, in lamps or candles, but none of these resources provided much light; they were also expensive, some were dangerous, and fish oil stank. The discovery that coal might be distilled to give off an inflammable gas dated from the seventeenth century and was another European innovation that Americans borrowed. By the 1790s, English entrepreneurs were attempting to commercialize the process. In 1802, William Murdock (1754–1839), an engineer with Boulton & Watt, installed gas lights in a foundry, and cotton mills soon began to build individual gas plants to replace their oil lamps. Not only were gas lights safer and cheaper; they were brighter as well. Early experiments indicated that, depending on the fixture burning it, gas might generate five to seven times as much light as candles. In 1811, a British writer termed gas "one of the greatest improvements of which modern times can boast." Others followed in Murdock's footsteps. Samuel Clegg (1781–1861), also of Boulton & Watt, and the German immigrant Frederick Winsor (1763–1830) developed and perfected the process of distillation and the cleansing of the gas. They also developed lamps, meters, and retorts for the new technology. In 1812, they established the London & Westminster Gas Light and Coke Co., and by 1815 they were laying gas main—made from recycled musket barrels—in London streets.[62]

American Beginnings

Americans were also experimenting with gas light in the early nineteenth century. In 1813, a tinker named David Melville (1773–1856) began to install a "pit-coal" gas-light system in a Rhode Island cotton mill. A visitor who saw Melville's system described "its superiority in every respect to any other artificial light." About the same time, Rubens Peale (1784–1865) introduced gas lighting derived from coal in the family's Philadelphia museum established by his father, Charles Wilson Peale (1741–1827). The motive, apparently, was not only to achieve better light but also to increase admissions; the Peales hoped the new technology might draw in the curious. Peale and his brother Rembrandt (1778–1860) had seen gas lighting on a trip to England in 1802 and could easily have kept abreast of English developments, which were regularly reported in the American press. Indeed, the Peales provide the first of several instances of Americans borrowing from British gas technology. The brothers also spread the new technology; in 1817, the *Boston Weekly Magazine* reported that a local manufacturer and artist—"Mr. Beath"—had visited the Philadelphia museum and come back to introduce gas lighting in his shop. In addition, Rembrandt Peale was instrumental in establishing the first commercial-gas establishment in America when the Gas Light Company of Baltimore opened its doors in 1816.[63]

There is precious little statistical information on the spread of gas lighting during antebellum times. Boston and New York followed Baltimore in 1822 and 1823, respectively. They probably also employed "mechanics" who borrowed from the British or the Peales or Baltimore. For potentially large users, such as factories and hospitals and a few of the wealthy, there were also small gas generators that could use rosin, wood, or coal for fuel until they were replaced by gasoline.

Everything about gas production was an exercise in chemical engineering. Producers needed to know the chemical structure of their fuel, the definition and measurement of candlepower, the proper loading and firing of retorts, methods of gas purification and testing, and much more. The *American Gas Light Journal* (AGLJ) was founded in 1859; it was one of the first technical publications to specialize in energy matters, and such journals played an important role in making and perfecting energy markets. The AGLJ routinely reported on and provided informed criticism of industry developments and in its first issue announced, "We wish to take every periodical newspaper published in Europe (in whatever language) on the subject of gas lighting." It routinely translated important articles from French or German. Its pages included advertisements by sellers of coal, engineering firms that constructed gasworks, and offers of the specialized equipment needed, such as meters, retorts, and exhausters. There were also listings for consulting chemists and engineers.[64]

By this time, some coal companies had begun to advertise the chemical composition of their coal, perhaps because most large gas companies seem to have employed self-described engineers and occasionally consulting chemists. Benjamin Silliman Jr. and the geologist Charles T. Jackson (1805–1880) consulted for the New Haven and Boston gas companies, respectively. The AGLJ did note, however, that smaller companies might be run by "common laborers" and even, in one case, an "intelligent Irishman." The AGLJ boasted its own coterie of scientific consultants, which included such well-known scientists as John Torrey (1796–1873), Wolcott Gibbs (1822–1908), James Renwick, and Joseph Henry (1798–1878). It also published much European research and technical information in articles such as "Practical Management of Gas Works."[65]

Initially at least, gas manufacturing had quality-control problems. If a company used high-sulfur coal and improperly filtered the gas, it might smell when burned, and the fumes could tarnish metal and destroy paintings and fabric. By the 1850s, these problems were being solved. Yields of gas per ton of coal seem to have risen as the importance of coal chemistry became clear and as companies replaced iron with clay retorts. These were a British invention of the 1820s that allowed higher heat, thereby raising yields. By this time, too, market incentives ensured that most plants were recovering and using or selling some of the various byproducts of gas manu-facture. Distillation of a ton of coal might yield as much as 1,350 pounds of coke, some of which could be burned as fuel and the rest sold. There was also tar, creosote, ammonia, and sulfur. Around 1856, the Englishman William Perkin (1838–1907), while working with coal-tar, accidently dis-covered mauveine, the first artificial dye. For some companies, revenue from byproducts constituted an important source of income. For exam-ple, in 1856, sales of tar, ammonia, and coke covered about 44 percent of the cost of coal to the gas companies of New York City.[66]

Byproduct recovery was far from perfect, however. The first method for removing sulfur used a wet-lime process that was effective but generated highly polluting wastes with an appalling smell. In 1857, the *Medical Times* reported on the "clouds of smoke" and the "deposition of a number of nau-seous and poisonous substances . . . sent forth into the atmosphere." Although new methods of removing sulfur soon appeared, they were slow to arrive in New York City, where the stench lingered on. In the late 1860s, the Board of Health, led by the distinguished chemist Charles F. Chandler (1836–1925), sued to require the new processes. Chandler, then at Colum-bia College, had worked for a gas company, had graduated from Yale's Sheffield School, and had founded the *American Chemist*. He went on to become president of the American Chemical Society and contributed to the development of safe kerosene.[67]

The spread of gas lighting reflected the gradual rise of an urban middle class. Census data indicate that by 1860, America had 392 "urban places" (towns with at least 2,500 inhabitants) containing about six million people—sharp increases from previous decades—while real per capita GDP was about 73 percent higher than it had been in 1800. An 1859 survey by the AGLJ showed that at least 237 cities had commercial gas-lighting plants, only nine of them built before 1840, 31 in the decade of the 1840s, and 172 in the 1850s. There were also an unknown number of private gas manu-facturers. Initially, most systems were small: Baltimore had no domestic customers until 1836, while the median number of customers in the 1859 survey was 254, and the total that year for all plants surveyed was 211,502. The slow spread of gas reflected its economics; although on a per-unit basis it was often cheaper than other light sources (table 1.4), customers faced substantial up-front costs that discouraged small users and those with modest incomes. The house needed to be piped at customer expense, which likely left out renters; moreover, before meters began to arrive in the 1830s, companies sold gas by type of fixture. The cheapest offered by the Balti-more Gas Company cost $12 a quarter. In 1850, when craftsmen might

Table 1.4. The Cost of Light in 1860

Illuminating material	Cost per unit ($)	Duration (hours)	Cost per hour (¢)	Candle-power	Cost per candlepower-hour (¢)
Star candle (lbs.)	0.25	44.03	0.688	1.00	0.69
Tallow candle (lbs.)	0.15	49.73	0.324	0.25	1.29
Coal oil (gal.)	1.00	59.33	1.69	14.00	0.12
Lard oil (gal)	0.30	40.60	2.96	9.00	0.33
Mfg. gas (Mft.³)	3.00	200*	0.015	14.00	0.10

Source: Charles Wetherill, "On the Relative Cost of Illumination in Lafayette, IN," *American Gas Light Journal* 11 (May 1, 1860): 230; and author's calculations.
*Gas burning at 5 ft.³/hr. for 200 hrs. = 1,000 ft.³ and costs $3.00.

make about $1.40 a day and clerks $42.00 a month, such expenses, alongside the high cost of gas, confined its use to the well-to-do throughout the antebellum decades. Later, when meters arrived, companies rented them. Moreover, gas manufacturing faced both economics of scale *and* high fixed costs—meaning that small cities with small plants and a small customer base were inherently high-cost and therefore high-price. Such a cost structure also ensured that companies simply would not pipe streets where either poverty or paucity of residents led them to expect few sales. Yet even in large cities, managers seem to have been slow to learn that price reductions, because they stimulated sales, might actually reduce unit costs.[68]

Still, by the Civil War, for those who could afford the initial installation cost, gas was clearly economic, as table 1.4 demonstrates. Its customer-weighted average price in the 1859 survey was $3.00/1,000 cubic feet (hereafter ft.³), and five cubic feet typically yielded 14 candlepower of light per hour making the cost a tenth of a cent per candlepower-hour. This was not always true, of course; some cities sold gas at $7.00/1,000 ft.³, which made it less economic than coal oil (kerosene).

A Lighting Miracle

But for those who installed it, gas was—quite simply—a lighting miracle. It cost far less than candles, gave better light, required much less labor, and was far safer. Consider the family of a clerk who might make $600 a year in 1860. Had the family budgeted $30 (5 percent of its income) for light that year and bought best-quality star candles (see table 1.4), it would have had 4,347 ($30/0.0069) candlepower-hours of light (in modern terms 54,642 lumen-hours)—equivalent to burning a 40-watt bulb about 118 hours, or 20 minutes a day. Switching to gas, the clerk could purchase 10,000 ft.³ of 14 candlepower gas for $30, which when burned at 5 ft.³/hr. would yield 2,000 hours of light or 28,000 candlepower-hours (about 351,960 lumen-hours): seven times as much light for the same money.[69]

The improvement in welfare from such a shift to gas lighting—like those that resulted from heating with coal in stoves rather than wood in fireplaces—are hard for us to grasp; nor are such gains in human welfare well captured in the statistics of GDP. But to the men and women of 1860, they must have seemed wondrous indeed.

A picture of gas lighting at the end of the antebellum era thus reveals that it remained an infant industry. The 211,502 customers of the survey amounted to about 5 percent of the 4.6 million white families in 1860 (and roughly 8.1 percent of the 2.8 million nonfarm households). But since many customers may well have been business firms, that is surely an over-estimate of the number of domestic customers that year. I estimate that gas consumption that year amounted to 2.7 billion cubic feet of gas, of which about 2.2 billion went to private customers and the rest to munici-palities (see tables 1.3 and A2.8). Some municipal gas plants had initially distilled wood or rosin or had imported English cannel coal, but by late antebellum times, most had converted to domestic bituminous coal. Frag-mentary data suggest that the yield at this time was about 8,000 ft.^{3}/ton, implying that gas production required about 343,544 tons of BCE.[70]

Gas was not the only fuel that might derive from coal. In 1846, the Canadian Abraham Gesner (1797–1854)—a geologist, chemist, and entrepreneur—first produced what he called kerosene from Nova Scotia asphalt deposits. He then came to the United States and in about 1853 ap-plied for a patent. In another example of nearly simultaneous discovery, the Scotsman James Young (1811–1883) had made a similar discovery in 1852, ensuring that kerosene would become embroiled in patent wars. Gesner also set up kerosene manufacturing works and found that bring-ing kerosene to market was altogether different from producing it in a lab; Gesner undertook many experiments to coax the maximum amount of oil from coal and to purify the resulting product. Competitors soon fol-lowed, and they, too, demonstrated the increasing value of scientific analysis in industry. One of these, the Breckenridge Cannel Coal Com-pany, relied on the analysis of Benjamin Silliman Jr. for its choice of coal to refine. Kerosene was as cheap as gas, was available where gas was not, and had no up-front costs. By 1860, perhaps a million families burned coal oil for light, consuming the energy equivalent of nearly 190,000 tons of coal (table 1.3), and they, too, experienced a lighting miracle, getting nearly six times more light for their money than they had with candles (table 1.4).[71]

But while the term "coal oil" lingered for decades, it would soon be-come a misnomer as petroleum quickly emerged as its source. In retro-spect, this becomes a central moment in energy history. Chemists could produce liquid and gaseous fuels from coal, and experiments with vari-ous processes would proceed well into the post–World War II years. But none of the processes yielded a product competitive with that derived from petroleum or natural gas, and as discoveries of these fuels eventually pro-liferated, coal would hemorrhage sales.

Conclusion

In 1860, when the bonds of union were beginning to fray, and even in 1865, when quiet descended on Appomattox, the coal age remained in its infancy. Americans still obtained far more energy from wood than from coal, while horses and mules provided more power than did coal-fired steam engines in manufacturing and mining. Railroads mostly burned wood, and it dominated fuel choices for homes and businesses. Led by anthracite, however, coal was thrusting.

While hard coal was abundant in 1800, it required three key developments to bring it to eastern markets. The rising relative price of wood was one; the second was entrepreneurs and tinkers and salesmen who—motivated by these price developments—innovated ways to burn the new fuel and marketed it aggressively. The third development was transportation innovation, as canals and railroads lowered its delivery cost to eastern consumers. These were very largely the product of the private energy market with modest assists from governments, which augmented markets by providing education and information. Yet exploitation brought conservation as well, for markets also rewarded efficiency. Initially, woodburning was simply the recycling of farm wastes; later, stoves conserved fuel. If high-pressure steam engines were profligate of fuel, varying the cutoff saved it. The hot blast dramatically reduced energy requirements in the production of pig iron, while long trains improved energy efficiency and gas manufacturers saved byproducts.

A characteristic feature of economic development has been the rising share of energy devoted to power, and this was true for antebellum America. In 1800, the BCE of horses and mules amounted to about 8 percent of total energy consumption. Water and wind power are unmeasured, but adding them might raise the share to perhaps 15 percent. By 1860, power from these sources, along with wood and the increasing use of coal for power in factories and mines, as well as on steamboats and railroads, amounted to about 31 percent of energy use.

As many historians have noted, the shift to newer forms of energy was not an unalloyed blessing, as citizens of smoky Pittsburgh could have attested. Accordingly, it is useful to imagine how else the antebellum world would have differed if there had been no coal or if it had been very expensive. Wood would have become much more expensive, driving up costs of all goods and burdening especially the urban poor, whose dwellings would have been more cramped to save on heat. Deforestation would have proceeded further, and wood plantations would have arisen near cities, which would have been smaller owing to the high cost of transporting fuel and other goods—for railroads and steamboats would have burned wood, too. One can speculate further, but it is clear that these energy transitions to coal brought improvements in the standard of living for everyone, improvements that are not adequately captured by such measures as real per capita GDP. In particular, the widespread use of coal stoves for cooking and heating and the cheapening and improvement of artificial light enhanced well-being in ways that must have seemed miraculous to the men and women of 1860 who remembered a world of smoky fireplaces, wooden logs, and the dim lighting of candles.

2

The Age of Bituminous Coal,
1860–World War I

Old King Coal was a merry old soul;
"I'll move the world," quoth he;
"My country's high and rich and great,
But greater she shall be."

. . . .

. . . .

"And it's dig," he said, "in the deep, deep earth,
You'll find my treasures better worth
Than mines of Indian gold!"

—American Farmer's Magazine, *1859*

By the use of these waste gasses it is estimated that
there has been a saving of about 600,000 tons of coal
in the production of a million tons of pig iron.

—Manufacturer and Builder, *1878*

As we have seen, on the eve of the Civil War, coal was by no means the
preponderant source of energy and power in the United States. Yet in five
decades—from Bull Run to World War I—coal came to dominate Americans' energy choices, much as wood had a century earlier, accounting for
roughly 71 percent of all energy use in 1910 (table A2.2). Production, which
had amounted to about 20 million tons in 1860, rose to 678 million in 1918
(table A1.4), while per capita consumption of coal, which had been near
zero in 1800 and about half a ton in 1860, rose to approximately six tons
during World War I. Neither gas nor oil were serious competitors during
these years. Gas was a regional fuel, whereas until the twentieth century,
petroleum production was small and its major product kerosene. Only on
western railroads and in rural kitchens did it contest coal for markets.

Use of all forms of energy per person rose during these decades as well,
reflecting the rise in living standards and the full flowering of the Industrial Revolution. In 1860, the United Kingdom was the largest economy
in the world, with the highest standard of living, although the United States
was a close second. By 1910, the United States had the world's highest living standard, and the American economy was larger than that of Britain
and Germany combined. American steel production, which had lagged
behind that in Britain as late as 1880, exceeded the combined production
of Britain and Germany by nearly 40 percent. As the men and women of
Teddy Roosevelt's time understood, coal was supplying the energy that

drove this enormous expansion, accounting for about 92 percent of American manufacturing energy use in 1910 (table A3.2).[1]

These decades also saw the increasing importance of technically trained engineers and scientists in nearly all aspects of production. The inspired tinkers did not disappear, but they no longer held center stage. By the 1880s, large users of coal, such as gas companies and railroads, were routinely performing chemical analyses of coal and other purchases, creating an expanding market for individuals with technical training, which encouraged the development of academic programs in applied science and engineering. By 1892, *Engineering News* concluded that 94 schools were conferring engineering degrees; a later survey found that the number of students in engineering had risen from about 3,000 in 1889 to nearly 9,700 by 1899. The American Society of Civil Engineers had been founded in 1852, the American Institute of Mining Engineers in 1871, the American Society of Mechanical Engineers in 1880, and the American Institute of Electrical Engineers in 1884. The transactions of these organizations, as well as those of trade organizations such as the National Electric Light Association (founded in 1885), provided a forum for the testing and diffusion of new ideas. Much of their content focused on using and saving energy. Entrepreneurship extended to publishing as well. Specialty journals such as the *American Gas Light Journal*, founded in 1859, along with newer entrants such as *Power* (1880), *Colliery Engineer* (1881), and *Gas Age* (1883), performed similar functions, while their advertising brought buyers and sellers together. These organizations and their members midwifed the transition to the age of coal from an age of wood.[2]

Market forces largely drove these events, although public policies supported the rapid exploitation of coal and other resources. Expanding energy consumption contributed to the rise in the standard of living: real GDP per head increased about 2.6 times from 1860 to 1918 or 1.63 percent a year. GDP is an imperfect measure of welfare, yet if rising energy use worsened pollution in some cities, it also contributed to a dramatic rise in life expectancy. While national figures are unavailable, data from Massachusetts reveal about a 16-year increase during these years. Use of more energy, more efficiently, contributed to these gains and to increases in comfort that are simply not captured in statistics. And while these were the decades when Americans began to worry about prodigal waste of resources, conservation also remained central to these early developments.[3]

We begin this chapter with the expansion and industrialization of mining. Section 2 focuses on steamships and railroads as a major source of coal demand and the central technology that widened geographic markets. The increasing demand for soft coal, the dominant source of factory fuel and power, and the gas-light era are the focus of sections 3 and 4. The position of coal as a domestic fuel in 1910 follows. The last section reviews the status of coal at its peak, about World War I, and briefly sketches the role that war played in reshaping fuel markets during the postwar decades.

The Industrialization of Mining

As we saw in chapter 1, anthracite was a high-cost fuel owing in part to geography because its location (only in eastern Pennsylvania) ensured that transport costs restricted its market. Geology also made hard coal expensive to mine. Seams were thin, and beds were deep, pitching, and faulted, while mining methods remained largely unmechanized until the twentieth century. Beginning in the 1870s, Pennsylvania began to document the human cost of hard coal: in those early years, annual fatality rates often exceeded five per 1,000 employees—a staggeringly high figure that gradually declined only well into the twentieth century.

> Think that each miner's fate
> Leaves a home desolate
> Then you may estimate
> How much the cost is.

Such fatality rates also contributed to the demand for improved technical education. Pennsylvania's mine law of 1885 (P.L. 218) required anthracite mine foremen to pass a state examination. In response, *Colliery Engineer*, which had supported the requirement, began assisting miners studying for the exam. This evolved into a full course and ultimately into the International Correspondence School.[4]

Companies were employing steam power at anthracite mines by the 1850s for dewatering and hoisting; in 1870, the first steam engine arrived for underground haulage, and electric trolleys arrived in 1887, followed by compressed-air-driven locomotives in 1895. In response to market pressures, use of breakers spread to size and clean the coal. Between 1860 and 1900, production of hard coal rose about 4.2 percent a year (table A1.4). Increased output resulted largely from rising employment as the number of workers rose from about 53,000 in 1870 to 169,497 in 1910, and despite the increased mechanization, daily output per worker grew slowly—about 0.76 percent per year from 1890–1918.[5]

As I noted in the previous chapter, early mining had largely been a cut-and-try affair involving little scientific or engineering knowledge; accordingly, by later standards, it was sometimes wildly inefficient, leaving much coal in the ground. The most visible wastes resulted from blasting and handling, which produced vast quantities of dust ("culm") that was unmarketable, and the breakers that sized anthracite for domestic consumers added to the piles. But this waste was simply gold that had yet to be discovered, and by the 1870s, market incentives encouraged many attempts to burn coal waste directly or process it into briquettes. By then, Europeans

who faced higher coal costs had been burning briquettes made from compressed coal wastes for decades, but coal was cheaper here, and despite
tests showing briquettes' heat content superior to coal, no one found gold
in American briquette making until the twentieth century—and not much
of it then.[6]

Waste of anthracite seemed especially worrisome, for when combined
with the limited reserves of that fuel, it raised concerns over the possible
exhaustion of supplies. In the late 1870s, the Pennsylvania legislature directed the state's geological survey to report on the problem, which it did
in 1881 and again in 1883 and 1893. The geologist Franklin Platt authored
the 1881 report, which did, indeed, find much evidence of incompetent
mining practice and noted the mountains of waste from breakers that generated unmarketable fine coal. By the 1890s, some companies were pumping it back into mines to hold up roofs that were being crushed as a result
of bad mining practice. But Platt also noted that "the question of waste is
to a large extent dependent upon the cost of labor, timber, transportation
and the price of coal." He explained that while current practice left thin
seams of coal, "without doubt [seams] as small as 15 inches will be worked
here to a profit . . . after the larger ones have been exhausted as to make
coal higher in price." Writing about the same time, the mining engineer
Charles Ashburner (1854–1889)—a graduate of the University of Pennsylvania and a member of the Pennsylvania geological survey—also noted
the enormous losses of earlier days. But he claimed that "better and more
economical methods" had recently been introduced. The first meeting of
the American Institute of Mining Engineers (AIME) was in 1871 and included a discussion of wastes in mining. Coal conservation, thus, had been
a concern of state and private market actors long before Progressives discovered the matter.[7]

Moreover, much of the waste reflected economics. Market prices provided important economic information, and they were signaling that it was
in no one's interest to mine poor coal—a lesson later Progressives would
find hard to swallow. Waste also reflected the usual mistakes of pioneering entrepreneurs, and market forces were working to reduce it. While
little came of these investigations at the time, they provide another example of Americans' concerns with energy efficiency and foreshadowed later
worries over resource shortages and successful conservation efforts.

By the 1870s, railroad cartels had come to dominate anthracite production; some marketed through affiliated sales agencies while others
employed wholesalers, but in neither case was consumer advertising common. Data on hard coal mine prices exist from 1880 on; relative to the
price of soft coal, these show no trend before about 1900, so supply and
demand must have increased roughly in tandem. A simple relationship is

$$\%\Delta Consumption = \%\Delta Population + y\%\Delta Income$$

Thus, the rise in consumption resulted because population increased about 2.1 percent annually, while per capita GDP increased about 1.66 percent, implying that y = 1.27. Anthracite was therefore a luxury good (y > 1) in the late nineteenth century, as rising living standards led to more than proportional increases in consumption. Yet around 1900, the price of anthracite began to rise relative to the price of soft coal, and its growth slowed to 3.3 percent a year from that year to a peak in 1917 as competition began to erode markets for hard coal.[8]

The price rise resulted in part from the great anthracite strike of 1902. There had, of course, been strikes in both soft and hard coal before, but they had typically been short or involved only a subsector of the industry. The 1902 strike dragged on for six months (from May 12 to October 23). Historians have emphasized its political importance: for the first time, the federal government—represented by President Roosevelt—asserted a public interest in labor disputes. Yet these same historians have largely ignored the impact of the strike on the long-term market for hard coal: coming in summer and fall, when companies and households normally built up inventories, it resulted in price spikes, shortages, and panic buying well into the winter. In Boston, retail prices, which usually fluctuated around $5.00 a ton, had risen to $8.00 by July, and papers were warning of a coal famine. The *Boston Globe* reported that consumers, in shifting to bituminous coal, were "learning that they could do without anthracite" and that some markets were "permanently lost." The resulting pollution, as households shifted to bituminous coal, was termed "soft coal eye." The *Chicago Tribune* reported a "big demand" for oil stoves. The *New York Herald Tribune* announced that "thousands of New Yorkers have been moved by the coal famine to use gas entirely for cooking and other domestic purposes for which anthracite has been used," and it observed that the "fruits of the anthracite strike [are] likely to be lasting." Similarly, *Scientific American* observed that many who used the new fuels "are certain to continue their use." Anthracite, in short, was sowing the wind.[9]

Compared to anthracite, soft coal seemed to be inexhaustible. In 1860, 23 states mined soft coal, and the list grew to 31 by 1890. There had been 335 bituminous mines in 1860, but with the iron industry booming and the railroads about to shift to coal, mining must have seemed like a good business to enter. "We may confidently anticipate . . . an unprecedented increase in the coal trade," one writer for the census observed in 1860; others apparently agreed, for there were 1,335 establishments engaged in mining bituminous coal in 1870, 2,943 in 1880, and 5,652 in 1902. By 1918, there were more than 8,000 commercial mines and thousands more "wagon mines" that the USGS derisively termed "country banks," as farmers short of money might go to a coal outcrop on their land from time to time and "mine a little cash." Employment in mining similarly skyrocketed—from about 53,000 in 1870 to 615,305 in 1918.[10]

Mining soft coal—like mining anthracite—was labor-intensive: the 1880 census indicates that wages accounted for more than 60 percent of the value of bituminous output that year, which must have focused companies on mechanization. As I have noted, steam power invaded mining before the Civil War, and soft-coal companies (as they had done with anthracite) began to employ steam locomotives for underground haulage. Although bituminous coal producers also experimented with compressed air for haulage, only in the late 1880s, with the advent of electric trolley haulage, did mules face serious competition. A compressed-air-powered coal cutter appeared in 1877, and electric cutting machines appeared in 1885 (fig. 2.1). The first USGS data for 1891 indicate that about 5 percent of soft coal was then mechanically undercut; by 1896, the figure had leapt to 12 percent, and by 1918, it was nearly 57 percent. Still, there is little evidence that labor productivity rose much until about 1890, so until then, the growth in output simply reflected more mines and workers. After that date, however, with the rise in mechanization, output per man-day rose more rapidly than in anthracite—about 1.4 percent a year from 1890 to 1918. Mechanization seems to have come with a price tag, however, for soft coal had fatality rates around two per 1,000 workers in the early 1880s, which doubled by 1910.[11]

Marketing of soft coal was sometimes aimed directly at large consumers (railroads, gas companies) who routinely tested for quality; otherwise, sales went through both company-controlled and independent selling agencies. Companies often advertised gas coal in the *American Gas Light Journal*. Otherwise, advertising was rare, and individual companies' market

Figure 2.1. A miner maneuvers an electrically powered coal cutter that will undercut the coal before blasting. The new power and machines raised labor productivity but also worsened workers' risks. (*Cassier's Magazine*, Jan. 1908, 367)

Figure 2.2. Bituminous coal production and price, 1880–1918. The expansion of production largely held prices down except for two shocks in the 1890s and during World War I. (Production is from appendix table A1.4; prices are from USGS, MR, 1918, 732, deflated by David-Solar price index from Carter et al., *Historical Statistics*, series Cc2, 1860 = 100)

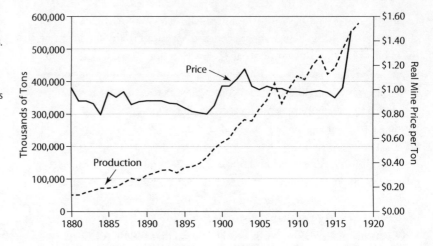

shares reflected their costs and freight rates. The first data on the mine price of soft coal appear in 1880, and adjusted for inflation, they are presented, along with production, in figure 2.2. As can be seen, output increased while prices in 1860 dollars remained around $0.85 a ton until 1897, when a boom drove up prices and wages as well, thereby locking in higher costs and prices of roughly $1.00 a ton. The rise in wages no doubt contributed to the incentives to mechanize from then on. In any event, even in the face of enormous increases in demand, the mine price of bituminous coal declined relative to that of anthracite until World War I. Moreover, the delivered prices of soft coal must have fallen even more sharply relative to those of anthracite as its mining spread to a growing number of states, reducing transportation costs and increasingly evicting hard coal from industrial markets.[12]

The ability of the soft coal industry to supply the enormous increases in demand after 1880 at stable prices deserves some emphasis, for patterns of industrialization would surely have differed sharply had prices risen substantially. The wide distribution of soft coal allowed industrialization to spread over the Northeast and Midwest while its energy density could support large firms and large cities. Had prices risen sharply with expansion, energy-intensive industries such as steel, cement, glass, and paper would have grown somewhat more slowly, as would overall industrialization, while more expensive coal would have reshaped location decisions of industrial fuel users and therefore patterns of settlement. Higher prices would, of course, have encouraged more use of wood, resulting in additional deforestation and additional energy conservation, both of which would have slowed the price rise.

Soft coal also brought clouds of smoke to industrial cities. Engineers understood that smoke was a form of wasted coal, but with hand firing it proved difficult to control. In 1890, the *Chicago Tribune*, which had inveighed against smoke for two decades, explained that "soft coal has of late years been so generally substituted for hard [coal] . . . that Chicago has become a very smoky city," and Chicago was not alone. As modern historians have shown, the "smoke nuisance" was a target of reformers in many large cities from the 1880s onward, usually with little success. Death rates in large cities exceeded those in small towns, and smoke pollution may have been one of the causes.[13]

While most early mines were small, in soft-coal mining as elsewhere, the post–Civil War years saw the emergence of large corporations. While none of the mines integrated forward into wholesaling or retailing, some of them came to be owned by major users that wished to assure supplies. Virtually from their beginning, railroads had surveyed route structures with an eye to resource availability, including, of course, coal. Increasingly, not only major railroads but also steel companies and electric utilities came to operate their own large mines. The first data—for 1913, when these "captive" mines produced 18 percent of all coal—show them to be much larger than average, and whether captive or not, with size came science and engineering.[14]

The desire to mine at scale required a long-time horizon, substantial reserves, and careful planning. Private surveys, whether by railroads with an eye to route structure or large coal companies, relied on consulting geologists to find and evaluate coal seams. These reports contained far more than just estimates of the quantity and quality of the coal and the nature of the beds, often including discussion of mining costs, transportation, and availability of markets. Mining itself required engineers to design an efficient layout and to take responsibility for the ventilation, explosives, safety, and increasingly sophisticated mechanical and electrical equipment. The entrance of larger firms that hired engineers probably accounts for the gradual rise in recovery rates that occurred in the twentieth century.[15]

Transportation

By the onset of the Civil War, the shift from wood to coal was under way on steamboats and railroads and from wind to coal on ocean vessels. With technical problems largely solved, energy choices now became mostly a matter of economics. By 1880, while wood continued to be a common fuel in rural areas far from most coal regions, the spread of mining and cheaper transportation had tilted the calculations toward soft coal.

Steamboats and Steamships

Steamboats accounted for about 5 percent of coal use in 1880. In New England and the Middle Atlantic states a good deal of this was anthracite: elsewhere it was soft coal. Overall about 86 percent of steamboat fuel came from the mines (table A2.4), and census writers and steamboat inspectors noted what seemed its profligate use—especially on western lines. Describing lower Mississippi steamboating, one writer found "the same wasteful practices here as on the upper lines." In fact, what most writers described as waste was—once again—the result of an economic choice. Critics urged abandonment of small, high-pressure engines without condensers, yet western steamers needed shallow draft, so space and weight were at a premium. As one writer acknowledged, the need for light draft "prevents the use of fuel saving machinery which is generally heavy and cumbersome." It is also likely that the economic efficiency of steamboat transportation rose, even if steam boilers and engines showed little improvement, and for the same reasons that productivity had risen in antebellum times: larger, better-designed boats. In one instance, for example, the shift from wood to iron construction after the Civil War raised the ratio of payload to draft by 23 percent. The conversion of fuel from wood to coal raised output relative to capital and labor and other inputs as well (total factor productivity), because it created usable space, and—since coal required fewer stops—vessels spent less time in port.[16]

Coal-fired steam power was also at the center of a great many innovations that dramatically improved the productivity of ocean transport in the decades between the Civil War and World War I. Coal raised total-factor productivity in ocean transport for the same reason it did so on steamboats. In addition, first iron and then steel increasingly replaced wood construction; the stronger materials allowed larger ships and more than proportionately increased cargo capacity. In few areas did a grasp of scientific principles have greater consequence than in ship design; as vessels became much longer, deeper, and larger, they dramatically increased cargo capacity relative to weight and fuel consumption, thus reducing costs. Whereas the *Savannah* had been 100 feet long and weighed but 350 tons, on the eve of World War I, the *Lusitania* was 785 feet long and weighed in at 38,000 tons. Nowhere is fuel efficiency more important than on ocean-going vessels, for every pound of coal reduced a vessel's payload. Screw propellers improved efficiency, and they dated from the 1840s, while the 1850s saw the compound use of steam (in which the exhaust from a high-pressure cylinder entered a lower-pressure cylinder), which squeezed more power from the coal. By the 1870s and 1880s, triple and quadruple compounding arrived, and steam pressures that had been 30 to 40 psi in the 1840s rose to 200 psi or more, again wringing more power from the coal. Consumption fell from 10 pounds per horsepower-hour in the earlier period to perhaps two in the 1870s, and in the 1880s, even more efficient steam turbines entered service. By World War I, freighters traveling about 10 miles per hour might burn 1.3 pounds of coal per horsepower-hour—down from 1.6 pounds a quarter century earlier. More important, between 1875 and 1920, coal consumption per hundred ton-miles of freight had declined from 17 pounds to just four.[17]

Collectively, these great changes dramatically reduced the cost of carrying passengers and cargo. For example, Douglass North (1920–2015) found that an index of American export rates, which stood at 362 in 1815, had fallen to 32 by 1910. This price collapse did not by itself cause the dramatic expansion of trade and migration that characterized the late nineteenth century, but it is hard to imagine such increases coming about in a world of high costs and wooden sailing ships. World trade grew rapidly—US merchandise exports (unadjusted for price changes) were nearly eight times as great in 1914 as they had been in 1860—and cheap transport helped make America a nation of immigrants. American coal sales benefited modestly from these changes. Bunker fuel rose from 5.7 million tons in 1906, the first year that data are available, to a peak of 11.1 million tons in 1917. More broadly, net coal exports, including bunker sales, accounted for about 5 percent of bituminous output that year.[18]

Yet coal's domination as the fuel for ocean shipping would be brief. In 1886, *The Economist* had reported on the maiden voyage of the oil-burning steamship *Himalaya*. Its fuel costs, the magazine pointed out, were a seventh that of coal burners, while oil also raised total factor productivity because "a great saving is also obtained in space and labor." In short, everything coal could do as bunker fuel, oil could do better—and sometimes cheaper. By 1914, about 15 percent of American fleet tonnage was oil powered; it would be 42 percent by 1920 while worldwide oil use also climbed. By 1940, American use of coal as bunker fuel had declined to fewer than 1.5 million tons.[19]

Railroads

The shift to coal by the railroads was just as dramatic and far more consequential for the coal industry. American domestic commerce expanded faster than did ocean-borne trade while railroading boomed, and—thanks to railroad competition—riverboating became a declining industry. About 31,000 miles of track had operated in 1860; by 1880, the total reached 87,000, on its way to 402,000 in 1918. Of course, locomotive and train miles grew even more rapidly.[20]

By 1880, the carriers also dominated in coal haulage, moving some 89 million tons that year, which slightly exceeded total production and amounted to about 30 percent of all tonnage the carriers hauled that year. With the spread of railroads and of coal production, its distribution became increasingly complex. Market competition—as shaped by mine location, production, and transport costs—and coal quality typically determined a mine's market area. A snapshot of these forces and of the railroads' importance appears in a survey of production and distribution of soft coal by state that the USGS published in 1915. In that year, the USGS reported soft coal mined in significant quantities in 27 states. Some 86 percent of production was shipped entirely by rail, with the remainder going by combinations of rail and water routes, and about two-thirds of production moved out of state.[21]

Transportation costs, of course, reflected in part fuel costs. As we have seen, by 1860 the carriers had solved the technical problems of burning coal, and with the economics increasingly favorable, they shifted rapidly to the new fuel. The 1880 census provides data on fuel choices by region of the country, and these are presented in table 2.1. As can be seen, coal dominated in the Middle Atlantic and the Midwest, while wood hung on only in parts of the South. By 1880, coal accounted for about 89 percent of all railroad fuel, and it accounted for 13 percent of all coal mined. By 1918 the railroads burned 138 million tons of coal—about one-fifth of all coal produced that year. Soft coal, that is, provided the power for the railroads, which returned the favor, creating nationwide markets for Old King Coal.[22]

By 1900, only about 3 percent of railroad coal was anthracite. Hard coal had certain attractions: roads that carried it to market also used it for fuel with the Delaware Lackawanna & Western creating the advertising character Phoebe Snow to emphasize its cleanliness:

Says Phoebe Snow about to go
Upon a trip to Buffalo,
"My gown stays white from morn to night
Upon the road of anthracite."

But anthracite's concentration in eastern Pennsylvania meant that except for local lines, its transportation costs were relatively high. Soft coal was not entirely without competition, however. Californians discovered oil in the 1860s, and with residual fuel oil a byproduct of kerosene refining, it was cheap, while local coal was of poor quality. So as early as 1880, the Southern Pacific was experimenting with oil to fire steam locomotives. Other lines also dabbled in the new fuel, and in the West especially, some found it economic. The carriers' long goodbye to coal began well before the diesel locomotive.

Table 2.1. Coal Use by Railroads, 1880

Region	Tons coal	Cords wood	Pct. coal*
1. NE	461,641	258,817	69.0
2. NY, PA, OH, MI, IN, MD, DE, NJ, DC	5,781,416	315,745	95.8
3. WV, KY, TN, MS, AL, GA, FL, NC, SC	435,268	358,687	60.3
4. IL, IA, WI, MO, MN	1,678,030	310,600	87.1
5. LA, AR, OK	2,764	12,491	21.7
6. ND, SD, KS, TX, NM, CO, MT, ID, UT, AZ, CA, NV, OR, WA	684,761	153,916	84.8
US Total	9,043,880	1,410,256	88.9

Source: Calculations based on Tenth Census of the United States, *Report on Transportation* (Washington, 1883), 574–581.
*Wood converted to BCE at 1.25 cords = 1 ton of coal.

The railroads' consumption of such staggering amounts of coal generated staggering costs as well, with fuel typically constituting around 10 percent of operating expenses. Accordingly, in the post–Civil War decades, the railroad technological community generated many efforts to conserve fuel. Boiler encrustation from hard water wasted prodigious quantiles of coal, and although companies experimented with endless additives, good solutions remained elusive until the twentieth century. Another potential road to better fuel economy was to improve coal quality. The Pennsylvania Railroad had tested coals since the 1850s, reporting the results, and the railroad press continued to publish many similar tests by other lines in the postwar years.[23]

Until circa 1900, the main trends in locomotive technology were toward larger size. This likely had little impact on their technical efficiency, but it did increase economic efficiency. Railroad managers grasped that the most economical way to haul freight was to employ a large engine and fire it heavily to pull a very heavy train. Accordingly, freight-car size rose from 10 to 50 tons by 1900, and—thanks to the availability of the air brake—trains grew longer. Heavy firing reduced the technical efficiency with which coal was burned, but heavy trains raised ton-miles per ton of coal. As the *Railroad Gazette* put it in 1894, "Less fuel is being used now per ton mile . . . but on the other hand . . . the efficiency of the boiler is less."[24]

Despite such apparent improvements, economic efficiency measured as ton-miles per ton of coal burned deteriorated from the 1880s into the twentieth century. Some thought that the carriers simply did not take the matter very seriously. Angus Sinclair (1841–1919), who had emigrated from Scotland and studied chemistry at Iowa State, had also worked as a fireman; he was also a recognized expert on railroad mechanical engineering. Sinclair recalled, "I began running [engines] in this country in 1876 . . . [and enginemen] . . . paid no attention to the saving of coal." In 1888, the *Railroad Gazette* also decried the "desultory" attempts to increase fuel economy and the casual and unscientific treatment of the topic by the Railway Master Mechanics Association.[25]

Perhaps, as well, changes in traffic simply swamped the fuel gains from longer and heavier trains. The decades after 1880 saw the proliferation of branch lines, where short, light trains were more likely. In addition, as the railroads became a network, freight-car switching became more common, and switchyards burned coal but produced very few ton-miles. The growth of networks, along with specialized cars, also resulted in much empty freight-car movement (empty cars amounted to 29 percent of all those moved in 1894) and motivated many proposals to pool equipment. But for whatever reasons, railroad fuel economy improvements would have to wait until the twentieth century.[26]

Factory Fuel and Power

American manufacturing came of age in the decades separating the Civil War and 1900. Output (value added) grew nearly tenfold over these decades, and coal was the fuel that fired the growth. The census estimated that by 1910, manufacturing used about 165 million tons of coal for fuel and power, 91 percent of it being bituminous, which accounted for nearly 40 percent of all soft-coal production at the time (table A3.2).[27]

Natural gas entered the energy picture during these years as well, but transportation difficulties confined it to areas near the wells. New discoveries, such as those around Pittsburgh circa 1884, led to floods of production in the local market and caused Pittsburgh to lose—briefly—its title of "smoky city." In Pennsylvania and elsewhere, gas and oil reservoirs often lay beneath multiple landholdings, each of which had the right to drill, but the resulting oil or gas belonged to whoever pumped it. This was the "rule of capture." Since gas and oil will flow to low-pressure areas underground, landowners had strong incentives to site wells near property boundaries and pump like mad to drain oil or gas from adjoining areas. Since no one could appropriate savings from conservation, the rule encouraged the hydrocarbon equivalent of a gold rush. One suspects that when Progressives decried Americans' lack of foresight in dealing with energy, some of what they were observing were the perverse results encouraged by that legal rule. Although the consequences of overdrilling were not well understood until about World War I, they included a loss of reservoir pressure and flooding, reducing ultimate recovery from the pool. By about 1890, the Pittsburgh boom was over, and field production declined sharply. Thus, common property rights ensured that markets, far from encouraging conservation, would reward waste.[28]

But some of the wastes associated with gas and oil production resulted because the process was new and largely unknown. The inability to predict production was a source of much waste; a well might yield a dry hole or be a gusher that overwhelmed storage. Expensive transportation also contributed to losses; by depressing local prices, it made much conservation uneconomic and led to flaring of gas.[29]

Typically, local firms might convert to the new fuel—until the gas dwindled—resulting in much commentary on the amount of coal that gas was displacing but also expressions of the widespread view that because gas reserves would soon vanish, that fuel could not be a long-term threat to coal. Still, nationwide production rose rapidly: gas output jumped from about seven billion ft.[3] in 1880 to 239 billion in 1890. By 1910, total sales amounted to about 509 billion ft.[3]—the equivalent of about 21 million tons of coal. Yet as table A3.2 points out, natural gas remained a minor source of fuel in manufacturing, where coal reigned supreme, and in no industry was coal more dominant than the production of iron and steel.[30]

Beginning in the 1820s, iron production had shifted from charcoal to anthracite, and it remained the dominant source of fuel in pig-iron production until about 1880. During these years, chemical analysis of raw materials became widespread, as did the role of scientific publications in diffusing information. In June and July 1874, the *Engineering and Mining Journal* published a table showing the fuel economy of a large pig-iron maker between 1855 and 1873, along with commentary and analysis. The *Transactions* of the American Institute of Mining Engineers then published a long paper by Lehigh University professor of metallurgy Benjamin West Frazier (1841–1905), demonstrating that one of the causes of poor fuel economy was the use of siliceous iron ores that required much limestone to smelt and, accordingly, much heat. The company must have paid attention, because its records soon showed a sharp drop in limestone consumption and, accordingly, a rise in fuel economy.[31]

This hard-won wisdom was not lost when iron making shifted from anthracite to bituminous coal, which resulted from the discovery of good, cheap coking coal about 1859 near Connellsville, Pennsylvania, along with the development of the railroad network. By 1880, these were driving anthracite from iron manufacture. Iron and steel are weight-losing commodities (the weight of raw materials exceeds the weight of final product), so economics drew the industry toward cheap ore and fuel. Accordingly, new fuel sources and iron ore discoveries around Lake Superior and in Alabama shifted the industry's geographic base from eastern Pennsylvania and New York, near the anthracite beds, closer to the iron and bituminous coal mines in the Midwest and South.[32]

These were also the decades when industry output simply exploded. Production of pig iron rose from 800 thousand tons in 1860 to 16 million tons in 1900 and 44 million in 1918. Blast-furnace fuel was coke, and increased economy in coke making appeared with the first byproduct ovens, which debuted in 1893. Iron masters also shifted to ores that required less limestone to flux, while combustion improved as companies increased blast temperatures from perhaps 600°F to as much as 1,500°F, and increased pressures from 5 to 15 psi. While such measures benefited fuel economy, in some measure, fuel saving was once again a means to an end: as one writer observed in 1869, "economy of fuel is an important object . . . because wages are high; the handling of the fuel . . . requires labor."[33]

During these years, production was becoming increasingly scientific as well, and when *Iron Age* reviewed the two decades after 1855, it noted that "the appreciation of the chemistry of iron smelting [had] developed economy in fuel consumption." Here, Andrew Carnegie seems to have pioneered; in the early 1870s, Carnegie hired a chemist to test iron ores and found that the supposedly "good [ore] was bad and the bad was good." In 1890, the distinguished English metallurgist and ironmaster Lowthian Bell (1816–1904), toured American iron- and steelworks and reported on his findings. He described the "excellently appointed laboratories" and emphasized the "familiarity of the managers with the scientific truths" that underlay steelmaking. Bell claimed that American private philanthropy had donated at least £5 million (about $24 million) to support higher education, and his findings were "proof of the wisdom" of their charity.[34]

Like many other industries, iron and steel production required energy not only for process heat but also for power. For decades, blast furnaces had employed waste gases to power steam engines that generated the hot blast, until about 1901, when the Lackawanna Steel Company began to employ these gases to fire much more efficient internal combustion engines. When US Steel built its vast Gary (Indiana) works about 1909, *Iron Age* claimed that "a crowning feature of this plant is that all the [waste] gas from the blast furnaces will be utilized." Waste gases not only generated the hot blast and ran the blowing engines; they also powered the gas engines that generated the plant's electricity. As appendix table A2.5 demonstrates, smelting a ton of pig iron required about 20 percent less fuel in 1900 than it had two decades earlier.[35]

The shift in the industry's basic products from cast and wrought iron to steel that began in the late 1860s with the advent of the Bessemer process also conserved fuel. Henry Bessemer described his process in a paper entitled "On the Manufacture of Malleable Iron and Steel without Fuel": it simply used a hot blast to burn out the carbon in molten pig iron. In 1896, the ironmaster John Fritz explained how the shift from puddled iron to Bessemer steel had reduced coal use: "in ten minutes [a Bessemer converter could produce] ten tons of steel ingots with a consumption of twenty hundredweight of coal. It will require a puddling furnace ten days . . . to produce a like amount of puddled iron and require about twenty tons of coal." Yet Nathan Rosenberg has argued that the Bessemer process exemplifies Stanley Jevons's argument that conservation can expand energy use, for by cutting costs, it made steel cheap, thereby enormously expanding its use. Bessemer steel also generated a byproduct that could be recycled: its slag might be high enough in phosphorous to be marketed as a fertilizer. Cement makers located near steel mills also found a use for slag, making it into "purzollan" cement.[36]

Later, especially after 1900, open-hearth methods that employed fuel-efficient Siemens regenerative furnaces supplanted Bessemer's process. Finally, as companies increasingly combined under one roof the major production processes—blast furnaces, Bessemer and open-hearth steel producers, and finishing mills—they could avoid reheating, thereby speeding up processes and raising capital, labor, and energy productivity. In this case, it was not a new fuel but rather conservation that raised total-factor productivity.

In the post–Civil War decades, coal came to dominate fuel for factory steam power, as well as for process heat. While the importance of steam for industrialization, and the reasons for its triumph, have fascinated historians, the focus here is on steam power as a source of coal demand.[37] An underappreciated byproduct of steam power was "free" heat as well, for companies often employed the waste steam to warm buildings. The horsepower of steam engines in manufacturing rose 18 times between 1850 and 1900—from about 450,000 to 8.2 million. By 1900, it constituted 81 percent of all primary factory power. In appendix 2 (table A2.6), these horsepower data are combined with plausible estimates of coal use per horsepower-hour and annual hours to generate an estimate of coal consumed by steam power in manufacturing, which amounted to about 61 million tons in 1900, or nearly a quarter of all soft coal burned that year. Similar calculations indicate that power in mining consumed another 12 million tons (see table A2.7). By then, however, electricity from utilities was beginning to contest the dominance of steam power. Because of this competition, along with improved efficiency, coal burned for factory power began to decline.[38]

The growth of steam for factory power continued to raise the share of energy devoted to power production. In chapter 1, I estimated that power production had risen from perhaps 15 percent of energy use in 1800 to about 31 percent in 1860. I estimate that the expansion of steam power in manufacturing and mining, along with the growth of railroads and the rise in electricity, resulted in about 48 percent of energy being devoted to power production in 1900. Well before the automobile, Americans' use of the power had begun to rise sharply.[39]

Figure 2.3 (derived from table A2.6) isolates the importance of efficiency gains in steam power between 1850 and 1920. Coal per horsepower-hour measures efficiency, and it declined from nearly 10 pounds in 1850 to less than two by 1920, and the trend in coal use per horsepower-hour in mining reveals a similar trajectory. On the latter date, Pittsburgh was still known as the smoky city. One can only wonder what its skies and those of

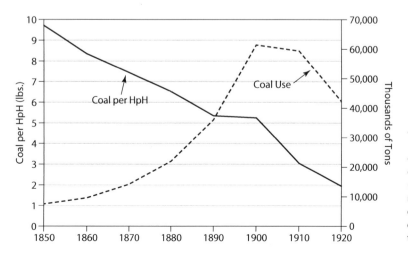

Figure 2.3. Steam-engine efficiency and coal use, 1850–1920. Gains in the efficiency of steam engines caused the usable power from coal to rise much more sharply than did coal use itself. (Chart data drawn from appendix table A2.6)

other major industrial cities might have looked like had the efficiency gains in steam power been less.

Some of the gain revealed in figure 2.3 resulted from better methods of operating and manufacturing steam engines that reflected the efforts of myriad tinkers and generated improvements such as more tightly fitting parts. Expansive working of steam spread and fuel savings resulted because steam engines and boilers grew in size. But while bigger equipment burned less fuel per horsepower, it still burned more fuel in total, focusing the interests of large power users on efficiency. George Corliss (1817–1888)—a "gifted mechanic"—responded to such concerns, developing an automatic variable steam cutoff that both economized on fuel and smoothed engine operation. Users who switched to Corliss engines reported fuel savings of 30 to 40 percent. Other, later, improvements included compounding and increases in steam pressure from perhaps 50 psi in 1860 to 200 psi by 1890. While this evolution was largely the outcome of intuition and empiricism, later writers would interpret these events in thermodynamic terms. Louis Hunter phrases the relationship between science and steam somewhat differently: "the impact of thermodynamics was less to change than to strengthen by correction and refinement the long standing trends" in steam engine design.[40]

Yet the greatest inefficiency of nineteenth-century factory power lay not in the boilers and engines that generated it but in its transmission through a maze of shafts and belts, so typically the horsepower delivered to a machine amounted to only about half that indicated at the steam engine. Efficiency gains in transmission would have to wait for the coming of electricity.

The Gas-Light Era

The expansion of the manufactured gas industry after the Civil War resulted in good part from the increasing urbanization and rising incomes that characterized the period, but technological progress that cut the cost of gas making contributed to the story. In 1900, the census counted 1,740 cities and towns with at least 2,500 inhabitants—up from 392 in 1860— and 877 of them had at least one gas company—up from 237 companies in 1859. Between 1860 and 1900, urban populations rose from six to 30 million people, while real income per person increased more than 90 percent. There were 212,000 city gas customers in 1859 and 4.2 million in 1900; 20 years later, customers would number more than eight million. As the market exploded, production of manufactured gas rose from 2.7 billion cubic feet in 1859 to about 68 billion in 1900 (see table A2.8), on its way to more than 150 billion in 1910, most of it for domestic use.

As I have noted, some of these gains resulted because gas became increasingly affordable. As late as the 1880s, companies typically charged a flat rate per 1,000 ft.[3], but declining block rates began to appear as companies realized such pricing might increase sales and, by reducing average costs, raise profits. By 1915, a survey revealed that 70 percent of companies employed some type of differential rate helping to bring down prices, so adjusted for inflation, average rates declined about 66 percent from 1890 to 1920, even as quality increased and meter charges disappeared.[41]

Widening the Market

Declining prices, as later chapters will underscore, were one of the reasons that manufactured gas began to drive coal from the kitchen. By the 1870s, articles in the *American Gas Light Journal* began to promote gas as a superior fuel for cooking that would be cheaper than coal, while a small market developed employing manufactured gas for space heating as well. When companies sold gas for lighting only, they had a very unbalanced load: a plant big enough to satisfy evening demands was too big for daytime use. Cooking and space heating were used by day, thereby increasing load factors and reducing average costs. The movement of gas into cooking was thus both a cause and an effect of declining costs and prices. Yet few gas companies seem to have done much to promote gas cooking at this time, preferring to let stove makers take the lead. As an important competitor of coal in the kitchen, gas would wait until the twentieth century.[42]

Domestic and commercial natural gas consumption about this time was of somewhat greater magnitude (table A1.6) although sharply limited in geographic scope. It was vastly cheaper, however. In 1912, when manufactured gas retailed for about 70 cents per 1,000 ft.^{3}, natural gas was selling at 26 cents for the same quantity (tables A1.6 and A2.10). Moreover, natural gas also produced about 95 percent more heat, making it a far cheaper fuel, and where available, it, too, began to drive coal from the kitchen.[43]

As the manufactured gas industry expanded, its consumption of coal and coke for fuel rose—to about 2.3 million tons by 1890 and nine million by 1920. The industry became more science-based as well. Investigation of the efficiency of retorts and furnaces had an economic payoff. Testing of coal became widespread, while investigation of gas determined its constituents and candlepower, as well as how to strip out contaminants. The American Gas Light Association (AGLA) began in 1873 and gradually evolved from a social organization into a forum for evaluating research, as was the American Gas Engineering Association. The *American Gas Light Journal* routinely reported scientific and technical developments. Not all companies seem to have been scientific-minded, however, for in 1874, the journal grumbled that "there are scores of huge and rich [gas] companies . . . that *employ no* chemist *or chemical analyst or advisor.*" About 1895, the AGLA set up committees on research and education, and in 1900 the Michigan Gas Association began to award fellowships in gas engineering at the University of Michigan. One motivation for such interest was the possibility of collecting and marketing byproducts. Coke was such a byproduct and had been marketed from the beginning, but by the twentieth century, the capture and sale of tar, ammonia, and a stew of other byproducts became economic for large plants that produced them in volume. By 1919, byproducts accounted for about 9 percent of industry revenue. Their broader significance was that coke provided increased competition with anthracite coal for domestic markets, while other byproducts became chemical industry feedstocks.[44]

The making and using of gas changed dramatically in the post–Civil War decades. While the production of traditional coal gas had become more efficient, employing ever-larger clay retorts heated to higher temperatures by regenerative furnaces that increased yields, far more important was the introduction of a new product: water gas. An outcome of European experimentation, this method passed steam over a high-carbon fuel, such as coke or anthracite, instead of gasifying the coal via destructive distillation. British companies had employed the new process since 1854; it was less capital intensive and used much less labor and fuel than coal gasification. To improve its illuminating qualities, producers mixed it with some liquid hydrocarbon, calling the result carbureted water gas. The new gas reduced fuel use and costs, while the rise of competition from the nascent electric light industry encouraged companies to pass the savings on to customers.[45]

Initially, established scientists and others in this country dismissed the idea of making gas from water as a delusion. In 1870, the *Boston Journal of Chemistry* referred to it as a "plague," and three years later, *Scientific American* equated it with perpetual motion. When it proved to work, established companies with an investment in coal gas technology fought it. Made by some processes, water gas might be high in carbon monoxide, and in Massachusetts, the coal gas interests succeeded in banning it for a time as a safety risk.[46]

Finally, in 1874, Benjamin Silliman Jr.—whom the reader met in chapter 1—and Henry Wurtz (1828–1910) decided to publish a private investigation of water gas they had conducted some years earlier—apparently to combat the misinformation being spread by the coal-gas interests. The report was a thorough and persuasive endorsement, and gradually water gas won out; it constituted 38 percent of manufactured gas in 1890, at which time a Massachusetts report demonstrated that it required a quarter as much coal per 1,000 feet of gas as did the older process. By 1920, water gas constituted two-thirds of industry output. Spurred by competition with electricity, companies increasingly enriched both types of gas to raise their candlepower, using oil and similar products. One ingredient, ironically, was naphtha, a nearly useless—and hence cheap—byproduct of refining crude oil into kerosene, the main competitor of gas.

Two results of this were sharp increases in the energy efficiency of gas making and a declining share of soft coal as a feedstock (table A2.8).[47]

A Revolution in Lighting

As we saw in chapter 1, in 1860, a dollar spent on gas light yielded seven times as much illumination as the same expenditure on candles. This was an illumination revolution, and it continued, for the lighting power of gas rose from about 14 candlepower in 1860 to 18 (that is 3.6 per ft.^{3}/hr.) by 1900 and often higher (table 2.2). Burners improved dramatically as well, with the most important innovation being that of Carl Auer von Welsbach (1858–1929), who patented a chemically impregnated mantle that gave a brilliant white light when heated. Although his early efforts failed, by 1894, American tests showed that gas rated at 3.6 candlepower per ft.^{3}/hr. in a test burner generated 20.25 candlepower per ft.^{3}/hr. employing a Welsbach mantle (and later as much as 30). The result was a dramatic improvement in the ability to transform fuel into light.[48]

In 1860, a ton of coal might generate 8,000 ft.^{3} of gas; by 1900, output had risen to 17,840 ft.^{3}/ton. The increased efficiency with which utilities could convert coal to gas resulted in a sharp decline in the price of gas from $3.00/Mft.^{3} in 1860 to $1.04 in 1900 (table 2.2), even as the Welsbach mantle increased the light productivity of gas from 2.8 candlepower per hour to 20.30.

Collectively, this amounted to yet another lighting miracle before the arrival of electricity, and here again, the effects on welfare are largely absent from traditional measures of the standard of living. The reader may recall from chapter 1 the clerical worker of 1860 who might have earned $600 a year and by spending $30—5 percent of his income—on gas costing $3.00/Mft.^{3}, could buy 10,000 ft.^{3} of gas, yielding 28,000 candlepower-hours of light, or about 351,840 lumen-hours. In 1900, if his income had grown at the same rate as real per-person GDP, he would have earned $1,159 in 1900, and $30 would now constitute but 2.5 percent of his income. But with gas at $1.04/Mft.^{3}, the same $30 of gas burned in a Welsbach

Table 2.2. The Gas Lighting Revolution, 1860–1900

	1860	1890	1900
	Gas	Gas	Gas/Welsbach
Mft.^{3}/ton BCE	8.0	12.85	17.84
Candlepower-hrs./ft.^{3}	2.8	3.60	20.30
Lumen-hrs./ft.^{3}	35.2	45.20	255.10
Lumen-hrs./ton	281,478.0	580,849.00	4,550,792.00
Price ($) per Mft.^{3}	3.0	2.00	1.04
Lumen-hrs. per $	11,728.0	22,619.00	245,278.00

Source: Yields and prices from chap. 1n70 and tables 1.4 and A2.8; other data author's calculations.

burner would yield 30 × 245,278 = 7,358,340 lumen hours—nearly 21 times as much light as in the early days of gas—and this for a smaller share of the family's income.[49]

Still, this remained a miracle of limited scope, for as I have noted, manufactured gas companies served only 4.2 million customers in 1900—about 26 percent of all households—while perhaps another 700,000 burned natural gas. The rest would have used kerosene lamps. These were also a great improvement over candles, and they, too, were improving, for in the 1890s, new lamps with Welsbach mantles appeared and were far brighter than old, flat-wick lamps.

Domestic Fuel Use

Table 2.3 presents a summary sketch of Americans' domestic energy consumption from 1800 to 1910. These data reveal that although coal was the dominant source of fuel by 1910, Americans still burned prodigious quantities of wood. Even so, per capita domestic energy consumption continued to decline, as hard and soft coal gradually pushed out wood.[50]

This long-term decline resulted in part from retreat of the fireplace in favor of more efficient stoves and furnaces. Production of stoves alone reached two million in 1870, and they must have been ubiquitous, raising comfort levels even as they diminished fuel use. The decline in per capita domestic energy use also reflected the impact of urbanization, for apartment buildings could heat more efficiently than detached dwellings. Moreover, by the 1880s, some cities had district heating in which a central source of steam supplied a range of buildings, which also economized on fuel.[51]

In 1910, domestic fuel was the one area of the economy where anthracite continued to hold its own. As the USGS put matters in 1911, "The

Table 2.3. Domestic Energy Use, 1800–1910

Thousands of tons of BCE	1800	1860	1910
Bituminous coal	59	3,634	65,562
Anthracite coal	—	2,648	41,485
Wood	18,000	95,760	63,173
Whale oil	0.297	1.3	—
Coke	—	—	3,637
Briquettes	—	—	179
Kerosene	—	(37)*	3,325
Gasoline	—	—	2,232
Electricity	—	—	2,393
Mfg. gas	—	(27)*	2,314
Natural gas	—	—	6,806
Total	18,059	102,042	191,019
Total tons per capita	3.4	3.2	2.07
Coal tons per capita	0.01	0.18	1.14

Source: Tables A2.9 and A2.10.
*Figures in parentheses for kerosene and manufactured gas are excluded from 1860 totals because they are included in the way bituminous coal is measured.

principal demand for anthracite . . . [is] restricted largely to domestic trade," by which it meant private homes, hotels, apartment buildings, and retail trade venues. Beginning about 1900, the mine price of hard coal had been slowly rising relative to that of its bituminous cousin (table A1.5), reducing its attractiveness, and it was predominantly an eastern fuel. In 1915, about 87 percent of sales were in New England and the Middle Atlantic states.[52]

Outside the Northeast, bituminous coal was shoving aside both wood and anthracite as domestic fuels, contributing to the smoke nuisance. Most families burned it in stoves, except for the wealthy urban and suburban middle classes, who increasingly had central heat, employing hot air, steam, or hot water. Hot-air furnaces had appeared before the Civil War; early models were little more than a stove in the cellar, in an enclosure that piped the heat via convection to one or more upstairs rooms. They were cheap to install but tended to dry the house out, and they were easily contaminated by furnace exhaust. Domestic steam heat owes its origins to the work of Stephen Gold (1801–1880), who designed a low-pressure system that included boiler, safety equipment, and radiators. Gold seems to have gotten the idea from the Nason & Walworth high-pressure system. With some assistance from Yale's Benjamin Silliman Jr., Gold formed the Connecticut Steam Heating Company and began marketing his system in 1854.[53]

No data exist to chart the spread of central heat until the census of 1940 revealed that by then, about 42 percent of the population had central heating. Like the stove before it, domestic central heating surely reduced exposure to indoor air pollutants, with attendant health benefits. Moving the furnace to the cellar transferred much of the coal dust and ashes there, too, as well as the smoke that resulted from firing and leaky pipes.[54]

Peak Coal and War

The economy-wide use of coal peaked during World War I. The European war that broke out in 1914 resulted in a flood of orders from the allies by 1915, touching off an economic boom. Industrial production more than doubled from 1914 to its peak in early 1920, and consumer prices roughly doubled as well. Coal output also rose, but by far less (table A1.4). Bituminous output rose about 37 percent from 1914 to its peak of 579 million tons in 1918—a level it would not exceed outside of wartime until 1970—while anthracite rose 10 percent, peaking at a fraction less than 100 million tons in 1917. Oil production, however, boomed, nearly doubling between 1914 and 1920 with no decline in sight (table A1.7).

Under the press of wartime demand, the mine price of soft coal, which had hovered around $1.00 to $1.15 a ton since 1900, took off in 1917, rising to $3.75 in 1920 (table A1.5). Lured by the prospect of such riches, companies piled in: the number of commercial mines jumped from not quite

5,800 in 1913 to 8,200 in 1918. Moreover, by 1916, supply shortages were erupting as well; *Railway Age* warned of a "coal famine" due to labor disputes and a lack of railroad cars. "America Faces Coal Shortage," a representative of the Council on National Defense warned readers of the *Chicago Tribune*. By 1917, in New York, the Interborough Rapid Transit briefly stopped subway service for want of coal, and some interurbans instituted a "skipped stops" schedule. In a sellers' market, quality deteriorated as well; in October 1917, the Columbus [Ohio] Railway Light and Power Company shut down the railway several times because there was so much dirt in the coal supply that it was growing grass. In November, Fuel Administrator Harry Garfield warned of a 50-million-ton shortage; by June 1918, his forecast had mushroomed to 80 million tons. "Today it is everybody's Business to Save Coal," *Electrical World* proclaimed as it reported on the US Fuel Administration's national conservation campaign; warnings of shortages continued well into 1920.[55]

The result was a brief, unprecedented federal intervention into energy markets. The Lever Act of 1917 led to the US Fuel Administration and gave the federal government power to fix prices and allocate production of coal along with somewhat less-stringent controls over other fuels. Anthracite and bituminous coal were rationed, and there were major campaigns to conserve coal, natural gas, and petroleum products.[56]

The shortages and government-determined market allocations had long-term consequences, for they introduced consumers to fuels such as coke and coal briquettes, while the conservation campaigns continued into the 1920s and later, with far-reaching results for coal. The wartime shortages also induced Congress to enact the oil depletion allowance and to allow producers to write off intangible drilling expenses against current income. These tax breaks proved long-lasting, and by encouraging more drilling, they contributed modestly to coal's postwar woes.[57]

The war also reinforced the Progressives' resource pessimism and market skepticism, and it shifted the focus to petroleum, for war now required oil for transport. Accordingly, forecasts of declining reserves raised concerns over national defense, reinforcing the concerns of market skeptics. In 1916, Mark Requa (1866–1937), soon to become head of the US Fuel Administration's oil division, told Congress that petroleum reserves might last but 28 years and excoriated that industry for its waste, which he blamed on competitive conditions. Oil shortages, he warned, might put us "at the mercy of the enemy." Other experts voiced similar concerns, which continued to worry policy makers during the interwar decades and after World War II, when they would support major interventions in energy markets.[58]

These events marked the apex of the age of coal and the beginnings of its long goodbye. The retreat of anthracite from its 1917 high was not simply a business-cycle readjustment, for the industry would never see that level of output again, while bituminous coal would not return to its wartime high for decades. The USGS caught the significance of these events. "The period from 1914 through . . . 1919 will . . . be viewed as a separate epoch in the coal industry," it noted, and the year 1917, in particular, was "the most momentous in the history of the industry."[59]

Conclusion

As this chapter and the previous one have shown, the rise of first hard and then soft coal to such prominence reflected not simply their abundance but the head-over-heals exploitation of the new fuels by entrepreneurs aided by an emerging scientific community. Like anthracite, soft coal was a better fuel than wood because it was more dense and less labor-intensive. In turn, the keys to the triumph of bituminous coal over anthracite were its wider geographic dispersion and greater ease in mining, along with modest productivity gains and a developing rail network, which made its delivered price increasingly cheaper than that of hard coal. The enormous increase in coal production at constant prices after 1880 contributed to the expansion of energy-intensive industries and technologies. Overall, energy use per person rose only modestly because so much of it replaced older, poorer sources of fuel and power, while domestic energy use per person declined until 1910. More energy, more efficiently employed raised standards of living and comfort for nearly all as travel became more common and more pleasant, cooking less burdensome, and homes less drafty and better lighted.

And although the Progressives of Theodore Roosevelt's generation depicted this vast expansion of energy use as rife with waste, the reality was somewhat different. In coal mining, much of what seemed wasteful to critics was, in fact, the outcome of an economic choice in the context of high wages and cheap energy. As engineers understood at the time, some coal was best left in the ground. Waste of all forms of energy also resulted from ignorance—not all who opened mines were skilled—while gas or oil production might outrun transport and storage facilities—a result of rapid, uncoordinated development. Those who decried these events were wishing for a world in which Americans would have stayed poorer longer. An additional source of waste characterized production of gas and oil—what I summarize as common property problems—and here again, some of the cause was ignorance, for the problem was little understood until World War I.

In energy consumption, the story reveals a continuing quest to conserve fuel—in iron-making technology, in the production of manufactured gas, in the expanding ubiquity of the domestic stove, and in the evolution of the steam engine. Sometimes, as with the Bessemer process, these efficiency gains were so large they increased rather than decreased energy use.

Yet just as Marx had argued that capitalism would foster its own destruction, so the science and engineering and entrepreneurship that helped midwife the coal age would lead to its undoing. The seeds of coal's decline and stagnation had long been germinating. By World War I, the continuing quest for efficiency, along with the discovery and rapid exploitation of seas of oil and gas, would begin to drive coal from hearth and home and from factory floors, while the industry's critics came increasingly to feel that they could improve on market outcomes. The details of these energy transitions, and of coal's attempts to fight back, are the subjects of the chapters that follow.

Part II

LOSING INDUSTRY

3

Soft Coal in Industry, 1900–1940
The Long Goodbye

The demand for coal is like a lump of tar in cold weather. It yields very little to sudden pressure [of high prices] but if the pressure is long continued [it] will flow out of shape and perhaps never return to its original position.

—*Frederick G. Tryon, in* Mineral Resources of the United States

This transformation of waste materials into valuable resources has been one of the most pervasive of all the effects of the application of science to industry.

—*David C. Mowery and Nathan Rosenberg,* Technology and the Pursuit of Economic Growth

In 1908, the Bituminous Coal Association happily predicted that consumption in 1940 would total 1.2 billion tons; seldom has a forecast been more wrong, for the actual consumption that year was 460 million tons. As late as 1918, bituminous coal had been the fuel of industry, with production at its peak of 579 million tons; but in the 1920s, it stagnated while the Great Depression brought disaster, with output bottoming out at 309 million tons in 1932. In 1923, the USGS took official note of the new state of the American coal industry. "Since the war . . . the market has ceased to expand," it observed. Bituminous coal consumption that year "was 75,000,000 tons less than might have been expected on the basis of prewar experience."[1]

The fall of coal resulted primarily from the same market forces that had brought it to prominence—the routine workings of a network of individuals and institutions focused on energy production and use and driven by the profit motive. More specifically, coal's vicissitudes reflected the confluence of three broad trends that predated World War I. First, the long-term rise in the industrial employment of scientists and engineers supplied the skills needed to save energy. Second, electric utilities continued to grow rapidly; their special incentives to save fuel made them a conservation laboratory that generated improvements with broad applicability. Finally, the discovery and rapid exploitation of enormous reserves of oil and gas, along with improved transport, led to sharply increased interfuel competition.

The fuel shortages and price spikes associated with World War I reinforced these developments, sharply focusing companies on energy conservation. War-induced immigration restrictions also put upward pressure on wages and therefore on the cost of labor-intensive fuels such as coal, which raised the payoff to conservation and fuel switching.[2] Other energy policies played only a minor and conflicting role in shaping coal's fortunes, while the great strikes of the 1920s that cut into domestic consumption of anthracite seem to have had little impact on sales of soft coal to industry.[3]

Section 1 below sketches the flowering of a network of scientists and engineers who innovated and spread new technologies that institutionalized energy efficiency and enhanced interfuel competition. The next section reviews innovations in the recovery of wastes from coal mining and coke production. The following section discusses the complexities of fuel substitution and conservation in manufacturing and electrical utilities. The last part of the chapter briefly discusses the consequences of these changes for urban smoke pollution.

Institutionalizing Energy Efficiency

By the twentieth century, what had begun as a trickle of technically trained individuals into industry became a flood. The number of engineers in the labor force grew 15 times between 1860 and 1910, and there were 25 times more chemists in the latter year than in the former. This profusion of technically trained individuals both caused and reflected the sharply expanded role of science in industry. David Mowery's data on the founding of industrial research establishments illustrate these trends. Mowery found that there were but 819 industrial research laboratories in the United States in 1919, most of which had been founded in the previous decade. After 1919, the number exploded, rising to 2,358 in 1936, of which 1,915, or 81 percent, were in manufacturing. These figures indicate a broad awakening to the value of industrial research and provide an early example of the institutionalization of innovation described by William Baumol and others, but as we will see below, much energy research involved engineering experimentation that was conducted outside of laboratories on the shop floor.[4]

The most important source of public-sector energy research was the United States Bureau of Mines. The bureau performed market-enhancing activities, working closely with industry in the service of resource conservation and taking over much fuel research from the USGS. In addition, the Bureau of Standards, state mining bureaus, and several universities all conducted energy research, much of it focused on coal. In the 1930s the Bureau of Mines reported that nationwide, 40 laboratories were conducting research on coal, and it is safe to say that fuel use and efficiency generally were important topics at many others. Thus, while few coal companies did much formal research in the interwar decades, the industry's allies in government and academia, along with users and suppliers, generated important improvements.[5]

Engineers, chemists, and others not only entered laboratories in increasing numbers; these years also saw a "passing of the baton" in the boiler room as technically trained individuals replaced a previous generation of managers and line workers who had worked by rules of thumb. What economists call "on-the-job learning" increasingly involved science and engineering, and although it was not formal research, it often pushed the technological envelope, with the results typically shared in technical journals. There were also consulting engineering firms that diffused new technologies and approaches.[6]

An Energy Network

These research and consulting organizations and the technically trained men and women who worked in industry formed a network—an "invisible college" in Derek Price's expressive phrase. This network had expanded throughout the nineteenth century and included men and women who worked in governments, scientific societies, universities, trade associations, and at scientific and trade journals. Their work produced a kind of collective invention of energy technology. The engineering and other technical societies generated a vast literature on nearly all aspects of energy. Some, such as the American Gas Association (AGA) and the American Society of Heating and Ventilating Engineers, even supported research laboratories. These decades also saw a flowering of trade associations—the Portland Cement Association (1902) and the Technical Association of the Pulp and Paper Industry (TAPPI, 1915), for example—which also undertook research and discussed and disseminated technical information. The membership of TAPPI included representatives of supplier firms (e.g., DuPont, Combustion Engineering), government researchers (United States Bureau of Standards and Forest Products Laboratory), and consultants (Arthur D. Little) as well.[7]

Journals that emphasized applied science and technology also multiplied rapidly. These included industry-specific periodicals such as *Coal Age* (founded in 1911) and *Blast Furnace and Steel Plant* (1912). Other technology and energy-focused journals such as *Power Plant Engineering* (1896), *Industrial and Engineering Chemistry* (1909), and *Combustion* (1919) spoke to nearly every business in the country. Their articles ranged from the highly technical to more commonsense pieces intended for readers with less formal training, and some journals included sections such as "Ideas from Practical Men" and question-and-answer sections. Thus, the individuals who populated this technological network included not only those who worked in laboratories but also practicing "shop floor" scientists, engineers, and many others in the invisible college.[8]

Technological convergence made such networks both possible and important. Chemical reactions and the laws of thermodynamics applied universally, so innovations at one firm might benefit companies far removed from that industry. Yet because raw materials and products differed in an infinity of ways, combustion and other engineering and scientific problems might differ among firms and industries (one expert referred to combustion as an "art"). Technological diffusion, like evolution, usually involved "descent with modification."[9]

War Focuses on Fuels

Thus, fuel economy was a plant with deep institutional roots; it grew rapidly during the hothouse years of World War I, sharply focusing attention on energy efficiency and powerfully reinforcing prewar worries over future scarcities. Speaking to the American Mining Congress in 1916, George Otis Smith (1871–1944), the longtime head of the USGS, had urged that coal be made a public utility. In 1920, lecturing to the American Iron and Steel Institute (AISI) on thrift in coal, he sounded an alarm. Echoing the Progressives of a decade before, he told the steelmen that "Coal is the shortest word we have to express national power and domestic comfort," and he went on to rehearse the by-now-standard list of the ways Americans wasted that fuel and to urge conservation on his audience. But the potential shortage of petroleum was even more serious, Smith warned. "The decade from 1910 to 1919 is best described as a transition from oversupply to over demand" for oil, he told the audience. Since energy experts imagined that gasoline had no substitutes, while coal could raise steam, oil should be conserved for "essential uses." Smith urged that "use of fuel oil as a substitute for coal should be discouraged," and he warned that "the use of gasoline to serve our pleasure cannot go unchecked."[10]

Smith's worries were not idiosyncratic. Two years later, the USGS estimated American oil reserves to be a little more than nine billion barrels—about a 16 years' supply at current rates of production. In a speech to the ASME that year, David White of the USGS publicized the reserve figures; he emphasized the need for enhanced recovery and reduction in such wastes as the use of oil to generate steam. These estimates of oil reserves were routinely recalculated during the interwar decades, invariably demonstrating that reserves were equal to roughly 15 to 20 years of production. The calls of Smith and White for conservation did not go unanswered. The American Society of Mechanical Engineers responded to wartime concerns by creating a fuels division in 1920. Thus, wartime developments provided a powerful stimulus to energy conservation, reinforcing existing efforts.[11]

Clearly, energy experts' ideas about resource conservation were evolving. As I noted in the previous chapter, national defense was an additional conservation motive and likely to be ignored by private enterprise even as oil and gas, which had once been dismissed as bit players, were now taking center stage. Moreover, experts' resource pessimism ensured that few saw this as the beginning of a long-term energy transition from coal to the new fuels. Rather, the scarcity of the new fuels, policy makers imagined, would make their day in the sun a brief one, and they needed to be conserved for "essential functions," with coal consigned to lower-value uses. Men and women had once spoken of "old KING coal," but in the interwar years, he was losing his throne and slowly becoming simply "OLD king coal."

Recovering Wasted Coal

In 1908, when Joseph Holmes informed the governors' conference on conservation that in the bad old days, anthracite mining recovered only 40 to 60 percent of the coal, he was drawing on findings that dated from the 1880s. But the increasing employment of engineers had steadily improved mining methods, increasing recovery. Indeed, bureau experts claimed that losses in both hard and soft coal mining had fallen from 65 percent or more in 1850 to about 35 percent by 1923.[12]

Practical Conservation

Much waste, however, resulted *after* the coal arrived at the surface, and here, too, recovery efforts long preceded World War I. Small sizes of anthracite were once of little economic value because they did not burn well in stoves and furnaces, so companies simply threw them away. Immense "culm" piles arose outside breakers, where some of them washed into nearby rivers. By 1918, however, wartime shortages and price spikes were generating a boom in recovery (fig. 3.1). That same year, a chemist at the Lehigh Valley Coal Company estimated that 175 million tons of anthracite culm were then available, and efforts to market it had been ongoing for decades.[13]

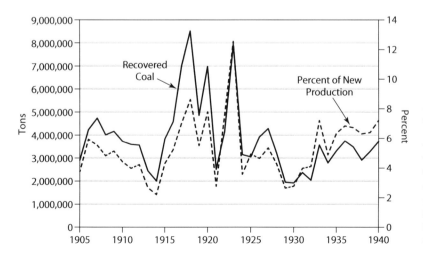

Figure 3.1. Recovered anthracite wastes from culm banks and rivers, 1905–1940. Innovations in combustion technology combined with expanding markets to encourage recovery of what had once been wasted coal. (Chart data drawn from appendix table A3.1)

As a writer for the census of 1900 put matters, "The refuse of today is a source of profit tomorrow," and waste on such a scale had long drawn entrepreneurial efforts to convert it into gold. Pennsylvania had commissioned a report on the wastes of anthracite in 1893, and for commission member Eckley B. Coxe (1839–1895), the culm banks were of special interest, for he also oversaw his family's vast coal properties. Coxe represented the new scientific breezes blowing through mining in the late nineteenth century. A scion of one of Philadelphia's first families and a graduate of the University of Pennsylvania, Coxe became a mining engineer. He was also one of the founders of the American Institute of Mining Engineers, where he had chaired an early committee on waste in anthracite, and he became the institute's president in 1893. In that year, Coxe obtained the first of a number of patents for inventing a stoker that would more efficiently burn the small sizes of anthracite. His invention employed a special traveling grate with air supplied at different pressures along the way; it could burn the smallest sizes of anthracite.[14]

Coxe's stoker appeared at a time when the new electric utility plants were looking for cheap fuel; markets married the two, thereby converting culm into cash. By 1897, an editor of *Coal Trade* remarked that "there are fortunes in these old dumps." Dredging of the Susquehanna and other rivers for coal also began during the 1890s. The USGS first took serious note of this new supply of coal in 1905. By that time, recovery amounted to 2.9 million tons a year, or about 4 percent of annual anthracite production (see fig. 3.1 and table A3.1). Five years later, the survey reported—with some overstatement—that the "unsightly culm banks which blotted the landscape in the anthracite region are fast disappearing." The marketing of small sizes and culm was, the author enthused, "an object lesson in practical conservation."[15]

World War I boomed this market for recovered coal, and, even on the eve of World War II, recaptured wastes continued to supply 6 to 8 percent of America's anthracite. By 1940, recovery had yielded 138 million tons of coal (table A3.1) that had once been considered irretrievably lost—contributing to the woes of underground miners and testifying to the power of market incentives, technological inventiveness, and entrepreneurial enthusiasm.[16]

Artificial Anthracite

While Coxe's grates might burn some anthracite waste, a number of inventors tried compressing it into briquettes, usually with a binder added, and they applied the process to bituminous wastes as well. Because bituminous coal was then usually sold "run of the mine" (i.e., unsized), wastes consisted almost entirely of dust, vast piles of which were strewn about the mines and transshipment points such as the Great Lakes docks. Briquetting was an old idea—one writer traced it back to 1594—but the first successful machine appeared in France in 1842. Briquettes had a number of advantages over coal: they promised better combustion, generated less dust, and required less storage space. By the 1860s, briquetting was a thriving industry in Europe.

American experiments with briquettes began about 1870, but none of them proved commercially successful. Briquettes were usually more costly to produce than soft coal and therefore likely more expensive unless sufficiently distant from the mines. Moreover, many processes used coal tar as a binder, and the resulting briquettes smoked, making them inferior to anthracite. Still, the USGS was enthusiastic. In 1907 it reported eleven manufacturers that had marketed about 67,000 tons of briquettes. The "rock of opposition," it gushed, "is being worn away" by the publicity given to wasteful practices.[17]

Companies produced industrial sizes of briquettes—some weighing as much as eight pounds—in addition to small domestic sizes. A number of railroads and industrial coal users experimented with them. The USGS, and later the USBM, studied them extensively, and both agencies promoted briquettes enthusiastically. One important finding was that because briquetting removed water, the product had higher energy content than the equivalent weight of coal.[18]

Although modestly popular for domestic uses, briquettes enjoyed little acclaim in the United States as an industrial fuel. In Europe, coal was expensive and labor cheap, making waste processing economically attractive, but in the United States, these conditions reversed, making briquettes typically more expensive than run-of-the-mine coal. By 1930, annual sales for all purposes amounted to only about 1.2 million tons; thereafter, production averaged a bit more than 800,000 tons a year through 1940.[19]

Fuel Substitution and Conservation in Manufacturing

By 1900, the United States had become the world's leading manufacturing nation, accounting for about 30 percent of world production, and by the late 1920s, its share surpassed 40 percent. The interwar years were also America's most technologically progressive decades. A host of new or improved products appeared, and factory electrification and mass production spread rapidly. Productivity (relative to all inputs) in manufacturing, which had grown at less than 1 percent a year from 1869 to 1919, accelerated to 3.8 percent a year from 1919 to 1937, beginning what one economist has termed America's Great Leap Forward. Energy productivity also rose sharply in manufacturing, contributing importantly, as we will see, to the overall productivity surge. The interwar decades also saw the beginnings of an energy transition that would eventually largely drive coal from the factory floor.[20]

Table 3.1 provides an overview of primary energy use by American manufacturers in the years between 1910 and 1940—along with the post–World War II decade for comparison. Two broad trends emerge: shares of oil and, especially, gas rose relative to those of coal, while primary energy efficiency—an index of manufacturing output divided by the BCE of coal, oil, natural gas, and purchased electricity—rose sharply in the interwar years. To avoid double counting, I exclude mixed and manufactured gas, as well as produced electricity, most of which were derived from coal. Indeed, their use was one of the reasons primary-energy efficiency improved. For similar reasons, I also exclude coke. Total coal use fell by 59 million tons from 1920 to 1940, and its share of all primary fuel fell about 32 percentage points even as manufacturing output rose 50 percent. Nor were these events an artifact of the Great Depression, for as indicated, the decline in coal's share of energy continued in the postwar years. The improved efficiency in the use of coal clearly reduced consumption, but as we will see, in the absence of such efficiency gains, and in the face of expanding supplies of gas and oil, coal use would surely have fallen even more. As the Red Queen told Alice, sometimes "it takes all the running you can do to keep in the same place."[21]

Gas and Oil Become Industrial Fuels

The growing use of oil and, especially, gas relative to coal has one obvious explanation: they were superior fuels. As one user put it in 1916: "Even at as great a cost per heat unit compared with pulverized coal, natural gas has material advantages. . . . To a considerable extent the same comparison applies to oil." As I noted above, war and postwar immigration restrictions handicapped coal because it was a labor-intensive fuel. Thus, in the 1930s, one of the motives that induced some lime kilns to switch from coal to natural gas was the reduction in labor requirements. In that industry,

Table 3.1. Primary Energy Use in Manufacturing, 1910–1954

Thousands of tons of BCE	1910	1920	1930	1940	1947	1954
Bituminous	151,123	188,836	196,780	138,182	210,141	176,716
Anthracite	14,036	13,328	9,168	5,013	8,518	—
All coal	165,159	202,164	205,948	143,195	218,659	176,716
Oil	4,588	16,195	31,687	31,024	50,130	58,039
Natural gas	10,737	12,840	29,617	39,296	69,060	251,801
Manufactured gas	—	(5,152)	(31,496)	(38,366)	(46,367)	—
Mixed gas	(8,579)	—	—	—	(39,409)	(78,809)
Purchased electricity	—	37,786	30,289	30,273	66,834	92,578
Produced electricity	—	—	—	(7,863)	(11,643)	(14,982)
Primary energy BCE*	180,483	274,136	329,038	282,155	451,050	579,135
Coal percent of primary BCE	0.92	0.74	0.63	0.51	0.48	0.31
MFG index (1910 = 100)	100	150	229	225	397	501
MFG/Primary (1910 = 100)	100	99	125	144	159	156

Source: Table A3.2.
*Totals exclude figures in parentheses for manufactured and mixed gas and produced electricity.

gas also yielded a superior product and an increase in the number of heats from two to three per day. Thus, switching the *form* of energy raised output relative to all inputs. While both oil and gases contested energy markets in manufacturing, in the interests of space, the focus here is on natural gas.[22]

By World War I, natural gas had a bad reputation as a boom-and-bust fuel. As I explained in chapter 2, much of the problem was that multiple drilling rights to a reservoir might touch off a gas rush that would rapidly deplete pool pressure and reduce recovery. An unhappy example occurred at McKeesport, Pennsylvania, during 1919 and 1920. The initial well that found gas on April 19, 1919, yielded 1.5 Mmft.³/day, and by January 1920, field production exceeded 60 Mmft.³/day. A year later, in January 1921, it had fallen by two-thirds. Such events had been occurring for decades, and by this time, the Bureau of Mines had repeatedly publicized the technical reasons for such rapid depletion, but because of the drilling incentives resulting from the rule of capture, effective solutions were slow to appear.[23]

The demand increases associated with World War I also contributed to the fears that supplies of oil and gas would soon run short, and in 1919 and 1920, the federal government and the gas industry jointly sponsored a gas conservation campaign. While the campaign focused mostly on domestic users, companies surely paid attention as well, and fears of gas shortages lingered throughout the 1920s. Writing in 1921, a Bureau of Mines researcher concluded that "the annual output [of natural gas] will never be very much more than it was in the period 1916–1920," and he thought a decline would soon set in. No doubt, he came to regret the prediction: annual production then averaged a bit less than 800 million ft.³; a decade later, it was double that figure, and at this writing is around 34 trillion ft.³ Probably because of such forecasts, companies initially focused on oil rather than gas as a possible coal substitute. Even here, however, supply worries were evident—thanks in part to the work of George Otis Smith and others noted above. Writing in *Paper Industry* in 1922, the consulting engineer John Ferguson concluded that "the merits of oil burning far outweigh those of coal burning if we consider only the present." But he warned that if a manager considered that "the future means a continual rising cost of oil . . . he will likely decide in favor of coal." Similarly, a review of fuel developments in 1926 noted that the "greatly increased consumption of fuel oil for power . . . has not taken place . . . due to the feeling that the supplies of low-priced fuel would be of . . . short duration." Foresight in energy markets, it appears, may not have been as uncommon as President Roosevelt had feared.[24]

A temporary dearth of new petroleum discoveries in the 1920s seemed to confirm pessimists' fears, leading President Coolidge to create the Federal Oil Conservation Board, which appeared in 1924. With George Otis Smith as an adviser, the board was reliably pessimistic. Its first report, in 1926, estimated that oil reserves equaled but six years of future supply. The board also toyed with the idea of reducing petroleum consumed in such "low value" uses as raising steam. These various worries over petroleum and gas supplies, with the implied need for experts to guide the market back toward coal, continued throughout the 1930s.[25]

While such worries were widespread, it is unlikely that they significantly retarded the shift from coal to gas and oil. In part, this may have been because those making the prediction soon got the reputation of the child who cried wolf. In addition, by the mid-1920s, new discoveries of oil were dispelling the clouds of gloom for all but the relentlessly pessimistic. In 1926, one trade journal reassured automobile manufacturers that there was "Plenty of Fuel for Oil Engines," while another asked skeptically, "What of the Oil Famine?" Moreover, risk-averse companies quickly learned to design flexibility into energy systems to avoid being locked into an uneconomic fuel. By 1930, one expert was noting that "modern burners are generally of the combination type suitable for the use of gas, oil, or powered coal," and soon, stoker coal would join the list, thereby sharply reducing the cost of hedging against fuel uncertainties and, accordingly, the lock-in risks of burning gas or oil.[26]

In any event, by the mid-1920s, fears of gas and oil shortages were gradually abating. Monroe, Louisiana, saw a large gas find in 1916, followed by a series of spectacular discoveries during and after World War I. These included the Texas Panhandle field (1918–1919), the Hugoton Field in Kansas and Oklahoma (1922), and the Kettleman Hills in California (1928), among others. Oil discoveries rose after 1926 as well, and in 1930, Marion (Dad) Joiner discovered the giant East Texas oil field that would ultimately yield more than five billion barrels. Far from simply being the result of resource abundance, these finds were socially created, reflecting the increasing importance of geologists and other scientists in gas and oil discovery and exploitation. For example, one expert estimated that before about 1905, knowledge of geology accounted for perhaps 10 percent of major discoveries, whereas in the period between 1930 and 1936, virtually all the large finds resulted from the application of scientific methods.[27]

Initially, development of most of the gas fields was slow owing to a lack of markets, for early pipelines rarely extended beyond 100 miles, and all leaked prodigiously. Such problems accounted for some of the waste and low-value use that characterized the industry. But the explosion in reserves after 1916, with its assurance that gas supplies would last for decades, justified long-term investments in pipelines, while improvements in pipeline technology so lowered transport costs that distant markets became economically attractive and entrepreneurs suddenly saw gold in what had recently been waste.[28]

Early pipe had been small—three to eight inches in diameter—and screwed together or bolted and sealed with rubber couplings; both methods were leak-prone. The volume of gas that a pipe can transport varies directly with the pressure and exponentially with pipe diameter. The development of alloy steels and seamless pipe production allowed larger pipes and higher pressures, while factory electric welding improved quality and sped up pipe production, thereby reducing costs. By the early 1930s, pipes were 24 inches in diameter, and pressures were more than 400 psi and rising—800 psi was "entirely feasible"—while the development of specialized trenching machines reduced labor requirements and sped up operations.[29]

Even with the best technology, however, transporting gas at higher pressure over long distances increased the number and cost of leaks, even as better markets improved the payoff to detection. In 1928, with industry aid, the Bureau of Mines concluded a study demonstrating the prevalence and economic cost of pipeline leaks. The bureau had long been concerned with the loss of gas from wells; in many publications, it had explained the importance of gas pressure for oil recovery and had promoted simple solutions—such as the use of mud-laden fluid—to prevent gas loss. Long-distance transmission, however, offered new opportunities for waste.[30]

In one of many similar instances, the bureau found a three-inch line 100 miles long leaking 600 million ft.^{3} of gas a year. It also found that most such leaks had simple, cheap fixes, and the bureau calculated the long-term economic loss from such wastage. With such incentives, companies sharply reduced transmission losses. Following bureau guidelines, Loan Star Gas, which transported gas at 300 psi pressure, determined that a hole 1/64" in diameter would lose 640,000 ft.^{3} of gas per year. The company developed a comprehensive program for detecting leaks. It estimated the cost per leak avoided at $14.59, and the average payoff to prevention was $115.[31]

Collectively, these innovations in pipeline transport sharply reduced costs, greatly increasing the range that it was economically viable to transport gas; accordingly, they disrupted a widening array of distant energy markets. Thus, a 900-mile, 24-inch pipeline from the Texas Panhandle transmitting at 425 psi to (say) Chicago might purchase gas at the wellhead for $0.06/Mft.^{3} that it could sell for an average of $0.28/Mft.^{3} and earn a 12 percent return on investment.[32]

The disruptions began in 1925, when Magnolia Petroleum built a 200-mile, 14-inch welded pipeline from Monroe, Louisiana, to Beaumont, Texas. In 1926, another company completed a line 170 miles to Baton Rouge and then extended it to New Orleans. In 1927, a 20-inch pipeline brought gas into Denver and other Colorado towns and cities. In 1928, a pipeline going east from Monroe, brought gas to Birmingham, on its way to Atlanta, which it reached in 1930. In the early 1930s, natural gas from Texas arrived in Chicago, Milwaukee, and Minneapolis, along with many other towns and cities in Illinois, Iowa, Kansas, Michigan, Minnesota, and Nebraska.[33]

Figure 3.2 captures the effects of these supply changes on the price of natural gas relative to soft coal. As can be seen, wellhead gas prices de-

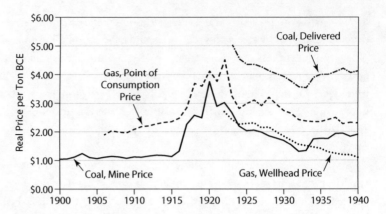

Figure 3.2. Prices of bituminous coal and natural gas for industrial use at points of production and consumption, 1900–1940. (Prices are from USBM, MR, and MY, various years, with gas converted to BCE. The inflation adjustment employs the David-Solar price index in Carter et al., *Historical Statistics*, series Cc2)

clined steadily and on a BCE basis were roughly equivalent to the mine price of soft coal up to 1933. Thereafter, while gas prices continued to fall, coal prices shot up despite coal's productivity gains from mechanization, because the miners' union engineered a 58-percent wage increase between 1933 and 1936. As the figure shows, the disparity between the average delivered price of coal (including freight transport) and gas widened after 1933.[34]

As a result, while use of natural gas for manufacturing and other industrial purposes (but excluding that used in the field or for carbon black) increased modestly between 1919 and 1923 (table A1.6), its use doubled between 1923 and 1929 and nearly doubled again between 1929 and 1940. For cement producers, gas had made up only about 9 percent of fuel use in 1909; the figure did not budge until 1929, when it jumped to 12 percent and by 1935 accounted for nearly 20 percent of fuel use. Iron and steel producers increased use of natural gas from 68 billion cubic feet in 1920 to 122 billion in 1939, while they cut in half the coal they burned. Chemical manufactures increased gas use from 12 billion to 305 billion cubic feet over the same period as they evolved into petrochemical companies, and they pared their use of coal as well.[35]

While this discussion makes the shift to natural gas seem like an almost-inevitable result of expanding supplies, that is not how matters appeared at the time to the men and women who were trying to sell gas. In fact, in the early 1920s, the recently formed American Gas Association developed a formidable research-and-marketing campaign to sell commercial and industrial gas that had no parallel in the coal industry. At this time, members of the association almost entirely sold manufactured gas—which was, because of its higher cost and lower heating value, a much tougher sell than was natural gas.

The AGA had begun to investigate establishment of a research laboratory in 1919 before it began to focus on industrial gas sales. A committee reviewing research on gas found much activity funded by state and regional gas organizations, as well as universities and governments;

nevertheless, the association established a central laboratory in 1925. Initially, most of its work involved testing domestic appliances.[36]

An industrial and commercial gas section of the AGA began in 1923; by 1927, it had established an annual budget of $100,000 for research. An additional $35,000 a year placed advertisements in 21 trade publications (e.g., *Iron Age*; *Chemical and Metallurgical Engineering*) and six college papers. It also published a journal, *Industrial Gas*, with a distribution list of 3,500 and was successful in getting trade journals such as *The Ceramist* and *Bakers Weekly* to reproduce its work. *Iron Age*, for example, carried a story by a member of the AGA informing readers of the importance of using gas firing for enameling. The AGA Industrial Section included a speakers' bureau that developed prepared talks for its members; it instituted short courses for industrial gas salesmen as well. The section also did market research surveys of industries likely to be particularly receptive to manufactured gas—ceramics, metalworking, baking—and it urged gas companies to develop rate structures that would make the sale. Because manufactured gas was high cost, companies worked with industrial equipment makers, who were often members of the AGA to improve combustion efficiency.[37]

Coal had nothing to match this. Between 1923 and 1927, industrial and commercial sales of manufactured gas rose 46 percent in comparison to domestic sales, which increased only 13 percent. Then, in 1927, the Natural Gas Association (NGA) joined the AGA, ensuring that they would also benefit from that institution's sales and research juggernaut. Thus, when natural gas began to appear in cities such as Denver, Chicago, Des Moines, and Minneapolis, the utilities that blended it with manufactured gas suddenly had a far better product to sell. In addition, they could rely on the research and marketing expertise that the AGA had been perfecting for nearly a decade.[38]

Research provided the key that would unlock many markets. In 1934, a representative of the AGA's laboratory explained how their work expanded gas markets by increasing conservation:

> The demand for lower costs has been partly met by research . . . [that has improved energy efficiency]. This type of research has taken several forms due to the almost infinite variety of heating operations practiced in industry. A wider use of radiant heat . . . the development of refractory insulating materials for furnace construction . . . better controls . . . [use of] synthetic and controlled furnace atmospheres; the use of direct gas heating in place of indirect heating; and the general shortening of time cycles have all contributed to reducing the Btus required for particular heating operations.[39]

Two years later the AGA's Committee on Ceramic Industries described some of the ways that new technologies were spreading the use of gas in ceramics. It had "sponsored a project in the application of gas radiant or hot tubes to porcelain enameling furnaces," and "the results . . . [were] so satisfactory that at least seven additional furnaces have been converted . . . from electricity." The committee also noted that in porcelain manufacture, furnace redesign had transformed "two old style muffle type kilns employing the direct fired principle," doubling kiln output while using less gas per hour. Another company that switched from coal to manufactured gas to fire vitreous enamel experienced lower costs and greater output with fewer rejections.[40]

At that same meeting, the Non-Ferrous Metals Committee reported that gas accounted for only about 10 percent of fuel employed in copper melting but "through the use of insulated refractory and carborundum linings, developed during the past year, and efficient burner equipment the gas industry is in a position to obtain the crucible melting business held by coal and oil." In iron and steel making during the 1930s, gas largely swept coal out of enameling, annealing, forging, galvanizing, and much else. Typically, it reduced fuel use, raised output, and improved quality. In 1940, a review in *Metal Progress* reported that "coal as an industrial furnace fuel has almost disappeared," while "gas remains the only fuel for all heating operations."[41]

These examples could be multiplied many times, but they should suffice to demonstrate that the spread of gas in manufacturing during these years was not simply a matter of resource abundance. Aggressive research into equipment design, insulation, and combustion, packaged with aggressive marketing, built on the natural advantages of gas and contributed to its conservation and its spread.

Improving Energy Efficiency

Energy efficiency in manufacturing also improved dramatically during the interwar years (see table 3.1) as primary fuel use per unit of output fell nearly 30 percent between 1920 and 1940.[42] This efficiency gain helps account for the economy-wide decline in fuel use relative to GDP in the two decades after World War I, since manufacturing accounted for about 30 percent of *all* American energy use in 1920 (tables 3.1 and A2.2). Moreover, because the share of manufacturing output in GDP changed little down to 1940, deindustrialization played no part in the economy-wide reduction in energy intensiveness. Nor was the decline simply the outcome of a diminishing importance of energy-intensive sectors, for the seven industry groups that accounted for the lion's share of fuel use saw their share of manufacturing value added increase between 1929 and 1939.[43]

Accordingly, most of the gains in fuel efficiency depicted in table 3.1 reflected improvements in production processes—a result of the application or perfection of ideas and technologies, many of which had originated before World War I, sometimes in Europe. Much of the improvement resulted from a kind of collective learning-by-doing as the technological network diffused ideas for improvements. Their rapid introduction after World War I reflected the influx of scientists and engineers, as well as the price spikes and fuel and labor shortages associated with the war, which focused attention on ways to save energy. Of course, all such efforts had to pass a market test. As one engineer put matters: "investment in heat recovery equipment is warranted only . . . [if] earnings on the new investment [could earn the company's] minimum rate of return." This market test was two-edged: while it winnowed out conservation that was unprofitable, the prospect of high returns proved a powerful spur to discover and implement energy savings that would pay off.[44]

Manufacturing companies used fuel to generate and transmit power throughout the factory. By the end of the nineteenth century, boilers had grown larger, operating at higher temperatures and pressures, and the Corliss reciprocating engine had achieved a high degree of efficiency. Yet by the 1890s, replacements for reciprocating steam engines were in the wings, one of which was the internal combustion engine. Before the widespread availability of cheap petroleum, such engines might burn coal gas, and they were much more efficient than reciprocating steam engines because they operated at higher temperature and pressure, thereby promising to save scarce resources. As a more efficient way to burn coal, gas engines proved enormously attractive to Progressive-era conservationists. In 1908, when Israel White admonished the governors' conference on the "insane riot of destruction and waste of our fuel resources," he cited the coal-gas engine as one remedy.[45]

Yet the coal-gas engine would never fulfill its advocates' dreams; although it was a technical success, it was an economic failure except where—as in iron and steel—large volumes of waste gases were available. Steam turbines, not internal combustion engines, became the dominant prime mover. Yet by the 1920s, for many manufacturing companies, all forms of internally generated power were giving way to purchased electricity.[46]

Factories Shift to Purchased Electric Power

For readers a bit unsure of the terminology of electricity production, the following sketch of the basics may be helpful. Thermal generation begins with coal or some other fuel burned in a furnace to generate steam in a boiler. The steam then powers a reciprocating steam engine or later a steam turbine. This power then drives a dynamo that produces electricity:

Fuel → furnace → boiler → steam engine/turbine → dynamo → kilowatts

Thomas Edison invented the first electric utility—the Pearl Street (New York) Station—which opened in September 1882, employing reciprocating steam engines powering six dynamos that supplied 600 kw of DC power to about 90 customers. Thereafter, the industry took off. Twenty-five years later, in 1907, the census reported more than 4,700 central stations. By this time AC power—which shipped economically over long distances—was replacing DC generation, while about 4,700 dynamos, driven increasingly by steam turbines, generated 2.7 million kilowatts. This rapid expansion continued throughout the interwar years and beyond, and the continuous growth of new and ever larger plants ensured that utilities would drive the technological frontier of power generation and conservation. Technological convergence, along with the institutional network noted above, ensured that utilities' new methods would spread to manufacturing and other sectors of the economy.

Fuel economy was central to the growth of these early utilities because conservation was profitable. As the chief engineer of the Boston Electric Illuminating Company described that company's motives for installing a new higher-pressure steam plant, "We were after dollar efficiency not thermal efficiency alone."[47]

Fuel and purchased power typically accounted for nearly half of central station costs compared to perhaps 2 percent for manufacturers, thus focusing the former much more sharply on energy saving. Because Pearl Street and other early companies had generated electricity mostly for light, they tended to have a low load factor (the ratio of average to maximum output) that resulted in expensive equipment sitting idle during daylight hours. This generated a strong incentive to develop markets for factory power that used daytime capacity, but because manufacturers could generate their own electricity, low prices for commercial power were crucial—reinforcing incentives for fuel economy. Finally, because utility rates were regulated, the sharp wartime rise in coal prices squeezed profitability, sharpening the focus on fuel efficiency. A result of these circumstances, as the Bureau of Mines put it, was that "the route to promotion was seen to lead through the boiler room and the best brains in the electric power industry were directed to squeezing more and more kilowatt-hours out of the same ton of coal."[48]

Sometimes utilities found it advantageous to sell steam, as well as electricity—a very early example of cogeneration. Factories that shifted to purchased electricity lost the use of waste steam for space heat, motivating central stations to market steam. In one imaginative contract, Quaker Oats shipped oat hulls through a 1,200-foot pipeline to the local utility, which burned them to supply Quaker with steam for processes and electricity for power.[49]

These various market incentives provided a sharp spur to fuel economy. In 1894, central stations required roughly 10 pounds of BCE to generate a kilowatt-hour. By 1902, this requirement had fallen to about 7.3 pounds. By 1940, fuel requirements averaged less than two pounds per kilowatt-hour (fig. 3.3; table A3.3). These gains resulted not only from transformative innovations but also from incremental technological and organizational

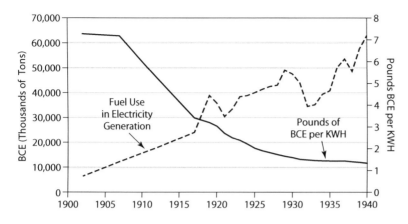

Figure 3.3. Efficiency gains and fuel consumption in electricity generation, 1902–1940. Dramatic improvements in utility energy efficiency spread to other steam users and underwrote the economy-wide surge in electrification during interwar decades. (Chart data drawn from appendix table A3.3)

improvements that reflected the increasing application of science and engineering to fuel combustion. These improvements were important for three reasons. First, they economized not only on fuel but other resources as well. Total factor productivity (output relative to all inputs) in electric utilities rose at an annual rate of 5.2 percent from 1899 (the first year data are available) to 1940, leading to a sharp decline in costs and prices. Second, rising energy efficiency contributed to the enormous expansion in electricity use during these decades and the prosperity that it helped generate. In 1929, *Electrical World* estimated that a 10 percent cut in electricity prices would expand sales 20 percent. Accordingly, with fuel making up about half of utility costs, an increase in efficiency of 1 percent would yield a 1 percent increase in sales. Third, as we will see, innovations at utilities diffused throughout manufacturing and elsewhere, improving fuel economy and saving resources there as well.[50]

Initially, the improvements in utilities' fuel efficiency depicted in figure 3.3 reflected the perfecting of reciprocating steam engines, but by the late 1890s, steam turbines were beginning to push them aside. Turbines were transformative: they were smaller, cheaper, and more improvable than any other prime mover. While initially less efficient than the best reciprocating engines, they soon surpassed the older technology, growing rapidly in size and efficiency. In 1903, General Electric and others were selling 5,000 kw turbines that required 23,500 Btu/kwh (1.8 lb. coal); a decade later, 20,000-kw turbines required about 14,400 Btu/kwh (about 1.1 lb.). By 1926, some units were as large as 200,000 kw and required 9,000 Btu/kwh (about 0.7 pounds).[51]

Turbines and boilers were, of course, part of a technological package, and the larger turbines inevitably required much greater power as well. The experience of Detroit Edison when it installed a new boiler circa 1911 demonstrates the kind of shop-floor innovation and information sharing that yielded fuel economies with economy-wide spillovers. The company replaced two older 520 hp. boilers with a new 2,365 hp. boiler that had been custom designed by Babcock and Wilcox with the aid of the utility's engineers. Engineers from nearby Michigan schools and from Stevens Institute of Technology assisted in the setup and testing. The boiler achieved the remarkable efficiency of 80 percent. As was common, the company shared its findings in technical journals. The likely motive for such sharing was reciprocity, as each company gained from the experiments of others. Looking back from the vantage point of 1935, the journal *Power* explained: "Cooperation between utility and equipment manufacturer explains much of the remarkable progress made during this 15-year period. And utilities, spared the handicap of competition, shared their findings to the limit. One would act as a laboratory for this; another for that."[52]

In 1895, maximum boiler pressure had been about 250 psi; by the mid-1920s, pressures averaged 400 psi with steam temperatures of perhaps 700°F. By 1929, a review described 1,200–1,400 psi as typical of new construction. Temperature increases were more moderate, being constrained by materials' limits, but in 1927, Detroit Edison began to experiment with a boiler that would deliver superheated steam to the turbine at temperatures of 1,100°F—a level the company historian describes as "totally unexplored outside the laboratory." The company's progress in fuel efficiency, he concludes, was "not the product of revolutionary invention but the persistent and patient company-wide inquiry on both method and material." Writing in 1925, the ASME Fuels Division observed that "to an increasing extent the [utility] boiler room has become a laboratory for steam generation." There were also experiments with combined cycle generation, with binary fluid systems employing mercury or diphenyl oxide (both have higher boiling points than water), but neither proved successful. In traditional steam systems, use of water tube boilers spread to accommodate the higher pressures, while boilers invariably included economizers (that used waste heat to raise the temperature of feed water). Superheaters and air preheaters also enhanced efficiency. These developments both reflected and encouraged major advances in metallurgy during the interwar decades. Better alloy steels, improvements in welding, and use of X-ray testing improved safety, reduced construction costs, and improved fuel economy. Joint meetings of the ASME and the American Society for Testing Materials helped spread the new methods among utilities and to manufacturing, mining, hotels, and other steam-using sectors of the economy. Journals such as *Power*, *Power Plant Engineering*, and other technical publications also diffused the new technology.[53]

Wartime developments reinforced longer-term developments in science and engineering that were driving these important changes in power generation. As the journal *Power* observed, "The post-war scarcity of coal and labor had intensified the development of fuel and labor saving equipment in the boiler room and engine room." With the increasing application of scientific principles, engineers could squeeze more work from

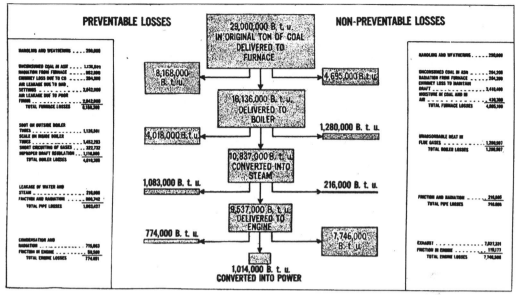

Figure 3.4. By the time of World War I, engineers had begun to compute heat balances to pinpoint how much heat was wasted and where it went. Here, out of 29 million Btus in a ton of coal, burned in a steam locomotive, only about a million become usable power. Some of the waste is inevitable (right-hand side) but preventable losses (left-hand side) amount to about 14 million Btus. ("Fuel and Heat Losses between Mine and Crank Shaft." William N. Allman, "Reducing the Fuel Consumption of Power Plants," *Railway Mechanical Engineer*, Feb. 1921, 88–91, 89. Public domain)

fuel. As one writer put it, by World War I, "alert engineers everywhere had learned that combustion was a chemical process." One result of this deeper understanding was that they increasingly began to calculate heat balances to improve efficiency (fig. 3.4). These were an accounting of the heat derived from burning fuel, and they invariably demonstrated that only a small fraction of it ultimately generated useful work, the rest being waste of one form or another.[54]

At Detroit Edison, the introduction of heat balances occasioned a changing of the guard from men without formal training to those with engineering skills. Alex Dow (1862–1942) was the company's plant manager circa 1911. Trained in both electrical and mechanical engineering, Dow was a pioneer in utility research and would go on to become president of the American Society of Mechanical Engineers. At Detroit Edison, he ran head-on into the power-room manager, an old-school, practical man. Dow explained to him the value of heat balances: "we should show how many heat units are accounted for by the electrical output, [and] how many by the losses in machines [etc.]." Unable to persuade the man, Dow replaced him with a younger engineering graduate from Cornell. Soon, the company's "power houses were on occasion cluttered with equipment and instruments of research and with men whose vocabulary was studded with scientific jargon."[55]

Such calculations, along with wartime pressures, focused attention on just how much heat—and money—higher temperatures and an uninsulated surface might squander (table 3.2). This was not a new idea. Rock wool, a waste product resulting from steelmaking, had been around for

Table 3.2. Industrial Fuel Saving from Pipe Insulation, 1918

Steam pressure	10 lb.	100 lb.	200 lb. + 100° superheat
Steam temperature (F)	240°	338°	400°
Loss from bare pipe (Btu/ft^2/hr)	409	802	1,737
Saved from insulation* (Btu/ft^2/hr)	333	676	1,525
Tons coal saved per 10,000 ft^2 per year	1,490	3,024	6,328
Value of saving @ $5.00/ton	$7,450	$15,120	$31,640

Source: "Conservation of Heat Losses from Pipes and Boilers," ASME Journal 40 (Nov. 1918): 918–922, table 3 and author's calculations.
*Insulation is 1" of 85 percent magnesia.

decades, as had calculations demonstrating its economic payoff. Use seems to have spread rapidly around World War I, however. In 1917, Rockwell Manufacturing claimed that insulating its steam pipes had saved about 51 tons of coal during the year. A few years later, a review of insulation among ceramics producers found that fuel savings ranged from 10 to 15 percent. New insulation materials also came on the market: in the 1930s, insulating refractories (firebrick) appeared, dramatically increasing furnace efficiency. Naturally, as one engineer from Johns Manville put it, "the matter of economical thickness of insulation . . . [was soon] the subject of extensive investigation."[56]

Much heat simply disappeared up the smokestack, so companies began to employ soot blowers along with water treatment to improve heat transfer. Laboratory analysis discovered ratios for CO_2, CO, and other flue gases that measured combustion efficiency. Soon, these laboratory instruments entered utilities' and manufacturers' boiler-room tool kits—with suitable instructions for users—while recordkeeping of meter results allowed continuous checks on efficiency. "With no other class of power plant equipment has the development been more marked and rapid than it has been with instruments," one writer claimed in 1940. When one utility installed CO_2 readers and draft gauges, it was able to cut coal use from 8–9 lbs./kwh to 6.1 lbs./kwh. Merrimac Chemical also metered flue gases and temperature along with coal use and steam flow, keeping records and computing weekly heat balances, thereby raising combustion efficiency and saving more than $16,000 a year. As a further check on efficiency, boiler rooms at Merrimac and elsewhere began to "sell" steam and electricity at cost to other departments, providing a check on their efficiency and that of the boiler room as well. By about 1930 automatic combustion control appeared.[57]

Organizational innovation also led to significant energy savings in electricity generation. For a utility with multiple boilers and turbines of varying ages and sizes, efficiency might vary sharply with the allocation of any given load among its equipment. Although a single station might strive to discover the most efficient allocation of power production among its boilers and turbines, the rise of multiplant companies, along with the grid, allowed companies to concentrate generation in low-cost plants. There were, of course, losses from transmission. But by 1899, the Southern California Electric Power Company was shipping power 81 miles at 33,000 volts (33 kv) with a line loss of about 10 percent—far less than transmission of only a few hundred feet through shafts and belts within a given factory. In the 1920s, losses from transmitting 300 miles at 220 kv were 10 percent. By the 1930s, companies were shipping power 200 miles at 287 kv with a line loss of less than 7 percent.[58]

Most of these improvements were not fuel-specific and likely had little impact on energy choices. The exceptions were use of improved stokers and, especially, powdered coal—which swept through utilities and other large energy users. The idea of an automatic stoker dates from the time of James Watt, and by 1900, a bewildering variety of types had emerged. What spurred their use was the combined desire to save labor and increase boiler size: "the mechanical stoker has, more than any other piece of boiler equipment, made possible the present size of boiler plants," an advocate announced to readers of *Combustion* in 1924. Developed for large users, stokers soon found a home at smaller firms as well. Stokers did away with the backbreaking labor in firing and, compared to hand firing, cut coal use by 15 percent. Typically, underfeed stokers were the most efficient; they also largely did away with smoke, and by the mid-1920s, they accounted for about 75 percent of sales. In a few instances, municipal smoke ordinances seem to have nudged companies to use stokers. In 1937, when public officials complained to one South Bend Indiana furniture company about its smoke, the company installed underfeed stokers that stopped the smoke and paid for themselves as well.[59]

Even as stokers were improving and increasing in their use, around World War I, the burning of pulverized coal began to spread from its origins in cement manufacture. Efforts to burn coal in this manner in England and Germany dated from the 1830s. In the United States, by about 1894, engineers at the Atlas Portland Cement Company had pioneered use of pulverized coal. One writer described the innovation as "an American device in labor saving utility." Cement makers' expertise in grinding materials gave them an advantage, and pulverization allowed them to substitute coal for more expensive fuel oil or natural gas. A survey in 1918 found that about 80 percent of cement making then burned pulverized coal, while it was also widely used in kilns at foundries and smelters, all of which found attractive the long flame that resulted from dust combustion.[60]

Kiln and furnace combustion were sufficiently similar that utilities and manufacturers also found the lure of pulverization attractive. Burning coal on a grate might leave perhaps 5 percent of the fuel unburned in the ashes; pulverized coal eliminated such losses, for it yielded almost complete combustion with the added benefits that it allowed the burning of low-quality coal and largely did away with smoke. Pulverized coal also reduced ash handling, for small particulates ("fly ash") went up the chimney—a problem Detroit Edison solved by applying an electrostatic precipitator. An added attraction was that with some of the properties of a liquid, pulverized coal was comparatively easy to handle, thereby reducing labor costs and allowing much more careful control of combustion. So beguiling was the idea that German companies experimented with pulverized coal as a substitute for gasoline in automobile engines.[61]

Despite such attractions, the differences between burning pulverized coal in a kiln and in a furnace were sufficient to retard its diffusion as a source of power. Because pulverized coal burned in suspension like a gas, proper combustion required large—and therefore expensive—furnaces, while the flame was so hot it could destroy furnace lining and, depending on the coal burned, sometimes left a molten slag. There were also problems involving coal moisture, particle size, and safety, for coal dust is highly explosive.

By the early 1920s, engineers were solving most of these technical difficulties. The breakthrough came with large-scale experiments undertaken jointly by Combustion Engineering, the US Bureau of Mines, and the Milwaukee Electric Railway and Light Company at the latter's plants from 1918 to 1920. The company's apparent motive was the need to burn low-quality Illinois coal—a result of World War I shortages and federal allocations. These large-scale experiments worked out many of the details of pulverization, the role of moisture content, furnace size, air supply, and much else. Reviewing the results, *Power* editorialized that "pulverized coal in the power plant . . . may be pronounced a commercial success." Like Detroit Edison's earlier experiments with large boilers, the experiment was widely publicized in technical journals and publications of the Bureau of Mines, demonstrating that pulverization might cut fuel consumption by 6 to 10 percent compared with stoker firing. A study published by the pulp and paper manufacturers confirmed these findings and demonstrated large labor savings as well.[62]

Other utilities soon followed Milwaukee's lead, and with the technology well-publicized, it spread to large manufacturers such as the Ford Motor Company, Allegheny and Bethlehem Steel Companies, St. Joseph Lead, and Lima Locomotive works. In 1923, a member of the National Electric Light Association reported that "the use of pulverized fuel equipment . . . has taken rapid strides and many new installations have resulted." In 1928, a survey found that pulverized fuel accounted for 17 percent of utility coal use. Its combustion technology also continued to progress rapidly. Initially, engineers had designed larger furnaces to ensure adequate combustion, but later experiments revealed that a design to increase flame turbulence could enhance combustion in smaller furnaces. Use of water-cooled walls also spread, reducing maintenance costs.

One historian ranks use of pulverized coal by utilities—along with Edison's lamp and distribution system, the transformer, and the steam turbine—as "one of the four fundamental technological development[s] that made low cost central station service possible."[63]

Companies also learned to save energy by choosing the proper type of coal. Coal is a complex mixture, and, as we have seen, some buyers had been trying to discover a rational way of purchasing it since the mid-nineteenth century. The USGS gave a boost to these efforts with the systematic study of coal chemistry and combustion that it inaugurated about 1904 and that the Bureau of Mines continued after 1910. Despite such studies, as late as 1916, one speaker told the American Institute of Mining Engineers that for many companies, the criterion for coal purchases remained "ocular inspection." This made little sense, he thought, unless the coal's intended use was as a museum exhibit, and he claimed that producers often simply threw away "dull" looking coal as worthless. In the face of rising coal prices, procedures were quickly improving.[64]

Classifications of coal by rank (from peat and brown coal to anthracite) and class (extent of impurities) were too general to be very useful. But much more detailed investigations of Btu content—as well as volatile elements, moisture, sulfur, ash, and other constituents—allowed large buyers such as companies and state and municipal institutions to make more rational fuel choices. Companies discovered that the best coal also depended on a host of boiler characteristics and that certain coals burned best with particular types of stokers. Not only coal's chemical composition but its size mattered as well, for not all sizes burned equally well with all stokers. The goal was to maximize energy output per dollar, subject to constraints such as availability, ash-fusing point, and labor costs. The Bureau of Mines and, later, the American Society for Testing Materials developed and explained the proper procedures for statistical tests and the need to employ identical tests among various coals. There were also instructions on the importance of random sampling (which proved difficult because coal might settle differentially during shipment) and the need for an adequate sample size to obtain a reliable result.[65]

By World War I, journals with a focus on energy or power were routinely publishing articles with instructions on how to buy coal. In 1919, one journal claimed that "coal analyses are rapidly becoming a necessity in power plants" and urged companies to consult Bureau of Mines publications on proper procedures, while a later writer asserted that "proper coal selection is the greatest single factor involved in efficient steam plant operation." By the early 1920s, large users of coal, such as utilities, employed complex specifications and sophisticated sampling when they purchased coal. Contracts would specify levels for Btus, ash, sulfur, and moisture, implementing price penalties for violations; for example, a contract might specify no more than 5 percent ash per ton with a penalty that rose as ash exceeded 6 percent up to 12 percent, at which point the shipment would be rejected. More rational purchasing encouraged the spread of washed and "de-dusted" (treated with oil), and sized coal as well, as producers strived to improve product quality in order to compete with gas and oil. Iron makers found that coal washing improved the quality of coke and reduced emissions. In a roundabout fashion, cleaner coal, which might reduce weight by perhaps 5 to 10 percent, saved energy in transportation as well. In such small ways do markets also economize.[66]

Such procedures, along with improved stokers and pulverization—all of which increased the efficiency with which coal was burned—likely helped stave off competition from alternative fuels. For example, the Hotel New Yorker, which had shifted from coal to oil to furnish steam in 1935, found the combination of price and dustless qualities of pulverized coal sufficiently attractive that it returned to coal in 1940. Still, coal's share gradually declined from nearly 90 percent of all utility fuel in 1917 to about 82 percent on the eve of World War II. Such trends concerned experts, who worried about future supplies of oil: "Not a barrel of oil should be used under boilers," George Otis Smith pronounced in 1925, and a year later, the Federal Oil Conservation Board also urged that oil be diverted to "higher uses." The market, it seems, agreed, for gas was the main completive utility fuel, and its use tripled between 1926 and 1936.[67]

Adding to coal's woes was the rapid growth of hydropower in the 1930s owing to such large federal projects as the TVA and Hoover Dam. These and other smaller federal projects accounted for about 4 percent of utility power generated in 1939, and they undoubtedly displaced some fuel. In the case of Hoover Dam, however, waterpower largely competed with oil- and gas-generated power in California and other Pacific states. TVA, which opened for business in 1934, was a more direct threat and resulted in a storm of complaints from the National Coal Association, which joined private utilities fearing competition from lower-cost federal power. Yet competition with cheaper federal power drove down private system rates as well, widening the market. The East South Central states, where TVA competition was greatest, saw total electricity production double between 1934 and 1940 while that from steam power tripled.[68]

Power Distribution in Factories

While the fuel economy of factory power production had steadily improved even before the advent of electricity, the same could not be said of power distribution within factories. Well into the 1890s, in most cases, a wildly inefficient maze of shafts, pulleys, and belts transmitted power from the engine to the machine. One careful study of 1897 found that half or more of power might be used in the transmission system. Another study found that 67 percent of power was consumed in transmission. In general, transmission losses were worse at large factories, where power needed to travel long distances. At Baldwin Locomotive, 80 percent of the power went into transmission. Another writer claimed to know of a factory in which transmission consumed 93 percent of the power! In such a situation, it is only a slight exaggeration to compare factory power distribution with the perfect Rube Goldberg machine that was immensely complicated but accomplished almost nothing useful. Finally, because friction was roughly independent of load, the share of power it consumed rose as engine load decreased. One author calculated that a steam engine and transmission system that delivered 45 percent of its power at full load could deliver but 28 percent at a three-quarters load.[69]

Companies experimented with various substitutes for the mechanical distribution of power. What finally won out, of course, was electricity. The change came in stages, and in the twentieth century, individual motor drive gradually triumphed. The motives for factory electrification were not energy saving, for as noted, fuel often accounted for only a small percentage of a company's costs. Rather, electrification was a boon because of its form value—the flexibility in equipment use and location that came from escaping the imperatives of mechanical drive. A result was that the shift to electricity often raised output perhaps 20 to 30 percent. Most engineers would have agreed with one writer who claimed in 1895 that this was "the most important advantage gained by the electric system." Modern writers concur that this was an important part of manufacturing's Great (Productivity) Leap Forward during these years.[70]

Still, the energy savings were enormous: when a company shifted to electric distribution, transmission losses within the factory essentially reflected those of electric motors, which were about 80 percent efficient. Westinghouse Air Brake found that in tests, the shift from steam engines to a turbine with electric distribution saved 30 to 40 percent of the fuel bill compared to mechanical power transmission. Another writer claimed that shifting from steam to electricity would cut steel mills' fuel bills in half. When the US Government Printing Offices switched to electric power distribution, the fuel bill fell 75 percent, and production rose 15 to 20 percent.[71]

The energy savings due to the shift of factory power to electricity thus include the gains from shifting away from mechanical transmission to electric drive within factories, as well as the energy savings from abandoning factory steam engines and purchasing electric power. The shift to purchased electricity reduced demand for coal (and other fuels) for two reasons: first, because utilities were likely more efficient than individual factories and, second, because (in 1917, for example) 45 percent of the utility electricity derived from waterpower. Thus, while utility electricity required 3.41 pounds of BCE per kwh (2.55 lbs./hph) in 1917, the *average* kwh produced by fuel and waterpower combined (table A3.3) required only 1.89 pounds of BCE (1.41 lbs./hph).[72]

Table 3.3. Fuel Consumption per Horsepower-Hour at Factories and Utilities, ca. 1917

Factory				
Reciprocating steam engine and mechanical transmission	Indicated hp 3.00 lbs./hph	Brake hp 15% loss	Transmission in factory 50% loss	Hp at machine 8.57 lbs./hph
Central station				
Coal and water-driven turbines	At buss bar 1.41 lbs./hph	Long-distance transmission 20% loss	Electric motor 20% loss	Machine 2.35 lbs./hph

Table 3.3 assembles these various concepts and assigns plausible values for coal or other fuel consumption and losses around World War I, all in horsepower-hours. The estimate for fuel per indicated horsepower-hour (ihph) in factory steam derives from table A2.6 and is for 1910. As can be seen, a steam boiler and reciprocating engine that burned BCE using 3.00 lbs./ihph required more than eight pounds of fuel to deliver a horsepower-hour at the machine. The average of fuel and water-powered utility turbines, however, required 1.41 lbs. BCE/hph, and with transmission losses of 40 percent, 2.35 lbs. BCE/hph at the machine.[73]

A further implication is that with the transition to electricity, *usable* factory horsepower at the machine rose much faster than did the indicated horsepower reported by the census. Writing in 1934, Harry Jerome calculated that average (indicated) horsepower per worker in manufacturing rose from 2.11 in 1899 to 4.86 in 1929, or by 130 percent. With the spread of electric transmission, horsepower at the point of production probably rose nearly three times, and this surely made a major contribution to the rise in total factor productivity during these years.[74]

While no one should imagine that these calculations are precise, what they suggest is surely true: the shift by factories away from reciprocating steam engines employing mechanical transmission of power to purchased electricity distributed to the machine increased usable power even as it conserved immense amounts of fuel.

Byproduct and Waste Fuel for Process Heat

Many industries used fuel not only for power but also in production processes. These industries included, for example, canning of foods, paper and cement making, chemicals, petroleum refining, and iron and steel production. Here, too, savings of primary energy were dramatic, especially in the interwar years, and they largely derived from the capture of waste heat and byproducts. "Nothing in the arts of manufacture is more indicative of economic efficiencies than the utilization of products that have been rejected as wastes or residues," a census writer observed in 1900. And nowhere were these efficiencies more dramatic than in the making of coke used in the production of pig iron. Coking constituted a major employment of bituminous coal, accounting for 15 percent of total production in 1910. Moreover, coke fired the blast furnaces that produced the pig iron that underwrote the Industrial Revolution. To Progressives, coke not only underlined coal's importance; it also symbolized the ways that Americans squandered it. In their 1909 report to the National Conservation Commission, two USGS researchers described the "enormous wastes" from American methods of making coke. A year later, a Bureau of Mines writer went even further: "probably in no other industry . . . is there so much waste as in the manufacture of . . . coke," he lamented. The process was famously polluting as well. The geologist Charles Van Hise recounted the "dense clouds of smoke" that made the "dull sky cheerless and unhealthful."[75]

In 1913, Berton Braley described coking to readers of *Coal Age*:

> The trees are black with dust and smoke,
> The grass is burnt and sere,
> The noxious gases from the coke
> Pollute the atmosphere,
> The valley droops as with a blight,
> There is no vivid green,
> And reeking ovens day and night
> Make desolate the scene.[76]

American coke making at this time used the beehive method almost exclusively. The name came from the dome-shaped kiln; in this process, workers placed coal in the kiln and fired some of it in a low-oxygen environment. The process, called destructive distillation, simply vented the volatile elements of the coal into the air, to "make desolate the scene," leaving the coke, which was almost pure carbon. And because beehive ovens used coal as a fuel, coke yields were low.[77]

The byproduct process, by contrast, used retort ovens to distill the coal at high temperatures (900°–1,200°C); it captured the volatile components, burning some of the resulting gas for heat, thereby saving coal. By 1900, Europeans had been using the process for decades. By any standard, it was technologically superior: in 1903, byproduct coking took 16 percent less coal than did the beehive process to produce a ton of coke (1.34 tons of coal versus 1.58). The volatile elements contained not only flammable gases but also ammonia, phenol, naphthalene, sulfur, benzol, and tar, among other things. And because the process captured the byproducts, it was far less polluting than beehive technology. While byproduct coking generated some fine waste, called coke breeze, producers burned it as fuel or sold it locally.[78] As Braley continued:

> For if we ceased to foul the air
> With murkiness and smoke,
> We'd have our rich byproducts there

And also have our coke,

It would not prove a huge expense

The profits would be fat,

But it might need some common sense

—And who could hope for that![79]

Common sense, embodied in the first byproduct oven, appeared in the United States in 1893. Using technology developed in Belgium in the 1880s, the Solvay Process Company erected it in Syracuse, New York, to produce the ammonia needed for soda making. Use by iron makers spread slowly at first; in 1900, byproduct coking accounted for only about 5 percent of production. The new ovens were more expensive to construct, and coal was cheap, while initially, markets for byproducts were thin. Writing in 1900, the USGS rather optimistically reported that "no sooner was the supply [of byproducts] available than the demand was created." In fact, markets for byproducts grew slowly until World War I sharply increased demand: the share of the byproduct process rose from 27 percent in 1913 to 65 percent in 1921. By 1940, the byproduct process accounted for 95 percent of coke production (table A3.4).[80]

Initially, byproduct coking conserved coal because—as I have noted—it used waste gases rather than coal to produce coke. Moreover, as the new process spread, increased use of cleaner, low-ash coal and better oven design resulted in further increased yields of coke per ton of coal. In addition, iron producers also reduced the coke needed in smelting pig iron. They increased insulation and use of scrap iron and discovered the importance of uniform sizing of coke in reducing fuel use. Some coke producers also began to save the heat normally lost when coke was removed from ovens and cooled (quenched). Instead of spraying it with water, companies began to experiment with dry quenching, which cooled by circulating inert gases in the coke, employing the heat elsewhere. Initial experiments revealed that the process saved 69 percent of the heat available from the coke. Together, improving yields from coking and reductions in coke needed to smelt pig iron shrank the quantity of coal needed to make a ton of pig iron by 23 percent (from 1.80 to 1.42 tons) between 1913 and 1940 (table A3.5).[81] If energy efficiency in coking and smelting had remained unchanged over this period, production of the 43 million tons of pig iron made in 1940 would have burned an additional 17 million tons of coal.

Perhaps even more impressive was the gradual substitution of scrap for pig iron in steel production. In 1900, steel production used about 24 percent scrap and 76 percent pig iron. But the gradual change from Bessemer to open-hearth steel production, which could use more scrap, reduced requirements for pig iron and therefore coal. By 1940, scrap constituted 49 percent of the 91 million tons charged to steel furnaces, or about 45 million tons. Modern figures indicate that scrap saved 74 percent of the fuel

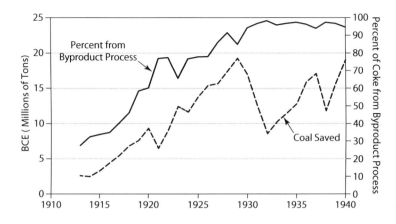

Figure 3.5. Percent of coke from byproduct processes and coal equivalent of byproducts, 1913–1940. The shift from beehive to byproduct coking and the rise of markets for the formally wasted byproducts saved large amounts of coal during the interwar decades. (Chart data drawn from appendix table A3.4)

required to make steel. Had pig iron been used instead, it would have burned an additional 47 million tons of coal.[82]

Yields of byproducts increased, too—from the Btu equivalent of about 20 percent of the coal used in coking in 1913 to about 35 percent by 1940— as companies purchased high-volatile coal for coking and substituted low-Btu blast furnace gas for coke oven gas to fuel the process. In the 1920s, companies increasingly began to sell coke oven gas to utilities and by 1940 were making up about a quarter of manufactured gas sold by utilities, thus reducing fuel use there. Calculations based on Bureau of Mines data (fig. 3.5; table A3.4) reveal that by 1940, byproduct coking produced gases and other byproducts that would have required an additional 19 million tons of coal had the byproducts been produced elsewhere.[83]

The economic importance of byproducts far exceeded the value of the coal that they saved. They provided the feedstocks that organic chemists and entrepreneurs would use to create a world of new products ranging from creosote to bubblegum and were the foundation of the new synthetic organic chemical industry.[84]

Many other industries found ways to conserve energy used for process heat. By the 1890s, cement production employed rotary kilns instead of older upright models. About 1899, the process caught the eye of Thomas Edison, who saw immediately that larger kilns would be more efficient. "The sixty foot kiln is a rotten proposition," he concluded, and his would be 150 feet long. Competitors followed Edison's lead, and by World War I, some kilns were 200 feet long. Companies also increased use of insulation and by about World War I began to capture exhaust heat from the kilns to generate power. By 1930, the industry had shifted from steam to electric power, which it purchased or generated using boilers powered by waste heat. These and other modifications reduced average primary energy use per barrel of cement by about 28 percent between 1909 and 1935. The combination of conservation and fuel switching combined to reduce coal consumption per barrel by 40 percent over the period.[85]

The development of boilers designed for waste heat along with bleeder turbines raised the payoff to conservation at many other companies that used process heat. Instead of using moderate pressure steam directly for process work and separately to power a reciprocating steam engine, firms discovered that they could employ higher pressures to drive a turbine and then bleed off some of the steam at a lower pressure for process purposes—another form of cogeneration. An engineer informed readers of *Paper Industry* that "the power required for the entire mill in most cases can be obtained from the steam used in the process of heating and drying." In such instances, as an editor of *Power* put it, "You Get Something for Almost Nothing." One company claimed that the modifications necessary to harness the process steam yielded a 100-percent return on investment. Sometimes companies might sell the surplus steam or electricity (fig. 3.6). So common were such schemes to squeeze power from process heat that in late 1929, *Power* devoted an entire issue to the topic.[86]

Where the uses of steam conflicted, some companies employed accumulators that stored steam ("thermic flywheels") for later use, or they purchased only the extra power required. Sometimes firms that generated large amounts of process heat, such as the Colorado Portland Cement Company, sold electric power to utilities. While such sales were modest in scale—about 539 million kwh in 1929, perhaps 0.5 percent of utility generation—they exemplify yet another way that markets reduced waste.[87]

Petroleum refining was another industry that made extensive use of process heat and became less energy intensive in the interwar decades, reducing primary energy input per barrel of crude about 30 percent. Some of these gains reflected improvements in technique that were specific to refining, but refineries were also adopting many of the same procedures that were saving energy elsewhere. By 1940, half of all refinery fuel came from waste gases and acid sludge. Moreover, the industry's center of gravity shifted away from the East and Midwest, where coal was cheap, to the Southwest, where it was not. Together, conservation and geographic shifts reduced total coal use from about six million tons in 1925 to less than a million on the eve of World War II, even as processed crude nearly doubled between 1925 and 1940.[88]

Many other examples suggest how widespread and varied were the techniques employed to reduce fuel costs. Waste-heat boilers spread rapidly among municipal gas utilities, saving fuel and labor. Sawmills burned waste wood; paper mills burned liquid and solid wastes and captured waste heat as well. In 1922, a small survey of firms in paper making found that "economizers" employing waste heat to raise the temperature of boiler feed water cut fuel use by about 13 percent and paid for themselves in a year. Ford Motor Company, which made window glass for its cars, discovered that briquetting the glass ingredients prior to charging its furnaces cut fuel use, increased output, and yielded a superior product. The company also

Don't Waste Exhaust Steam!

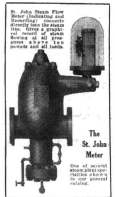

St. John Steam Flow Meter (Indicating and Recording) connects directly into the steam line. Gives a graphical record of steam flowing at all pressures above ten pounds and all loads.

The St. John Meter

One of several steam plant specialties shown in our general catalog.

Sell it to heat nearby buildings. Look over your own situation. Perhaps you can do as many others have done—get enough revenue from your exhaust steam to pay the entire cost of all of the power generated.

We have been designing and installing underground steam distribution systems for 40 years. More than 400 cities, villages, industrial plants and groups of buildings are heated by steam distributed through underground mains,—many of them by exhaust steam formerly wasted.

Write for Bulletin No. 20-PP. "Community Heating" of building groups; Bulletin No. 158-PP. "Adsco Heating" for individual buildings.

AMERICAN DISTRICT STEAM COMPANY
GENERAL OFFICES AND WORKS
NORTH TONAWANDA, N.Y.
Offices
New York   Chicago   Seattle   St. Paul

Figure 3.6. In the 1920s, calculation of heat balances encouraged companies to find uses for previously wasted steam by marketing it locally to supply heat to businesses and residents. ("Don't Waste Exhaust Steam!" Advertisement, *Power Plant Engineering*, Jan. 1923. Public domain)

developed a complex procedure for choosing among the various forms of waste and purchased gases differing in price and Btu content to minimize fuel costs.[89]

The Smoke Nuisance

Conservation and the substitution of natural gas and oil in manufacturing caused use of bituminous coal to fall about 27 percent in that sector during the interwar years (table 3.1), while improvements in combustion resulting from the spread of stokers and pulverized coal caused it to burn more cleanly as well. Accordingly, smoke pollution from manufacturers must have fallen much more than 27 percent. Nor did the shift to purchased power simply move the source of pollution, for as we have seen, utilities were leaders in fuel economy and in the use of pulverized coal and stokers. Along with similar developments on railroads and in households, these market-driven gains resulted in a decline of soft coal use by about 20 percent nationwide between 1920 and 1940 (table A1.4)—while its more efficient combustion must have yielded somewhat larger reductions in smoke.[90]

As we have seen, company motives for fuel switching or conservation were largely private economic gain as revealed by the many technical articles that invariably judged fuel conservation by its impact on the bottom line. Thus, the spreading use of automatic combustion control, stokers, and pulverized coal by large companies might nearly eliminate smoke—if not other forms of pollution—as did the switch from soft coal to gas or oil. Companies, however, saw these as added advantages of policies chosen on economic grounds—the frosting on the cake.[91]

St. Louis was virtually the only city to develop effective smoke-control procedures before World War II. In 1937, it began to require inspection and proper installation of new coal-burning equipment by large users. This did, indeed, reduce smoke. But St. Louis discovered what was becoming apparent in other major cities: large fuel users were no longer the primary cause of the smoke nuisance—for many of them had apparently already cleaned up—but rather the culprits had now become households. Thus, in Salt Lake City, between 1920 and 1929, the share of smoke from domestic sources rose from 20 percent to 85 percent. In Chicago, by 1937, domestic fuel consumption accounted for 67 percent of smoke. Increasingly, urban householders who wished to vanquish the smoke nuisance would need to look in the mirror.[92]

Conclusion

In the half century before World War II, a gale of creative destruction incited an energy transition that economized on fuel and began to sweep coal from the factory floor. Because this was a market-driven energy transition, the incidence of conservation and fuel switching was highly uneven across processes, firms, and industries, for companies undertook only profit-enhancing changes in energy use.

The gale of conservation had been rising for some time, for efforts to salvage anthracite wastes and to briquette coal dust had long attracted innovators. In the twentieth century, however, scientific and engineering expertise increasingly drove fuel technology. As more and more chemists and engineers entered large corporate laboratories and shaped shop-floor experimentation, they looked to improve fuel efficiency; and spurred on by wartime developments in energy and labor markets, they found much to do.

Innovations at electric utilities, especially, drove the technology of steam power with results that spread throughout industry and beyond. The shift to purchased electricity in manufacturing, along with its substitution for mechanical drive, contributed to the rise in productivity while reducing use of all forms of energy per unit of output.

Contributing to coal's woes, use of natural gas spread rapidly—the result not simply of enormous expansions in supply but also of investment and innovation in pipeline transport, while the AGA's research and marketing campaigns also helped spread use of the new fuel. Substituting gas for coal often increased energy efficiency; it also increased productivity because it improved product quality and sped up production. Use of gas and oil would surely have spread more widely but for the efficiency gains in coal consumption represented by such innovations as pulverization and improved stokers. Coal did not disappear from industry before World War II, but it began what would be a long goodbye.

Together, energy conservation and the spread of new sources of energy and power were important contributors to the spectacular rise in manufacturing productivity during the interwar decades; they fueled the arsenal of democracy and helped propel Americans' standards of living in the postwar years. These developments were entirely unanticipated by the Progressives who had assembled at President Roosevelt's Council of Governors in 1908, for none of them grasped the power of markets either to economize or develop new resources, and their resource pessimism discounted the role of natural gas and oil in the future. By the 1920s, their heirs, such as George Otis Smith, found these market developments unsettling. Oil and gas remained the scarcer fuels, and as their importance increased, so did warnings of impending doom. Yet while the Progressives and their descendants might have liked to slow the transition from coal to the newer fuels, the development of combustion technologies that avoided fuel lock-in ensured that their warnings would have little effect on companies' fuel choices in the interwar decades.

4

Railroads

Fuel Substitution and Conservation, 1885–1943

Eternal vigilance is the price of fuel economy.

—Railway Age

The extension of oil burning rests entirely upon the supply and costs.

—*H. J. Small, Southern Pacific Railroad, 1896*

As we saw in chapter 2, throughout the late nineteenth century, railroads and coal mines were like a good marriage: each partner improved the other. By providing cheap land transport, the carriers widened the market enormously for coal, while the mines, in turn, served up the cheap fuel that widened the market for rail transportation. By the early twentieth century, railroads were the largest single user of coal, about 104 million tons of it in 1909—nearly a quarter of all American coal consumption that year and about two-thirds as much as the entire manufacturing sector.

Yet by then, this marriage was beginning to come apart. After 1900, coal faced increasing competition from oil in the western United States (burned not as diesel fuel but as a substitute for coal in traditional steam locomotives), and by then, the carriers were starting to improve fuel economy as well. Fuel efficiency more than doubled from its low in 1904 until 1940, and as oil's share of fuel rose to 18 percent, railroad coal use collapsed, falling from nearly 136 million tons in 1917 to fewer than 80 million in 1940, before a brief wartime resurgence (tables 4.1 and A3.6). In short, coal's loss of the railroads did not begin with the advent of diesels about the time of World War II; that was simply the last chapter of a long story.

The railroads were also losing ground in the transportation market to trucks, automobiles, barges, airplanes, and pipelines, none of which burned coal. Short-haul passenger transportation virtually evaporated between 1920 and 1940. By the latter date, passenger miles on intercity buses amounted to nearly half those on steam railroads, while cars undoubtedly stole more of the carriers' markets. Similarly by that date, intercity truckers' 46 billion ton miles of freight added up to about 12 percent of the railroads' total. While not all of these gains came at the expense of the railroads, some did, and, accordingly, they cut into coal markets, too. After World War II, of course, these new modes of transport were irrelevant to the carriers' coal demand as the railroads rapidly shifted to diesel power.[1]

Table 4.1. Railroad Energy Consumption and Fuel Efficiency, 1889–1943

Year	All fuel (tons BCE)	Percent coal	Efficiency* 1918 = 100
1889	30,308,103	100.0	92
1890	34,973,437	100.0	85
1896	43,710,755	100.0	83
1900	62,479,964	100.0	81
1904	87,922,794	98.1	75
1909	108,754,990	95.5	75
1915	130,007,363	93.8	77
1920	150,480,650	92.6	99
1925	134,161,663	89.4	105
1930	113,818,577	87.5	110
1935	84,033,341	85.5	108
1940	95,711,088	83.5	125
1943	150,841,353	81.5	176

Source: Appendix table A3.6.
*Data are freight revenue ton-miles and passenger miles weighted by 1918 prices, relative to BCE.

In short, fuel developments on the railroads reflected the exploitation of newer, superior sources of energy and the quest for greater efficiency—the same forces that were reshaping fuel use in manufacturing. The railroad technological community was international in scope and quick to exploit oil as a new potentially more profitable fuel. But the railroads' story differs in several respects from that of manufacturing companies. Unlike manufacturers, the railroads' quest for improved fuel efficiency was motivated in part by a regulation-induced profit squeeze, as well as the rise in competition from trucks and automobiles. Fuel was also a far more important cost for the carriers than for most manufacturers, averaging about 5.6 percent of operating revenues in 1929 compared to 2.6 percent of the value of output in manufacturing at that time. Finally, because firemen and enginemen had a greater say in locomotive fuel economy than did shop-floor workers in manufacturing, the carriers paid greater attention to workers' roles in fuel economy.[2]

This chapter begins with the origins of railroads' experimentation with oil fuel in the West in the 1870s and then turns to the complex beginnings of conservation in the early twentieth century. The next section traces the spreading use of oil after about 1900 and is followed by conservation developments, especially during the 1920s and 1930s. The last section provides an overview of the railroads' retreat from coal as World War II approached.

The Railroads Discover Oil

In 1884, Thomas Urquhart, superintendent of the Grazi & Tsaritsin Railroad in southern Russia, delivered a paper entitled "On the Use of Petroleum Refuse as Fuel in Locomotive Engines" to the (British) Institution of Mechanical Engineers. Urquhart, whose education consisted largely of on-the-job training, had worked on the construction of the Grazi & Tsar-itsin and, when it was finished, joined that company and began to experiment with liquid fuel in 1874. By 1884, he was successfully powering locomotives by burning oil. Urquhart described his system in detail and with diagrams. As with the purchase of any fuel, buying the oil required knowledge of its technical characteristics—in this case temperature (because volume was temperature dependent) and specific gravity. The oil was pumped into tenders that Urquhart had modified to contain a tank in the coal bin. He emphasized the need to strain the oil, to drain off water, and, in winter, heat the heavy oil, which he accomplished with a warming coil in the tender tank. Flexible connections linked the tender to the locomotive, and for firing, an injector placed low in the back of the firebox used steam to atomize the oil. Urquhart installed a brick arch in the front and above the fire to keep the flame from the boiler tubes. He also detailed how to start the fire and the proper methods of firing and driving an oil-fired locomotive.[3]

The Diffusion of Oil-Burning Technology

In Russia, oil was economic because it was a residual—in Urquhart's words, "refuse." That is, it was the byproduct when crude oil was refined for kerosene (or later, gasoline). While oil refuse was abundant, without entrepreneurs such as Urquhart, it might have been thrown away, somewhat like the wood from land clearing in the United States. But Urquhart saw opportunity in the refuse: "The advantage in favor of petroleum . . . per truck-mile is . . . 57 percent over bituminous coal," he informed his audience. Oil from the fields around Baku yielded about 30 percent kerosene and 70 percent refuse, so the company could obtain the latter at 21 shillings to the ton, whereas coal, which had to travel 420 miles, sold at 27 shillings.[4]

Urquhart was by no means the first to try to burn oil in a locomotive. There had been experiments beginning in the 1860s; in 1882, the *Railroad Gazette* reviewed these efforts, all of which, it noted, were failures, for "at present prices, the heat producing power of a dollar's worth of coal is so much in excess of the heat producing power of a dollar's worth of petroleum." Nor were the editors sanguine about oil's future as a locomotive fuel, for even if it should somehow become economic, supplies were so limited that a large-scale shift by the railroads to oil would immediately run up its price.[5]

Despite the *Gazette*'s gloomy conclusions, Urquhart's work made a splash—perhaps because he had demonstrated that, in some places at least, oil could be economic. Charles Cochrane, president of the Institution of Mechanical Engineers, later noted that Urquhart's paper "had commanded a great deal of attention all over Europe and America." When Urquhart's work was reprinted in *Engineering* (London), it caught the eye of Theodore

Ely, general superintendent of motive power on the Pennsylvania Railroad, who charged the company's director of research, Charles B. Dudley (1842–1909), to look into the matter. Dudley was a PhD chemist the railroad had hired in 1875 when he established the first corporate laboratory in the United States and went on to help found the American Society for Testing and Materials.[6]

Dudley traveled to Russia, meeting Urquhart and investigating his procedures for burning oil. In December 1886, Dudley wrote Ely a long letter detailing Urquhart's system. "The technical part of the problem of petroleum burning . . . is solved," Dudley summarized; "it is now simply a commercial problem."[7]

The Pennsylvania experimented with the new fuel in 1887, modifying locomotives and conducting extensive tests. The *Railroad Gazette* noted that "it has taken considerable experiment and modification to adapt Mr. Urquhart's plans to American locomotives," but the company mechanical engineer A. S. Vogt had nothing but praise for the technology of oil burning. "It is capable of the nicest regulation," he noted. But with the delivered prices of coal and oil at $1.65 a ton and $0.025 a gallon, respectively, "an engine hauling the Day Express . . . would consume $6.25 worth of coal and $10.80 worth of oil." In 1888, Dudley presented these findings in a lecture to the Franklin Institute in which he sang the praises of liquid fuel. He presented figures on the relative prices of coal and oil at which the later fuel would be economic, calculated on an energy-equivalent basis, but he estimated that the other benefits of liquid fuel would justify paying a 13-percent price premium for oil. Nevertheless, like the *Gazette*, Dudley saw little future for large-scale use of oil by American railroads. He pointed out that the Pennsylvania alone then used the coal equivalent of nearly half of total American oil production.[8]

Even before Dudley published his findings, word had gotten out—"we have been flooded with letters . . . in regard to our experiments," Theodore Ely noted—and oil firing continued to interest the Pennsylvania's mechanical officers well into the twentieth century. In part, this was because oil seemed a solution to the urban smoke problem, but its economics remained a barrier. Moreover, as one official candidly noted, "the Pennsylvania is a coal carrying road so that even if the cost of oil and coal were equal we should use coal."[9]

But if expense vetoed the use of oil for the Pennsylvania, other railroads rejected the *Gazette*'s claim that it would never become economic, for crude oil production was exploding. It also moved west, where coal was sometimes of poor quality or had to travel long distances to market and was, accordingly, very expensive. California crude oil production dated from the 1860s. It was only 325,000 bbl. in 1891 but it then soared, rising to four million in 1900 and 73 million in 1910.[10]

Pioneering in the West

As California oil production expanded, the state's economic situation re-
sembled that of southern Russia. Little good quality coal was available; oil
was abundant and valuable because it would yield kerosene, but far from
major industrial centers, there was little market for the residual, which, as
Urquhart described, was refuse. Just as geography had initially sheltered
markets for anthracite, California geography provided the opening
through which oil poured into western railroading. The Central Pacific
quickly saw opportunity—another instance of the market turning waste
into gold. The company seems to have experimented with oil as a fuel in
1879 on the locomotive Young America and in the 1880s on its steamers
that plied the waters of San Francisco Bay. Again, the choice of fuels depended for the most part on their relative prices. The US Bureau of Mines
calculated that the energy content of a ton of bituminous coal was roughly
equivalent to that of 168 gallons (four 42-gallon barrels) of oil. On this basis, oil was economic if its cost per barrel was less than 25 percent of the cost
of a ton of coal, both including transport cost delivered to the locomotive
tender. But regional differences in fuel quality, along with combustion and
other efficiencies that favored oil, led companies to employ varying trade-
off ratios. For example, coal for the Central Pacific steamers was so poor
that a ton was equivalent to only 100 gallons (2.38 bbl.).[11]

Large-scale oil firing of steam locomotives in the United States began
on the Southern Pacific (SP) and Atchison, Topeka & Santa Fe (Santa Fe)
in the middle 1890s.[12] In 1895, with coal selling at $4.50/ton in San Fran-
cisco and oil delivered from Los Angeles going for about $0.75/bbl., the
SP began converting suburban engines in San Francisco to oil. The Santa
Fe began experimenting with oil in 1894, and its annual report for 1896
notes that "more than half the locomotives in use in Southern California
have been converted into oil burners." By circa 1900, both the SP and the
Santa Fe were equipping all locomotives in California (and in the case of
the SP all those west of El Paso) to burn oil.[13]

The payoff to such conversions was considerable. In 1908—the first year
for which such figures are available—SP locomotives averaged not quite
60,000 miles a year. Coal burners consumed about 4,576 tons of that fuel,
while locomotives burning oil consumed about the equivalent of 3,642
tons, where the company reckoned 4 bbl. of oil equal to a ton of coal. With
coal costing $3.79/ton and oil at the equivalent of $2.30/ton ($0.575/bbl.),
conversion to oil saved the SP $8,966 per locomotive-year. Company annual
reports indicate that it then cost the SP about $500 to convert a locomo-
tive from coal to oil, so the marginal return on such an investment was
nearly 1,800 percent per year.[14]

Companies such as Southern Pacific broadly adopted the technology
Urquhart outlined in 1884, later modifying it in the light of experience. A

rubber hose initially transferred oil from the tender to a locomotive, but companies soon found flexible couplings more reliable. Initially, mechanical engineers followed Russian precedent and placed the injector in the back of the firebox. But by about 1904, experiments by the SP indicated that placing it at the front resulted in more complete combustion and did away with the need for a brick arch to protect the boiler tubes. There were endless experiments with proper brickwork to protect the remainder of the firebox from intense heat.[15]

Locomotive tenders had a tank inserted in the coal bin and took on oil just as they took on water. While the typical fuel was a heavy residual, depending on cost, companies sometimes simply burned unrefined crude. In either case, because the oil was heavy, as Urquhart had instructed, it required heating, sometimes from a coil in the tender and sometimes—a new procedure—from direct injection of steam into the oil.[16]

This simple technology accounts for the comparatively low cost of converting a steam locomotive from coal to oil fuel or vice versa. It sharply reduced company fuel risks at a time when coal and oil prices might fluctuate sharply and energy futures markets were nonexistent. Thus, carriers simply reconverted locomotives to coal when relative prices dictated. "When it [oil] rises above a certain price, as was the case in June and July, we simply take the burners off and return to coal," an SP spokesman informed the *Railroad Gazette* in 1897.[17]

Many writers on energy transitions emphasize how long they take, and the explanations for such a glacial pace are often technological lock-in or, alternatively, the failings of the new technology. Such arguments sometimes adopt a national or global perspective—a level of aggregation that obscures the fact that many smaller transitions can, in fact, occur with great rapidity.[18] The use of oil firing by western railroads provides a case in point; switching between oil and coal remained common. Oil use on the Rock Island went from none at all in 1906 to 4 percent of its fuel in 1911; but a run-up in oil prices reduced oil to about 0.2 percent of the total in 1913. Some years later, when the Northern Pacific discovered huge deposits of lignite on its land, the company promptly converted locomotives from oil to that fuel. These rapid, often partial, conversions to oil that were sometimes followed by reconversions to coal paralleled the actions of manufacturers who designed energy systems that would avoid being locked into a particular fuel. They suggest that markets may be more flexible than the technological lock-in thesis suggests.[19]

Longer-term supplies of oil were also a worry. In 1911, Eugene McAuliffe (1866–1959) was fuel agent for the St. Louis–San Francisco and had recently founded the International Railway Fuel Association (IRFA). He cited the United States Geological Survey, suggesting that the United States might run out of oil by 1935. With railroaders, as with manufacturers, the shortages and inflation of World War I exacerbated these worries, and by

1920 McAuliffe was concluding that "a complete return to coal fuel is now imminent."[20]

Companies responded to these worries in several ways. Such concerns probably explain why they never pursued major changes in locomotive or tender design to enhance oil-burning efficiency. Price and availability worries also encouraged company decisions to invest in oil lands and were behind the Southern Pacific's venture into oil shale. With aid from the USBM, the SP set up a shale oil refinery near Elko, Nevada, circa 1919, but it seems to have been a failure, and the company sold it off in 1921.[21]

Worries over oil's long-term availability, combined with the lure of potential savings, also led carriers to experiment with powdered coal. That fuel promised advantages to railroads, as it did to utilities and manufacturing companies, such as less ash removal, longer locomotive runs, less smoke, fewer sparks, and access to a wider variety of coals. Beginning in 1914, several carriers experimented with burning powdered coal in locomotives, discovering that it yielded fuel savings but tended to burn out parts of the firebox, while pulverization itself was also expensive. The IRFA followed the developments closely, and by 1919, the verdict was in: powdered coal yielded favorable heat balances but not cash balances. Experiments with briquettes yielded similar results; they were superior to coal but not economically feasible.[22]

Conversion to oil also required the development of standards for fuel purchasing and new ways of firing locomotives. As Urquhart had instructed, the SP set standards for oil temperature, purity, water content, flash point, and gravity, and other carriers all had similar procedures. Firemen on a coal-burning steam locomotive needed strength and stamina before the introduction of the automatic stoker, as well as much skill to adjust the steam output to the engine driver's needs. Firing with oil did away with the job's strength requirements, but as Eugene McAuliffe informed other fuel agents, a skilled fireman was even more important with oil than coal. "Every time the engineer changes the throttle or [the] reverse lever, the fireman must change the fire," he noted. Firemen had to sand the flues to remove carbon as well, and it is worth noting how this procedure came to light: "An old carbide can filled with sand, for the use in case of fire was accidently overturned . . . and spilled some sand near enough the firebox door so that the suction drew some of it . . . through the flues [removing considerable soot]. . . . The effect was so noticeable on the steaming qualities of the engine that sanding engines was adopted as regular practice." Firemen needed to understand as well that the proper oil temperature for efficient combustion depended on its viscosity and specific gravity. Heating light oils to the temperature required by high viscosity—heavy crude, for example—caused the oil to flash too soon, leading to less-efficient combustion.[23]

Thus, carriers' decision to burn oil in steam locomotives required only modest, reversible changes in equipment and practices, and it retained the use of firemen, perhaps even enhancing their position, all of which surely facilitated its diffusion. The contrast with the spread of diesel locomotives after World War II is instructive. They were a disruptive technology, changing nearly everything about railroading, as the historian Maury Klein has pointed out, and the result was a far more difficult transition.[24]

By 1900, a few western railroads had responded to the availability of cheap California oil by converting some of their locomotives to burn that fuel. Absent additional discoveries, in all likelihood, that is as far as oil fuel would have spread. In 1904, coal accounted for 98 percent of all railroad fuel nationwide, with oil constituting most of the rest (see table A3.6). But in January 1901, wildcatters found a sea of oil at Spindletop, near Beaumont, Texas. The first well—the "Lucas"—initially produced 100,000 bbl./day! Texas production rose to 28 million barrels in 1905 before collapsing for a time and then booming again. Oklahoma oil production rose from 139,000 bbl. in 1903 to 52 million in 1910. Mexican production went from nearly zero in 1903 to almost 40 million bbl. in 1916. Soon after the "Lucas" came in, the SP and Santa Fe testing laboratories pronounced the Texas product perfectly acceptable for locomotive fuel. Their tests were widely publicized, as was the experience of an early adopter, the Gulf, Colorado & Santa Fe, with burning the new crudes. For western railroads, oil suddenly began to look economic over a much wider territory. By August 1901, the *American Railroad Journal* was cautiously predicting "a revolution in locomotive practice in a large and important section of the country."[25]

The SP and Santa Fe quickly expanded use of oil-fired locomotives beyond California. The Kansas City Southern, International Great Northern, Texas & Pacific, Fort Worth & Denver City, and many other southwestern railways immediately began to convert motive power on some divisions to burn oil. By 1910, 4.5 percent of Union Pacific locomotives were burning oil. Around 1910, as California oil became available in Washington and Oregon, the Northern Pacific and Great Northern began converting locomotives on western divisions to oil. By 1918, most of the major carriers that operated in 19 states west of the Mississippi burned oil in at least some locomotives. Except under special conditions, oil firing remained rare in the East.[26]

One reason oil firing failed to spread more widely was its changing economics. In 1913, William Burton (1865–1954), a researcher at Standard Oil of Indiana, patented a process for cracking residual oil to obtain more gasoline; suddenly, fuel oil was no longer refuse. Thus, instead of disposing of residuals as a leftover at whatever price they could get, oil companies could now calculate whether the residual oil was worth more when refined into gasoline or sold as fuel. Several other methods for cracking followed Burton's pioneering work, and the process spread rapidly. It is hard to overstate the importance of this innovation. In the absence of cracking, oil companies would have had to pump much more crude to meet surging demand for gasoline, and the resulting cheap refuse might have allowed oil to invade eastern railroads. Thus, cracking was simultaneously an oil conservation measure and a boon to coal producers.[27]

Conservation Beginnings

The reader will recall from chapter 2 that even though fuel conservation was a concern of nineteenth-century railroaders, they were not very successful at it. Table 4.1, derived from table A3.6, picks up this story in 1889 and carries it into World War II. As can be seen, efficiency (the value of transportation output relative to fuel use) declined until about 1909 and then rose sharply but unevenly.[28] Capacity utilization influences this measure of fuel efficiency, and for both passenger and freight service, especially during the 1930s, decreasing capacity utilization offset technical improvements. Overall, energy efficiency increased at an annual rate of two percent a year between 1909 and 1940. During World War II, however, such excess capacity evaporated, as is reflected in the sharp efficiency gains.[29]

This story begins early in the late nineteenth century, for while western carriers were experimenting with oil, nearly all carriers were experimenting with policies to economize on fuel. The Pennsylvania Railroad evaluated numerous substances that promised to save fuel and reduce the smoke nuisance in terminals. These included schemes to improve firebox draft, to add lime or salt to the fuel, or to use a "liquid composition" that augmented oxygen. All were "absolutely worthless," as one Pennsylvania engineer described a "smokeless fuel compound" he had seen tested. Some companies inaugurated fuel departments because fuel use depended in part on locomotive firemen. In the 1880s, the Pennsylvania Railroad pioneered an incentive scheme that rewarded careful firing. Although the system initially received glowing reviews, it was finally scrapped because of difficulties in the measurement of efficiency. Furthermore, some enginemen and firemen on coal trains had begun stopping along the road and shoveling fuel from the cars to the tender.[30]

Although most of the Pennsylvania's early efforts to improve fuel economy focused on the firemen and enginemen, studies from its test department revealed that the problem was broader. One 1908 report noted losses from overloading tenders and from firing too soon in the engine house. A test also suggested the importance of train delays. The company discovered that a standard H-6a freight locomotive coupled to thirteen air-braked cars burned 248 pounds of coal an hour simply standing still.[31]

In 1914, the Pennsylvania's Association of Transportation Officers raised the possibility of creating a separate fuel department. The committee wrote to a number of other carriers inquiring into their efforts at fuel economy, and it sent two officials on a tour to study such departments at the B&O, Chicago & North Western, Rock Island, and several other railroads. After much discussion, it rejected the idea of a fuel department, preferring to rely on its traveling engineers instead.[32]

Such was the state of railroad fuel concerns circa 1908, when Eugene McAuliffe, then general coal agent for the Rock Island, founded the IRFA. The association contributed to improving fuel economy in several ways; it scrutinized train dispatching, locomotive maintenance, signaling, and other aspects of railroading. Like the technical associations discussed in chapter 3, it also provided a forum for the discussion of new practices, disseminating the results through publication of its *Proceedings*, which were also reported in trade publications such as *Railway Age*.[33]

Public Pressures

Congress inadvertently gave these nascent efforts a powerful shove when it passed the Hepburn Act in 1906, giving the Interstate Commerce Commission (ICC) effective power to regulate railroad rates. In 1910, after long hearings, the commission—blind to the inflation of that day—denied the carriers' request for a general rate increase. But while railroad rates were held down, costs were not. Through strikes and arbitration, the railroad unions began concerted campaigns to increase their pay. As a result, their inflation-adjusted earnings rose sharply. In addition, while coal prices fell steadily compared to an index of other prices from 1900 to 1916, they *rose* relative to freight rates, contributing to the financial squeeze that reduced the carriers' net income from about 26 percent of operating revenue in 1911 to 21 percent in 1914. Rock Island's president, C. F. Richardson, summed up matters to the traveling engineers in 1910: "the increased cost of operation . . . makes it necessary to practice the strictest economy, and I believe that one of the greatest opportunities . . . in reducing the cost of operation lies in fuel economy." Thus, on railroads, it was not the wisdom of farseeing policy makers that would encourage fuel conservation, as Progressives had imagined, but rather the blunders of poorly informed regulators.[34]

Increasing public pressure to reduce smoke also played a greater role in motivating fuel economy on railroads than it did in manufacturing, as city after city enacted ordinances to reduce pollution. In 1913, a representative of the St. Louis–San Francisco told the traveling engineers that St. Louis had adopted a "very stringent" smoke ordinance. Part of the carriers' difficulty was their political unpopularity: a representative of the Erie complained that "in my territory, we are continually getting arrested for making smoke." And to at least some critics, there seemed to be a simple cure for railroad smoke. W. L. Robertson of the B&O noted that the carriers in Washington, D.C., were doing everything possible to reduce smoke because Congress was threatening to require electrification. Still, public efforts at smoke control largely reinforced policies the carriers were pursuing for their own ends.[35]

Here again, the railroads focused initially on their firemen and enginemen. The Lehigh Valley experimented with "thick" instead of "thin" firing and claimed astonishing increases in fuel economy. The Buffalo, Rochester & Pittsburgh also began to urge firemen and enginemen to save coal. On some runs, according to the company, ton-miles per ton of coal rose nearly one-fifth. Encouraged by these results, in 1912, the BR&P adopted a broader approach, publishing an instructional booklet aimed at engine dispatchers, hostlers, train dispatchers, boilermakers, machinists, and agents. In 1909, the Rock Island wrote to the Pennsylvania inquiring about means of smoke prevention and, around the same time, established a committee on fuel economy.[36]

In 1912, the Chicago & Alton launched a fuel-economy campaign, again placing most emphasis on the engineer and fireman. In 1914, the Northern Pacific set up an instruction car with a somewhat broader focus. It taught the principles of fuel economy to division officers, trainmen, and shop engineers, as well as firemen. At about the same time, the Union Pacific also employed a traveling instruction car that showed movies on fuel economy.[37]

The Pennsylvania officials who toured the Chicago Great Western in 1914 reported that the company ranked firemen and enginemen according to their fuel efficiency and that the men were "considerably interested in seeing their names at the head of this list." Since coal was extraordinarily heterogeneous, the Great Western ensured that locomotives always burned fuel from the same mine so that the men could develop expertise in its use. That company also emphasized poor locomotive maintenance as a cause of wasted fuel.[38]

Other major carriers pursued similar strategies, but until World War I, contemporaries concluded that these efforts had little effect. The Pennsylvania's officials who inspected other carriers penned a generally skeptical assessment of their fuel organizations. Although a few fuel departments, such as that on the B&O, had proved their worth, the officials noted that most lacked adequate supervision, and poor performance rarely led to penalties for trainmen. In 1915, L. G. Plant, fuel supervisor on the Seaboard, came to a similar conclusion: "the fuel department has failed to establish its position within the organization of the railroad [and] has not been successful in arousing any widespread interest in fuel economy." Yet as table 4.1 shows, fuel efficiency had begun to increase after about 1909. The origins of these gains reflected early twentieth-century improvements in motive power.[39]

Improving Locomotive Technology

During the nineteenth century, railroaders satisfied the need for more powerful locomotives by the simple expedient of building them bigger. There were many advantages to size, and the locomotive growth continued into the twentieth century. For example, about 1911, the New York Central replaced 60 of its 2-8-0 locomotives, each of which supplied 45,000 pounds of tractive effort, with 26 much larger 2-8-8-2 "Mallets," with 67,000 pounds of tractive effort. The new power not only reduced overtime by 80 percent but also cut 10 trains per day each way and raised ton-miles per ton of coal by 54 percent. Such equipment began the carriers' twentieth-century turnaround in fuel efficiency.[40]

Yet size constraints were beginning to change the trajectory of locomotive development, causing builders to look elsewhere for increases in hauling capacity. The result was a slew of modifications in locomotive technology that raised power without commensurate increases in size. Average tractive effort (pulling power) rose about 46 percent between 1916 and 1935, and, as a byproduct, fuel efficiency increased as well. The improvements included booster engines, brick firebox arches, and thermic syphons (which increased the flow of water through the firebox), along with many of the same devices manufacturing plants were installing at this time, such as superheaters, feedwater heaters, and automatic stokers.[41]

As in manufacturing, railroad technology increasingly benefited from the application of science to combustion problems, and many of these improvements resulted from the systematic study of locomotives begun by Dr. William Goss (1859–1928) at Purdue University's locomotive testing plant in 1892. Goss's work at Purdue, and later at the locomotive testing plant at the University of Illinois, along with similar work done by the Pennsylvania Railroad at Altoona, allowed engineers for the first time to apply thermodynamic principles to locomotive design to derive heat balances and to measure the relative importance of various energy losses. The result by the 1920s was what Lima Locomotive works called "superpower."

One beneficiary of superpower was the Texas & Pacific, which in the early 1920s introduced 15 new 2-10-4 Texas-class locomotives. The company compared their performance with its older 2-10-2s working under similar conditions. The new engines carried steam at 250 psi compared with 200 for the older locomotives, and they came with feedwater heating, superheat, boosters, thermic siphons, limited cutoff, and larger grate area. The Texas locomotives used 35 percent less fuel and hauled 24 percent more freight than had the 2-10-2s. Typically, the largest savings from the new power came from reductions in repairs and train crew wages, with fuel economy as frosting on the cake.[42]

A Conservation Lobby

Wartime pressures soon provided a powerful spur to carriers' interest in
energy efficiency. The Rock Island had organized a fuel committee circa
1910 and transformed it into a fuel department in 1916; a year later, it was
claiming to have cut fuel use 12 percent. By 1917, worries of coal shortages were widespread. The Northern Pacific stepped up its economy drive,
and the New Haven began a similar campaign. The New Haven claimed
that the program reduced fuel use by 14 percent between 1916 and 1917,
thereby saving one million dollars. The Air Brake Association chipped in
with a calculation that poor brake maintenance caused the railroads to
waste six million tons of coal a year simply to maintain brake pressure.
Finally, in 1918, when the US Railroad Administration (USRRA) took over
the carriers, it established a fuel conservation section, headed by McAuliffe, as part of the division of operation.[43]

The conservation section raised the status of existing fuel committees,
transforming them into departments, and it established new ones. Such
departments took a broader view of fuel economy than had the older committees. They included fuel inspectors, engineers, accountants, and stat-
isticians. These were staff organizations, usually run by fuel supervisors
or superintendents who reported to the superintendent of motive power,
or sometimes directly to the general manager or head of purchasing. Fuel
departments had to justify their existence, thereby providing a powerful
institutional "conservation lobby" for fuel efficiency.[44]

The USRRA fuel conservation section also rated the fuel performance
of individual carriers and urged laggards to improve. It took over the 1918
annual meeting of the Railway Fuel Association and turned it into a gigantic rally for efficiency. Although the association then had fewer than
200 members, the 1918 meeting attracted about 1,600 railroad men, who
were exhorted to go home and conserve. In 1920, *Railway Age* claimed that
"the work of the Fuel Conservation Section [of the USRRA] has advanced
the status of fuel conservation further than would ten or fifteen years of
normal pre-war development."[45]

The general inflation of costs during wartime also heightened the pressures on the railroads to look for any savings they could find. Net income
fell from about one-fifth of operating revenue to near zero in 1921, and it
never attained its prewar levels throughout the 1920s and 1930s. The worsening cost-price squeeze reinforced the efforts of the USRRA to establish
fuel departments; a 1921 survey of 50 large railroads found that 30 had some
form of fuel department, and most of these were not even a decade old.[46]

The Spreading Use of Oil

For western carriers, fuel economy involved not only conservation but
fuel switching as well, for access to cheap oil promised sharp reductions in

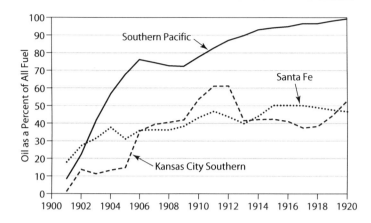

Figure 4.1. Percent of oil use for three western railroads, 1900–1920. The Southern Pacific is the percent of locomotives burning oil; other lines are the percent of fuel that is oil. (Data collected from various company annual reports)

fuel costs coupled with increases in productivity. The first available data for the SP show oil burners constituting 8.5 percent of the total of all locomotives in 1901 (fig. 4.1); by 1903, the total amounted to 42 percent, on the way to almost complete conversion. Some other western carriers seem to have made nearly as rapid—although less complete—conversions; the Kansas City Southern went from burning no oil in 1900 to 60 percent oil fuel by 1911, while the Santa Fe began oil use about 1896, and by 1910, oil constituted roughly 40 percent of its fuel.

Advantages of Oil

Firing with oil brought many ancillary benefits; accordingly, the desire to improve transport efficiency contributed to the substitution of fuel oil for coal. Because of its greater energy density and the higher combustion efficiency with which it burned, oil facilitated heavier loads and longer runs. The ability to regulate oil fuel more closely than coal also enhanced speed and hauling capacity. When the Missouri, Kansas and Texas Railway (MK&T) began conversion to oil about 1920, it found that the new fuel facilitated longer passenger runs and required fewer intermediate fueling stops between engine stations as well. Oil firing also played an important role when the railroads introduced art deco–styled streamliners in the 1930s, because many companies "streamlined" steam locomotives, while oil firing yielded faster schedules and longer runs. Of course, the SP steam streamliners were oil fired, but so was the Milwaukee's Hiawatha, which ran between Chicago and the Twin Cities. And when the Northwestern inaugurated streamliner service on that route, it converted two steam locomotives from coal to oil.[47]

Oil was easier to handle than coal, its movement, according to one estimate, costing only a quarter as much per ton. More precise regulation of fuel allowed harder working of locomotives and thus greater trainloads or higher speeds or both. About 1907, the Santa Fe reported that a train would go 60 to 70 percent farther on a ton of oil than on a ton of coal. In 1919, tests by that company found that an Atlantic-type (4-4-2) locomotive that could haul a 578-ton train when coal fired could pull 750 tons burning oil. There was no wastage with oil, while coal simply fell off the top of the tender (and oil was less likely to be stolen, as well); there were no ashes to clean out, making for faster turnaround and therefore greater locomotive utilization. Oil produced no cinders to start fires in forests or in passengers' clothes or to foul ballast, and it might be less smoky as well. Thus, just as the shift from coal to gas sometimes raised manufacturing fuel efficiency and overall productivity, so some of the railroads' fuel efficiency gains depicted in table 4.1 resulted from the shift to oil fuel, and it raised labor and capital productivity as well.[48]

In some instances, these ancillary benefits of oil were important in shaping fuel choices. While state law required the New York Central and the Delaware & Hudson to burn oil in the Adirondacks to reduce fire hazards, the Milwaukee voluntarily chose oil burners to reduce fire risk from operation in the Idaho Forest reserve. The Northern Pacific's early use of oil no doubt reflected the fire danger that resulted from burning coal in long snow sheds. The Southern Pacific had been burning down an average of 1,770 feet of snow shed a year, which dropped to 563 feet when it began the switch to oil, and after 1916, locomotive sparks disappeared as a source of shed fires.[49]

The Key Role of Costs

Still, these ancillary benefits seem to have played only a minor role in most companies' fuel choices. Lines with access to better coal and lines that were less well situated near oil fields burned little or no oil, while those that used oil did so selectively, burning it on some divisions and coal on others, the balance shifting with changes in costs and availability. The Santa Fe's experience in the early 1920s was typical. About that time, roughly half the company's locomotives burned oil (see fig. 4.1), but its use was sharply localized: in California and Arizona, as well as in parts of Kansas, the choice was oil; east of Kansas City and west of Newton, Kansas, coal ruled, but by 1941, it accounted for only a quarter of that company's fuel use. There is no reason why all western carriers— indeed, all railroads—could not have converted to oil with the speed shown by the SP had economics so dictated. Table 4.2 depicts the extent of oil firing on western lines as of 1941. The ancillary benefits were available to all carriers, yet companies displayed wide differences in oil use. Such behavior, and the almost complete failure of oil firing to penetrate eastern and southern railroading, suggests that while oil may have been a better fuel, its ability to defeat coal was largely a matter of relative costs.[50]

Table 4.2. Extent of Oil Burning by Steam Locomotives, Selected Western Carriers, 1941

Railroad	Percent of oil burning*
Atchison Topeka & Santa Fe	75.3
Chicago & Northwestern	7.0
Chicago, Burlington & Quincy	5.3
Chicago Milwaukee St Paul & Pacific	4.6
Chicago, Rock Island & Pacific	54.3
Great Northern	47.5
International Great Northern	100.0
Kansas City Southern	69.0
Missouri, Kansas & Texas	99.8
Missouri Pacific	13.7
Northern Pacific	2.7
St. Louis–San Francisco	40.8
St. Louis & Southwestern	97.9
Southern Pacific	93.3
Texas & New Orleans	100.0
Texas & Pacific	100.0
Union Pacific	19.7
Western Pacific	73.5
Average of all roads	45.9

Source: "Coal Equivalent of Other Fuels and Power," *Railway Mechanical Engineer* 116 (Nov. 1942): 487–489, table 1.
*Percent of gross ton-miles with oil-burning locomotives.

Shoveling on a Little Less Coal

As World War I came to an end, railroads were discovering many ways to get more power from their coal and oil. They had begun to invest in superpower locomotives, and they modified a host of operating practices to enhance efficiency. Often at the behest of their fuel departments, they modified personnel management to encourage conservation. Accordingly, the next two decades witnessed sharp gains in energy efficiency.

Improving Plant and Equipment

The new locomotive technology discussed earlier contributed to numerous other changes in plant and equipment and operating practices that led to major energy savings. Calculations revealed large savings in fuel, capital, and labor costs from running longer, heavier trains. Accordingly, freight-train weight increased gradually in the interwar years, from an average of 667 tons in 1920 to 676 in 1938 before it skyrocketed under the pressures of war. A representative of the Southern Pacific concluded that "a large part of any saving [in energy] that is made can be directly charged to an increase in the average tons per locomotive."[51]

The improvement in locomotive efficiency and the use of larger tenders (as well as the shift to oil as a fuel) also allowed the carriers to increase the length of locomotive runs in the 1920s and 1930s. So, too, did the loss of short-haul traffic, which—of course—raised the average length of haul.

Freight hauls rose to as much as 250 miles; for passenger trains, runs of 500 miles became possible. Longer hauls conserved fuel because they meant less idle locomotive time, and average mileage per locomotive increased, accordingly, from about 24,000 in 1920 to 31,000 in 1941. But the major gains from long hauls were the savings in capital and labor that resulted as companies required fewer locomotives and terminal facilities and fewer employees to operate them, while energy conservation was a byproduct. Thus, in 1924, the Southern Pacific replaced 50 older "Mikados," ten-wheelers, and "Atlantics" on the run from Los Angeles to El Paso with 25 new 4-8-2s that made the complete 800-mile run with stops only for supplies.[52]

Other improvements in plant and equipment also generated fuel economy as a byproduct. Reductions in grade or curvature could raise speed, thereby increasing line capacity and reducing the need for helper locomotives and crews, as well as saving energy. Thus, when the Central Pacific shortened its line from Ogden to Lucin, Utah, reducing the grade and curvature and cutting the distance from 147 to 103 miles, it chopped fuel costs in half.[53]

The increasing use of block signals (and later centralized traffic control) also saved fuel, although that was not the driving force in their spread. One study of the replacement of manual with automatic signals on 566 miles of road provided detailed estimates of both the costs and the benefits of the investment. Fuel savings accounted for only a quarter of the total gain and were overshadowed by reductions in overtime and telegraphers' wages.[54]

As investigations revealed the cost of stopping long, heavy freights, companies urged train dispatchers to give priority to heavy trains and to sidetrack the lighter ones. They were also warned not to sidetrack heavy trains before steep grades and not to use them to drop cars at local stations. Companies also began to substitute "form 19" train orders, which did not need to be signed by the trainmen, for "form 31" orders that did, thereby avoiding stops.[55]

Maintenance of locomotives also improved, as it did for stationary boilers, for similar reasons: studies demonstrated that plugged flues and encrusted superheaters greatly reduced power and fuel economy. Research demonstrated that as little as one-sixteenth of an inch of boiler scale could reduce locomotive thermal efficiency by 12 percent. As a result, more companies began to treat boiler water. Here again, however, benefits were far greater than simply the saving in fuel. When the Great Northern installed treatment facilities on 1,100 miles of line, it found that the main benefits were a reduction in repair costs and a saving in time over the road, thereby increasing locomotive productivity. In 1923, one expert estimated that 800 treatment facilities were in use. A decade later, the number of plants had risen to more than 2,000, and they treated about 40 percent of all boiler water.[56]

Companies also focused on fuel economy in terminals, one motive being the continuing pressures to abate smoke. In the 1920s, these motives led the Cleveland, Cincinnati, Chicago & St. Louis Railway (the Big Four); the Chesapeake & Ohio; the Grand Trunk; and the Great Northern to begin direct steaming (use of a stationary boiler to supply steam to locomotives) in roundhouses. Since stationary power plants were typically more efficient than locomotive boilers, this saved fuel directly. One carrier reported reductions in coal consumption of 800 to 1,000 pounds per locomotive turned. By 1930, reports indicated that there were 24 direct steaming plants then in use.[57]

Organizing for Fuel Efficiency

Energy savings also accrued from changes in personnel management. Carriers' fuel organizations often took the lead, employing fuel accounting to pinpoint potential economies. The practice improved the ability to monitor the performance of many employees who normally worked with little supervision, and companies began to maintain much more careful daily reports of train and locomotive movements. Such data could help isolate the impact of poor firing, improper train loading or dispatching, or shoddy maintenance. Thus, on the New Haven, in the mid-1920s, an internal company memorandum attributed improving fuel economy not only to "quality of fuel, improved condition[s] of power, [and] increasing use of coal saving devices" but also to "continued attention to fuel performance records."[58]

Tests made in 1920 demonstrated that poor firing might increase coal use by a fifth. As a result, the Southern Pacific, the Erie, and other carriers began to improve the training and supervision of firemen. The St. Louis–San Francisco and the Chicago Great Western employed traveling engineers and firemen to provide train crews with instruction and criticism.[59]

In the 1920s, several carriers instituted fuel-efficiency campaigns. The idea was to stimulate a rivalry among employees to see who could save the most coal. These campaigns were similar to the safety campaigns then sweeping through railroading. They were largely absent from manufacturing, reflecting the comparative importance of individual workers and lower-level managers in determining fuel efficiency on railroads. They almost invariably demonstrated large savings in fuel. The experience of the Central Railroad of Georgia was typical. Its efforts began in 1921, and the program included competition, commendations, and prizes. The campaign stressed proper coaling, assigning regular crews to locomotives, improving locomotive repair, reducing stops and slow orders, and giving train rights to the heavier trains. On-time performance increased, and engine failures and overtime declined, while coal consumption fell about 14 percent a year.[60]

In 1923, the IRFA instituted a prize essay contest on methods of fuel saving. It got nearly 2,000 responses, of which more than three-fourths came from engineers and firemen. The association and several carriers also put out fuel-economy posters and sponsored efficiency contests. About the same time, as part of the company's cooperative plan, the Baltimore & Ohio *Employees' Magazine* printed a steady stream of cartoons and slogans emphasizing the need for employee involvement in fuel savings, as did many other large carriers. The Texas & Pacific even urged on its employees with verse:

> Let all the employees start fighting the duel
> On the old T. of P. of saving the fuel.
> It can't be done by just a few,
> But Oh! what a saving when we all come thru.[61]

The Retreat from Coal, 1889–1940

While one may wonder how to weigh the relative contributions of better locomotives, fuel departments, poetry, cartoons, and cheap oil in the retreat from coal, collectively, they resulted in dramatic declines in railroad demand for that fuel in the years before World War II—long before the advent of diesels finally administered the coup de grace. As in manufacturing, the decline in coal use was largely market driven. But for railroads, regulations were a more important factor because ICC rate restrictions put a premium in efficiency gains, while the USRRA pushed institutional innovations.

Table 4.1 indicates that railroad transportation output rose about 64 percent relative to fuel use between 1909 and 1940 before skyrocketing under the press of wartime. The rise of oil firing added to the coal industry's woes, as did the carriers' dabbling in electrification and the appearance of the diesel. Coal use peaked in 1916 at about 136 million tons—which was 93 percent of all fuel burned. By 1940, coal consumption stood at 80 million tons, and it amounted to only 80 percent of fuel that year.

Regional Variations

A focus on the economy-wide changes in fuel use obscures sharp regional differences (fig. 4.2). By 1925, although oil accounted for about 11 percent of the fuel burned by steam locomotives nationwide, it constituted 28 percent of all such fuel in states west of the Mississippi. Thereafter, as western production of crude exploded, railroad use of oil increased; by 1940, well before diesels had any impact, oil had driven coal from nearly 40 percent of the railroad fuel market in the West. On eastern lines, however, liquids made up less than 1 percent of all fuel at that time and only 10 percent as late as 1950. Thereafter, the stampede to diesels began. Thus, only outside the West was the railroads' transition from solid to liquid fuel synonymous with the shift to diesels.

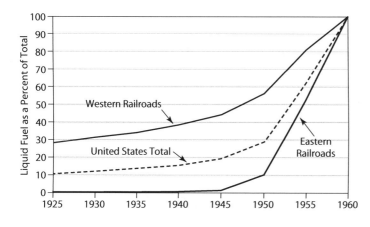

Figure 4.2. Liquid fuel as a percent of all locomotive fuel: US total and eastern and western regions, 1925–1960. (Chart data drawn from appendix table A3.7)

The Impact on Western Coal Mining

For western coal, this was a disaster. Across the nation, bituminous coal production fell about 20 percent from 1918 to 1940; west of the Mississippi, the collapse was far greater. Wyoming coal production fell 40 percent from its 1920 peak to 1940; in Kansas, the decline from its peak in 1918 was 38 percent. For these states, where strip mining had expanded during the interwar decades, the collapse in underground production was even greater. In Oklahoma, soft-coal output fell 66 percent from 1920 to 1940, while in New Mexico and Texas, coal mining virtually collapsed, falling 72 and 73 percent, respectively. Overall, western production was cut in half, and much of this relative collapse resulted from oil burning by western carriers; had they continued to fire with coal, the decline in western production would have been about 20 points less. Western coal interests, therefore, experienced a preview of what would be the industry's future nationwide.[62]

Conclusion

The gradual transition from coal to oil by western railroads reflected not only the abundance of cheap oil but also the carriers' entrepreneurial efforts, drawing on the work of an international technological network. The rise in fuel efficiency during these decades was similarly market-driven—reinforced by the need to escape the regulatory profit squeeze that began before World War I, as well as pressures to reduce smoke. These same motives spurred organizational changes in railroading, such as company fuel organizations, the IRFA, fuel accounting, and efficiency campaigns. Because railroads were interacting technological systems, the quest for fuel economy sometimes led to savings in capital or labor, raising total factor productivity, while in other instances these were the goals, with fuel economy as a side dish.

The carriers' performance during World War II provides the best glimpse into the ways these changes improved Americans' well-being. Had the railroads' fuel efficiency remained at 1918 levels of efficiency, in 1943, it would have taken 65 percent more fuel to move the same amount of freight. As the railroads carried the arsenal of democracy, there were no domestic coal shortages in 1943, as there had been in 1917, and one of the reasons was the railroads' greatly improved fuel efficiency.

Part III

LEAVING HOME

5

Coal Departs the Urban Kitchen, 1900–1940

The introduction of gas . . . stoves in the last few years is . . . an inestimable boon to housewives. . . . They will do everything a coal stove . . . [can do], often quicker and better. They are far cleaner and easier managed. . . . There is no bringing in coal or carrying out ashes.

—Good Housekeeping

New England's captains of public policy organized their preparations yesterday for a "war to end war" with the anthracite industry.

—Boston Globe

The National Commercial Gas Association (NCGA) created the character Nancy Gay in 1914 to promote the virtues of manufactured gas. In magazines and a pamphlet titled "The Story of Nancy Gay," the association explained that she had nearly broken up with her sweetheart, George, because he was so old-fashioned as to want a coal stove. True love finally won out, however, for Nancy persuaded George of the wonders of the all-gas kitchen—including, besides the range, a gas light, water heater, iron, waffle iron, and pancake griddle—and the reader learns "what gas provides in the way of domestic service and therefore contentment and happiness." Nancy so won over George that he wrote his parents—who lived in "Old Fashionedville"—urging them to convert to (manufactured) gas as well.[1]

Whereas previous chapters have emphasized the importance of technology and supply developments that drove American energy transitions, here we focus on demand and, especially, marketing, for if exploitation was as important as abundance in creating energy supplies, marketing was central to the creation of energy demands. The primary focus of this chapter is on hard coal because it was the first and most serious casualty of the assault by competitive fuels. The emphasis is also on fuels for cooking in urban households, for that is where coal's rout began and was most complete. Coal's loss of rural cooking markets to kerosene is the topic of the next chapter, and the battle of coal, oil, and gas for America's basements comes in chapter 7. We begin with a review of domestic coal markets at the time of World War I, and the next section traces the market penetration of substitutes for coal before the widespread use of natural gas. The focus then shifts to the marketing campaigns of gas companies before

Figure 5.1. Anthracite peak and decline, 1900–1950. Chart data collected from (USGS/USBM, MR and MY for various years)

World War I; the impact of World War I and labor disputes on anthracite sales is the topic of the following section. We then examine the national marketing campaigns for gas and gas stoves that began about the time of World War I, and the final section demonstrates how completely gas had routed coal from America's urban kitchens on the eve of World War II. The chapter ends with some conjectures and conclusions.[2]

Anthracite at Its Apogee

It seems unlikely that hard-coal producers paid any attention to Nancy Gay, for in 1914, their future must have seemed bright. Yet for anthracite, output in 1917 would prove to be its highest level ever (fig. 5.1). And while Progressives worried that the nation might soon run out of this wonderful resource, it was not supply constraints but rather the lack of demand that would lead to the eclipse of the hard-coal business. Nancy Gay, in short, was a warning—a canary in the coal mine, one might say.

Markets for Hard Coal

As chapters 1 and 2 demonstrated, anthracite coal and America grew up together. But by World War I, lower-priced soft coal was increasingly supplanting anthracite for industrial uses, and in much of the country, it was the dominant domestic fuel as well. Although anthracite's comparative cleanliness made it attractive to households, as we have seen, geography and transportation costs limited sales to the comparatively affluent and largely confined it to a narrow region of the country. The major markets were eastern Pennsylvania, New York, New Jersey, and New England—which accounted for about three-fourths of production in 1915, while cheap water transportation allowed modest sales in the central and lake states as well. Elsewhere, Americans cooked and heated mostly with soft coal or wood, but increasingly, coke, briquettes, manufactured (and, where available, natural) gas were also contesting its markets.

While thousands of companies produced soft coal, an informal cartel dominated anthracite production. Eight railroad-controlled coal companies typically accounted for about three-quarters of output, the remainder coming from a competitive fringe of around 100 independents. Essentially, anthracite was not branded until the 1930s, and soft coal was rarely branded even then. Coal companies and retailers sold hard coal by size, while bituminous coal might also be had as "run of the mine." In the case of anthracite, domestic sizes (e.g., "lump," "chestnut," "stove") were for home cooking and heating. A second size group of hard coal, termed "steam coal" (e.g., Buckwheat #1), was smaller still and sold to apartment buildings, utilities, and other large users that had equipment designed to burn them. All the sizes were produced jointly, and all cost the same to mine, but market prices roughly reflected size. Thus, in 1918, the average mine realization for chestnut and stove coal ranged from $5.87 to $6.64 per net ton. Steam sizes, essentially a byproduct of production for the domestic market, sold for less—from $3.48 to $3.55 per ton—because competition with bituminous coal governed their price. While such prices for steam coal did not cover average cost, they were a byproduct that contributed to revenue to help cover fixed costs.[3]

Companies marketed hard coal using a variety of wholesale and retail arrangements and rarely engaged in advertising or sales promotion. Accordingly, they were ill-prepared to counter the marketing campaigns that would soon come from the gas utilities. The railroad-controlled companies announced their "circular" prices about April 1 of each year. Companies discounted prices in the spring to encourage households to spread purchases through the months; otherwise, they usually remained fixed during the year. Prices charged by independent companies reflected market forces, however, sometimes selling at a premium or discount from circular prices. Soft coal also came to domestic markets through a bewildering combination of arrangements with wholesalers and retailers, and it, too, was discounted in the spring. For either kind of coal, transportation by water or rail or both might account for 18 to 20 percent of its retail price in cities close to producing areas, but in faraway Chicago, transport costs amounted to 35 to 40 percent of anthracite's market price. Wholesaler and retailer costs of coal included transport fees, while their markups similarly reflected market conditions. Taken together, transport costs and dealer margins ensured that the retail price of coal was usually more than twice its price at the mine (tables A4.2 and A4.3).[4]

Glimpses of the Future

In the years before World War I, when Progressives worried about natural-resource waste and scarcity, coal, especially anthracite, was a major concern. The reader will recall that in 1907, Chief of the US Forest Service Gifford Pinchot had served up a Malthusian vision of increasing anthracite scarcity to a popular audience, claiming that reserves would last only 50 years. This was a widely shared view. Anthracite was becoming "more and more a luxury," a writer for the USGS claimed in 1911, for he thought "prices must advance with the increasing cost of production." The maximum output, he concluded, would likely be about 100 million long tons (2,240 lbs.) followed by decline. A decade later, the US Coal Commission saw these problems in similar terms.[5]

In retrospect, it is easy to see that these fears of shortages were overblown. While the forecast of a maximum output of 100 million tons of anthracite proved quite accurate (fig. 5.1), the Malthusian explanation confused natural with man-made scarcity. Output per man-day in hard coal did indeed stagnate in the early decades of the twentieth century, and with rising wages, this led to sharply increasing costs. As a result, the mine price of anthracite, which had exceeded that of bituminous coal by 43 percent in 1900, was nearly triple that of soft coal in 1930 (table A4.2). Yet this productivity stagnation seems to have reflected company lethargy more than natural scarcity, for beginning in the mid-1920s, as the sting of competition induced companies to invest in better technology, labor productivity in hard-coal mining began to rise, and in the 1930s, the fuel's relative price declined.[6]

Thus, the collapse in sales of hard coal resulted not from natural shortages but rather from coal's increasing vulnerability to competition. As a high-priced fuel, hard coal could not defend against lower-cost alternatives that might duplicate some of its advantages. Because one of its selling points was convenience (less dirt; fewer clinkers), anthracite was susceptible to attack from even more user-friendly fuels. While soft coal was dirtier to burn, it was cheaper, which provided some protection from competition. In 1930, anthracite sales were off their wartime peak by about 30 percent. By then, the combination of rising prices and innovations in other fuel markets had put hard coal on the road to oblivion. By 1940, sales were half their peak level.

The Rise of Substitute Fuels

The boom in coal demand associated with World War I marked the end of an era of growth. Because soft coal was the fuel of industry, the war sharply boosted sales. Less affected by war, anthracite production grew only about 1.5 percent a year in the decade ending in 1917. Yet for both fuels, substitutes had been nibbling away like mice at domestic markets for years.

Solid Fuels

Along with hard and soft coal, a number of other solids were available for buyers of domestic fuels, and households of varying incomes and tastes no doubt balanced price against fuel characteristics. Except in East Coast cities, most households used cheaper bituminous coal for cooking, space heating, and hot water. Thus, while Boston households with moderate incomes used no soft coal around World War I, in Cincinnati, Columbus, and St. Louis similar families burned no anthracite at all.[7]

In addition to direct competition with bituminous coal, by 1900 anthracite faced increasing competition from a number of other fuels. Figure 5.2 presents data on domestic consumption of anthracite and some of its important competitors (imports of anthracite are not shown). The USGS reported in 1907 that "for domestic purposes, coke and gas, the products of bituminous coal, are competing more and more with anthracite in the markets of the larger cities and towns." Part of the reason for this increased competition was, as I have noted, anthracite's high cost. The first systematic data on retail coal prices began only in 1913; and, as table A4.3 demonstrates, from that date on, hard coal's cost per ton far exceeded the energy equivalent cost of soft coal, even as the price of manufactured gas fell relative to that of hard coal.[8]

Most coke that sold to households was a byproduct of the destructive distillation of soft coal to produce manufactured gas. Nearly pure carbon, it was smokeless, cleaner to handle and burn than bituminous coal, easier to light than anthracite, and had roughly the same heating value. Its disadvantages were that it required more tending than did anthracite and, because it was lighter, took more bin space. This "gashouse coke" had long been available for domestic fuel, typically selling at one to two dollars a ton below anthracite in local markets near the gas plant. As production of manufactured gas spread, coke sales increasingly contested coal markets. A second source of coke came from the byproduct coking process used by steel companies. Byproduct coke began to penetrate the domestic fuel market when steel-company demand lagged, because producers needed to operate ovens full-time. Byproduct coke also required consumer education, for it was harder and more difficult to light than the gashouse product. In 1923, domestic coke from all sources amounted to about 2.7 million tons (fig. 5.2); thereafter, sales took off, peaking at nearly 12 million tons in 1933. Most coke production was in the Northeast; a 1930 Bureau of

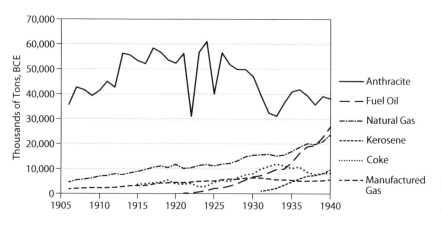

Figure 5.2. Domestic anthracite and some of its competitors, 1906–1940. (Chart data drawn from appendix table A4.1)

Mines survey found consumers used 90 percent of domestic coke in the north-central or Middle Atlantic States, where it competed directly with anthracite in both cooking and heating.[9]

Burning coke as a domestic fuel meshed nicely with Progressive ideals about reducing waste of natural resources. Because the coking of coal generated not only gas and coke, but also other byproducts such as ammonia and creosote, the Bureau of Mines explained that its use would "save many valuable by-products that are wasted when [bituminous] coal is used directly in the raw state." Finally, like anthracite, coke was smokeless. In the years before World War I, as urban Progressives began campaigns to pass city smoke ordinances, the bureau urged that this, too, made coke an attractive fuel, and it was far more widely available than was anthracite. Yet domestic use of coke gradually faded after the mid-1930s (fig. 5.2). It usually cost more than soft coal, and its smoke pollution benefits were hard to market, so it also fell victim to superior fuels.[10]

Like coke, coal briquettes warmed Progressive hearts, as well as their hearths and kitchens; the divided use of briquettes between heating and cooking is unknown. As discussed in chapter 3, briquettes resulted from entrepreneurial efforts to cash in on the vast piles of coal dust by adding a binder and compressing it into something much like modern charcoal briquettes. Because the producer price of briquettes typically exceeded that of anthracite, their natural market was near a cheap supply of fuel yet sufficiently distant from the mines to be protected from coal competition by high transport costs. The Great Lake states of Wisconsin and Michigan fit this profile because their docks contained vast stores of slack, so here, briquettes probably cut into sales of both hard and soft coal.[11]

Manufactured and Natural Gas

For cooking, manufactured gas was the most important of anthracite's competitors, although the urban poor, along with farm families and those in small towns without gas, might burn kerosene. The availability of gas for cooking turned all forms of solid fuels into inferior goods; rising incomes led to a reduction in their use. In the 1930s, a survey of family income and fuel use indicates that every 10 percent increase in income reduced the fraction of families cooking with solid fuels by about 8 percent, and this relationship no doubt obtained earlier.[12]

As we saw in chapter 2, by 1900, every large city and many small towns had a coal gas plant that derived its product mostly from bituminous coal, enhanced by the addition of liquid hydrocarbons (e.g., naphtha). A second source of manufactured gas came from the byproduct ovens that supplied coke for iron making. Here, the product was coke, with gas the byproduct, and, as the process steadily replaced beehive coking, a flood of byproduct gas appeared. As steel companies economized on its use, large surpluses became available and were marketed through gas utilities for domestic purposes. In this way, greater efficiency in the making of iron and steel increased competition among domestic fuels. As electricity increasingly drove gas out of illumination after 1900—and as economies of scale and technological change reduced gas prices while hard-coal prices rose—utilities increasingly marketed gas, sometimes at special discount rates, for hot water, room heating, stoves and radiators, and especially cooking.

Expensive transit tethered natural gas close to production areas until the late 1920s. In 1920, 36 percent of the 2.6 million domestic customers resided in Ohio and accounted for a third of all domestic consumption. Western Pennsylvania, western New York, California, Kansas, and Oklahoma were other important consuming states. Where available, natural gas was cheap—selling to domestic users at $0.38/Mft.^{3} in 1920, when its manufactured cousin (which contained roughly half the energy per cubic foot) cost $1.09. One result was that households used natural gas in prodigious quantities: 102,000 cubic feet per domestic customer in 1920 compared with about 37,000 cubic feet for *all* customers (including industrial) for manufactured gas. As a regional fuel, natural gas had largely driven hard and soft coal from Ohio's urban kitchens by 1920. Elsewhere, however, anthracite markets remained sheltered. I have calculated anthracite consumption for states and regions that accounted for 83 percent of the total in 1916. Natural gas consumption in those states amounted to about 6.4 million tons of coal equivalent in that year, and by 1940, that total had risen only to 6.6 million tons. Before World War II, natural gas competed mostly with manufactured gas, soft coal, and other solid fuels but not anthracite.[13]

Economics of Gas Cooking

The energy content of manufactured gas implied that 47,636 ft.^{3} of gas was the equivalent of one ton of soft coal, and this is the basis for the coal equivalent data for gas in figure 5.2. Using such figures to compare the cost of *using* gas or coal is misleading, however, for they ignore the efficiency with which the fuels burned. Because gas burned with greater efficiency than did coal, it was a far cheaper competitor of coal than a simple comparison of prices and energy content might suggest.

A careful study conducted by the US Bureau of Mines and Ohio State's Department of Home Economics shows that in cooking, about 105 ft.^{3} of manufactured gas could do the work of 28 pounds of hard coal (table 5.1). This implies that in cooking, 7,500 ft.^{3} of gas (not 47,636 ft.^{3}) was equivalent to a ton of anthracite. These efficiency considerations made gas cooking a potent competitor. Manufactured gas would be economic if the cost of 7,500 ft.^{3} was no more than a ton of coal, and this was often the case, while natural gas was clearly a bargain. Moreover, cooking with gas would displace (28 lbs./day × 365) = 5.1 tons of hard coal a year. (A similar, later,

Table 5.1. The Cost of Cooking: Gas vs. Coal, 1917

Price of Fuel	Natural gas (ft.^{3}) ($.31/Mft^3)	Manufactured gas (ft.^{3}) ($.91/Mft^3)	Bituminous coal (lbs.) ($6.50/ton)	Anthracite coal (lbs.) ($9.57/ton)
Breakfast				
Fuel Use	10	18.2	10.1	9.6
Btu	10,000	10,000	132,000	125,714
Costs (¢)	0.31	1.7	3.45	4.84
Lunch				
Fuel Use	24	43.4	11.5	11.0
Btu	24,000	24,000	138,000	131,429
Costs (¢)	0.74	4.2	3.75	5.3
Dinner				
Fuel Use	22	40.0	7.8	7.4
Btu	22,000	22,000	93,600	88,571
Costs (¢)	0.68	3.6	2.5	3.5
Totals				
Fuel Use	56	105	29.4	28
Btu	56,000	56,000	363,600	345,714
Costs (¢)	1.73	9.5	9.7	13.6
BCE*	4,000 ft.^{3} = ton	7,500 ft.^{3} = ton		

Source: Department of Home Economics, Ohio State University, *Kitchen Tests of Relative Cost of Natural Gas, Soft Coal, Oil, Gasoline and Electricity for Cooking* (Columbus, 1918). *Calculated for bituminous coal.

study of Indiana housewives found that cooking burned 6 tons of coal a year.) Thus, a family shifting from cooking with anthracite to gas would cut fuel consumption, reduce its energy bill, and improve the quality of life.[14]

But coal or wood stoves not only cooked; they also provided heat (whether it was wanted or not), and many contained a reservoir for hot water. These joint services complicate any assessment of how much coal gas cooking displaced. For example, a family that bought a gas stove and a coal water heater that would use 500 pounds of coal a month would save but 2.1 tons of coal a year. The point remains, however, that the advent of gas cooking sharply reduced coal demand, even if we cannot estimate the amount with great precision.[15]

The aggregate of these substitutes depicted in figure 5.2 was the energy equivalent of about 38 million tons of coal in 1930 and 75 million in 1940. These fuels competed with each other and both hard and soft coal in domestic markets for cooking and heating fuel. While their success or failure reflected the attractiveness of the price-quality package that each represented, utilities that sold gas and coke did not leave such consumer choices to chance. Rather, they created marketing campaigns that left coal producers in the dust.

Gas Markets Modern Cooking

As we saw in chapter 2, manufactured gas companies had begun to promote cooking by the 1870s, for it helped balance their load, thereby reducing average costs. Soon, the rise of electric lighting—enthusiastically promoted by utilities and equipment makers—provided an even sharper spur. This was not because electric light was lower in cost; indeed, with the advent of the Welsbach mantle in the 1890s, gas light became substantially cheaper and initially gave the electric industry a case of the jitters.[16] Electricity, however, gave a warmer, softer light than the Welsbach mantle and had other advantages as well. The gas-light era began to recede.[17]

Convenience, Cleanliness, and Leisure

Gas, in short, needed a new market, and it was not a simple task to change households' cooking habits. Domestic fuel choices embodied not only the physical investment in cooking equipment but the hard-won expertise in the use of a particular fuel and stove as well. Just as the rise of coal— especially anthracite—a century earlier had required considerable consumer experimentation and learning, so families again had to discover how to use substitute fuels before they would shift to them. The incentive to learn was the possibility of cheaper or easier cooking, while information came from suppliers' promotional efforts and from articles in women's magazines, which had long sung the praises of gas stoves, stressing themes of convenience, cleanliness, and leisure. Thus, Mary Warner (dates unknown) penned an enthusiastic assessment of (manufactured) gas cooking in *Good Housekeeping* in 1894. Recent price declines, she thought, had stimulated its use, and she advocated it especially for "dwellers of steam heated city flats" (because, of course, they did not need the heat from a coal stove), and she described the "immense saving of time, labor and dirt" that resulted from switching to gas. In 1902, Ellen Murdock (dates unknown) pointed out to readers of that magazine another aspect of gas cooking that must have appealed to many housewives: "The woman with a gas stove can economize a good many steps and many minutes by arranging a number of pantry things and cooking aids close by her stove. When one uses coal this cannot be done: dirt and ashes would keep things constantly dirty." Coal stoves were also balky and difficult to use (fig. 5.3). A home economics text of 1910 devoted eight pages to the complexities of coal stoves.[18]

Ellen Murdock's observations point to another benefit from switching to gas: it reduced indoor air pollution. I suggested in chapter 2 that the shift from fireplaces to stoves for cooking and heating must have reduced indoor air pollution, with special benefits to women, who were home more than men and in the kitchen much of the time. Yet as the complaints suggest, coal stoves were dirty: they smoked, and feeding and emptying them invariably resulted in the escape of coal dust and ashes. Large numbers of modern studies of domestic coal use in China link exposure to products of combustion to many illnesses. And while the pollutant levels in China are probably higher than Ellen Murdock experienced, the shift to less-polluting

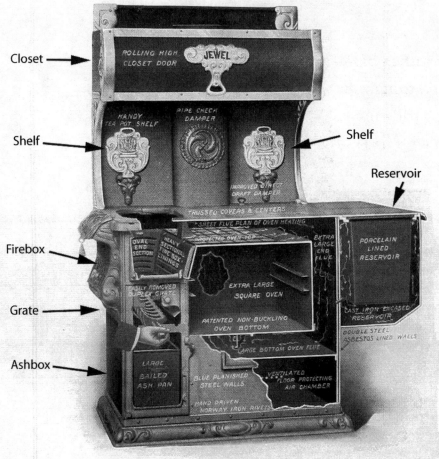

Sectional view of a Jewel Steel Range which shows plainer than
words the construction and special features

Figure 5.3. Coal cookstove, 1911. Coal stoves were massive compared to gas; they were dirty, time consuming, difficult to fire, and difficult to regulate. To heat one pot required heating the entire stove. (*Jewel Stoves, Ranges and Furnaces*, cat. 91, 1911–1912, 10. Public domain)

fuels, such as gas, must surely have improved the health of American women a century or more ago.[19]

Indeed, the comparative ease of gas cooking even provoked the muse in Lulu Eastman:

Yes that's the secret of success in weary woman's ways—
Use gas stoves bringing nights of ease and long and restful days.[20]

A gas stove purchased instead of a coal stove costing the same amount would show no change in the national income statistics, but like Ms. Eastman, many other women must have enjoyed the "nights of ease and [comparatively] long and restful days" that arrived courtesy of such a purchase.

An array of other gas-powered kitchen appliances also seemed to promise nights of ease and long and restful days to housewives, as well as burgeoning sales to gas companies; for Nancy Gay's all-gas kitchen included gas flatirons, waffle irons, and much else. In 1909, however, *Good House-keeping* advised that electric flatirons were far superior to those powered by gas, and by then, about a third of urban households had electricity. Similarly, electric toasters coffeepots, and washing machines were beginning to appear, and they, too, shouldered out gas. Yet for hot water and cooking, gas was cheaper, and having been used for a long time, it had little to fear from electricity.[21]

Selling Gas Stoves and Gas

Early gas stoves were often flimsy affairs, little more than hotplates, but by 1900, they came with optional pilot lights for the oven and in an array of sizes, some nearly as massive as coal or wood cookstoves. Producers tried to ape the advantages of the older equipment as well, offering gas stoves with hot-water reservoirs and hybrids that featured gas along with an option for either wood or coal. Before World War I, gas-stove manufacturers advertised in journals aimed at wholesalers and retailers such as *American Artisan* or *American Gas Journal* but rarely in newspapers or consumer-oriented magazines, leaving this, apparently, to the gas companies.[22]

The economic structure of gas production was far more conducive to advertising than was the coal business. In chapter 1, I noted that as coal became available for domestic use, its superiority made it easy to market, so dealers simply sprang up. Producers had little connection to dealers, and this separation remained intact into modern times. Because producers of manufactured gas were also inevitably retailers, they did not face the coordination problems that plagued coal producers, wholesalers, and retailers. Moreover, gas producers were usually gas monopolists; therefore, they could capture the payoff to local sales efforts. The industry also supported an assortment of national and regional organizations that organized and publicized sales work. The American Gas Association (AGA), which could trace its origins back to 1873, and its regional relatives (the Ohio Gas Association, the New England Gas Association, and others) were the most important for the manufactured-gas industry. The Natural Gas Association (NGA) spoke for that part of the industry, while the National Commercial Gas Association pushed both fuels. *Gas Age* and the *American Gas Journal* supplied cheerleading and information for both types of gas. Coal had no such institutional support for domestic marketing.[23]

While gas producers had been marketing their fuel for cooking since the 1870s, use spread slowly because of its expense. But rising incomes and declining prices widened markets. Real GDP per person rose about 76 percent between 1900 and 1940. Gas companies were experimenting with declining block rates as early as the mid-1880s, and by the early twentieth century, they seem to have become common. Inflation-adjusted prices of manufactured gas had declined throughout the nineteenth century, and between 1913 and 1925, gas prices fell 27 percent compared to soft coal and 59 percent relative to the price of anthracite (table A4.3).[24]

Producers grasped the opportunity. By this time, most municipal gas companies had moved into appliance sales as well, and they launched imaginative and powerful marketing campaigns. Since cooking was predominantly women's work, companies began to hire women in their marketing programs. In 1906, the American Gas Institute (predecessor of the AGA) polled its members on their various methods of getting new business, and the predominant focus was on expanding gas use for cooking and water heating—the core of the all-gas kitchen dear to the heart of Nancy Gay. The replies provide a glimpse into their sales practices. Nearly all employed newspaper advertising, sometimes in foreign-language papers as well. These ranged from the bland to snappy one-liners that played on themes such as modernity and comfort: "Might as well make your own shoes or weave your cloth as use a coal range. Cook with gas." "Who is afraid of the hot weather with a gas range in the kitchen?"[25]

A theme that runs through company responses was the need for good, efficient, honest service; a bad reputation was disaster. Several companies even taught meter-reading classes for customers. Bulk mailings were widely employed as well. Most gas companies also sold appliances, and the Battle Creek (Michigan) Gas Company sent out a mailing informing its recipients that "the price of one cigar a day would buy your wife a range." One Grand Rapids (Michigan) company sent service workers to check on existing appliances—and to gather information on customers who might be receptive to additional equipment. Most companies hired women "solicitors," who made house calls and provided advice, home-cooking demonstrations, and free items, such as gas waffle irons or horse blankets emblazoned with "Cook with Gas." L. C. Graham of the Winona (Minnesota) Gas Light and Coke Company explained why companies favored women for these jobs: "We find lady canvassers are better than men for selling gas ranges. It is possible for them to get in closer touch with the ladies and analyze the situation better and follow up what a man would think a poor prospect and turn it into a sale."[26]

Manufactured-gas companies often marketed stoves and heaters at cost, or sometimes at a loss, and might throw in free installation. They also sold on time, and the Bedford (Indiana) Heat and Power Company even provided 5,000 cubic feet of free gas for stoves bought in March. Since cooking with gas was so different from cooking with coal or wood, companies needed to educate consumers. Accordingly, many offered gas-cooking classes and sometimes managed to wangle them into high school home economics classes. Because gas was expensive, they provided tips on economical use. For those whose stove lacked pilot lights and were tempted to leave the burner on, companies instructed "matches are cheaper than gas." There were special classes for "colored" students and for adult cooks.

And there were endless contests: the person writing the best ad might get a free range, or there might be cash for the "lady baking the best loaf of bread, or cake in a gas range." The Bridgeport (Connecticut) Gas Light Company had women demonstrators in the office baking pastries. Some companies that sold gas, electricity, and transit advertised on their trolleys. The Butte (Montana) Gas Light and Fuel Company's offering read, "Everybody works but mamma, 'cause she uses a gas range." Mamma's view of this assessment is not recorded.[27]

As the market for coal gas expanded, utilities producing it inevitably generated an increasing amount of coke. Thus, in indirect fashion, every new gas range in the kitchen resulted in more fuel to compete with coal elsewhere. Coke that was a byproduct only had to cover the additional cost of shipping and handling, but to make the sale, companies had to employ many of the techniques they used to market gas. Here again, the sales material included a strong dose of information, for as I have noted, coke needed to be handled and burned differently from either anthracite or bituminous coal. Rome (New York) Gas and Electric made a virtue of coke's light weight, advertising it as the perfect fuel for "dainty women." The company also emphasized the cleanliness of coke: "We deliver coke in white canvas bags loaded on a white wagon with the driver in white canvas overalls. In muddy weather the wagon is washed . . . every trip. All the coal wagons are painted black [like] a funeral procession." Albion (Michigan) Gas Light successfully increased coke sales by arranging with local hardware stores to donate a quarter ton of coke with each gas stove sold. In some towns, grocery stores sold coke on commission and offered free samples from the gas company. Fort Dodge (Iowa) treated its coke and gas as complements, not substitutes, offering gas stoves with a side-arm heater to burn coke. In 1913, booming gas demand faced Detroit City Gas Company with the need to double its coke sales, which they accomplished by a stepped-up campaign featuring the usual mix of billboards, trolley ads, and discount coupons.[28]

Natural Gas Takes a Different Approach

The natural gas industry pursued domestic sales somewhat differently. Its retail pricing policies had long contributed to shortages and waste that plagued the industry. Well into the 1890s, many companies charged customers a fixed sum per appliance. In Muncie, Indiana, the charge for cooking stoves was $0.05 per burner per month, and such rates were typical. Since a flat fee meant that costs were independent of the quantity of gas used (economists would say that its marginal cost was zero), such pricing led to vastly expanded use. A very low flat, but metered, rate might accomplish a similar result. Cheap natural gas ensured its wide use for space heating as well as cooking, but it led to supply instability, hindering companies' ability to attract business, for very cheap gas resulted in waste and peak-load shortages and contributed to exhaustion of supplies. Paradoxically, then, natural gas producers faced the problem of too much use by existing customers, yet they found supply worries hindered efforts to attract new users.[29]

When natural gas utilities shifted to metering, consumers howled, but the result was often lower bills—and lower consumption. In 1906, companies formed the Natural Gas Association. From then on, and as a division of the AGA after 1927, the industry campaigned for higher, metered rates to ease peak-load problems and preserve supplies. By 1913, the Bureau of Mines claimed that "the majority of cities now supply [natural] gas through meters." In Humboldt, Kansas, however—which sold gas at $0.20/burner/month—"half the lights in town were left burning all night." Producers also stressed the need for improved economy by consumers in order to soothe the pain of rising rates. In going after new business, natural-gas distributors employed many of the same marketing arrangements pioneered by makers of artificial gas, adapted slightly to their particular needs. Kansas City's gas company referred to its appliance sales operation as the "gas saving department." And since natural gas was cheap, utilities could not sell appliances as loss leaders, which made allies out of local appliance dealers. Instead, they might offer a month of free gas and free service that emphasized economy of operation.[30]

These private campaigns for efficiency meshed nicely with Progressive worries over dwindling natural gas supplies. By 1919, natural gas shortages had already forced Indianapolis to return to manufactured gas, and it was widely assumed that more cities would follow. To postpone that day, the United States Fuel Administration inaugurated an efficiency campaign that I briefly noted in chapter 3. It issued pamphlets for schools and homeowners on reducing waste in gas cooking, and at war's end, the Bureau of Mines took over the program, working with the industry on a broad spectrum of conservation measures. Publications by the bureau, as well as by the consulting engineer Samuel Wyer (1879–1955) and the Bureau of Standards, complained that most existing natural gas stoves wasted 80 percent of the fuel they burned. They used too little air; the flame was too low; gas pressures were too high; and flat cooking tops diffused heat improperly. Yet with gas prices low, consumers had little incentive to conserve. To encourage efficiency, the conservation committee of the NGA distributed to customers a half million copies of one bureau publication on home use of gas. Such actions led to important improvements in cookstoves and supported distributors' efforts to raise retail prices as well.[31]

Well into the 1920s, concerns with resource exhaustion supported these programs to increase efficiency, as well as rates high enough to ensure stable supplies. By about 1930, a combination of massive gas finds, along with innovations in pipeline transport, boosted these efforts, transforming natural gas into an increasingly lethal competitor of coal for heating as well as cooking throughout an increasing number of states.

Wartime and Labor Disruptions

The rise of substitute fuels largely accounts for the decline in anthracite, but disruptions resulting from World War I, along with two massive strikes in the early 1920s, further battered domestic markets for hard coal. These events resulted not only in major price spikes but also in shortages—coal was sometimes unobtainable at any price—and in quality deterioration. High prices and shortages encouraged competitors to enter new geographic markets and encouraged consumers to experiment with alternative fuels, thereby speeding up learning. Public policies also encouraged consumers to experiment with new fuels. Although the bituminous and anthracite sectors each experienced disruptions, the focus here is on anthracite consumption, where the war and strikes had long-term consequences.[32]

Anthracite Goes to War

In 1917, shortages of bituminous coal in the East led the US Fuel Administration to allocate a disproportionate share of 1918 anthracite production to eastern states. This surely accelerated consumer learning about alternative fuels in those midwestern and western states that received sharply diminished supplies. The federal government also did its part to speed learning about alternatives: the Bureau of Mines continued to publicize its investigations into the economical use of anthracite and its substitutes, while the US Fuel Administration also chipped in with *Fuel Facts*, which it emphasized was for "SAVERS" of fuel. The war immensely expanded byproduct-coking capacity as well, leading that industry to push more strongly into domestic markets for manufactured gas and coke in the postwar years. Koppers, for example, began to market coke in New York City as early as 1919.[33]

Wartime shortages eroded coal quality as well. Individual anthracite producers had developed standards for size and impurities in coal and would condemn shipments exceeding the limits, but the Federal Trade Commission discovered that the volume of condemned shipments dropped sharply during the period of shortage in 1916. It seems unlikely that this reflected an outbreak of quality control, since households complained that their coal contained so much stone and slate that some termed it "fireproof."[34] After 1916, demand for gas skyrocketed in Baltimore, and per capita use doubled between 1916 and 1922. "Many new homes are built without a coal range in the kitchen so that gas alone is used," Johns Hopkins University Professor of Economics Jacob Hollander told the United States Coal Commission. "Gas water heaters are also coming into common use," he observed.[35]

Appendix table A4.4 presents sales by state of domestic anthracite for 1916 and 1921, two "normal" years. The data demonstrate that, as noted above, its use was concentrated in New England and the Mid-Atlantic states, which collectively accounted for more than 80 percent of domestic use in 1916, and that the wartime changes were important. Anthracite sales had been growing slowly for some time, but they declined about 5 percent from 1916 to 1921. Moreover, the greatest decline was in those central and northwestern states where distance had made anthracite expensive and marginal before the war, states that had experienced the greatest wartime shortages. Accordingly, by 1921, New England and the Mid-Atlantic states accounted for 85 percent of sales. It seems clear that anthracite was in trouble long before the great strikes of the 1920s.

Strikes in the 1920s

Two immense strikes—one in 1922 that lasted from April 1 to September 2, and another in 1925–1926 that dragged on for 170 days—hastened the shift away from anthracite. In the 1922 episode, the industry followed wartime precedent and instituted its own geographic allocations; as in wartime, these disproportionately favored eastern consumers. With domestic anthracite everywhere scarce and expensive, entrepreneurs saw their chance. Imports of hard coal, much of it from Wales, jumped from virtually nothing to 234,000 tons in 1922. Thereafter, they would range from that figure to as high as 800,000 tons (most of which went to New England), despite a two-dollar-per-ton tariff applied in 1932. The rise of imports was one manifestation of a revolt against domestic anthracite in New England. Citizens there remembered the long strike of 1902, and the region collectively seemed determined to escape the cycle of strikes and shortages that resulted from dependence on American producers. Massachusetts appointed a Fuel Administrator with "wartime powers," who promptly urged consumers to shift to soft coal. In 1923, Boston's municipal buildings switched from coal to coke for heating.[36]

The 1922 strike also appears to have reduced product quality. In the summer of 1923, the Bureau of Mines took samples from anthracite stocks at Massachusetts retailers. It found some samples of domestic sizes that contained as much as 46 percent ash.[37]

Even before the 1925 strike, the *Boston Globe* reported that "a very large number of New Englanders have switched from hard to soft coal." Late that year, *Coal Age* noted a "bitter anti-anthracite campaign in New England." The moving force behind the efforts to reduce anthracite use was the New England Governors' Council and, especially, Massachusetts governor Alvin Fuller (1878–1958) and the head of the council's fuel committee, John Hays Hammond (1855–1936), formerly of the US Coal Commission. Remarkably enough, Hammond at least urged the federal government to stay out of the way, apparently believing that it might interfere with New

England's efforts to punish producers by weaning the region from anthracite. Aside from a publicity campaign featuring a "war to end war" with anthracite, the council's most important work was to provide information on the availability and use of alternative fuels. It publicized the efforts of the West Virginia Smokeless Coal Operators (who marketed a low-volatile bituminous coal) to gain a foothold in New England markets. When those operators opened an advertising booth to display their wares on Boston Common, Governor Fuller inaugurated the festivities by shoveling the first scoops of coal.[38]

The Bureau of Mines and the council also tried to educate consumers on the advantages and techniques of burning soft coal, as well as coke and briquettes. In a 1923 report that sounded like an advertisement for the smokeless coals, the bureau concluded: "The 'smokeless' Pocahontas [coals] . . . are higher in heat value and usually contain less ash than anthracite; and as a general rule they can be bought considerably cheaper . . . the purchaser actually gets almost twice the amount of available heat for his money."[39]

By December 1927, with anthracite at $16.50 a ton and on its way to $18, the Massachusetts Special Commission on the Necessaries of Life underlined the bureau's claim that, adjusted for heating value, the cost of smokeless coal was about half that of anthracite. An assist in these efforts to shift New England away from anthracite came from the Interstate Commerce Commission, which established new, lower joint-freight rates on coal from West Virginia to New England. Rising anthracite prices again attracted imports, including hard coal from Wales and coke from Scotland. Massachusetts coke sales from all sources jumped from about 270,000 tons in 1924–1925 to 500,000 in 1925–1926.[40]

With local variation, similar events played out in New York, Philadelphia, and Chicago, among other cities. Noting the increasing availability of coke, the *Chicago Tribune* editorialized that "it is the consumer's chance . . . [for] independence." Indeed, while consumers in that city used about twice as much soft as hard coal around World War I, by the mid-1930s, they used about five times as much. In New York, the state and city began an educational campaign in 1925 to explain to households the proper way to burn soft coal. The *New York Times* reported that Pennsylvania byproduct coking plants were stepping up production for New York markets. Since the 1922 strike, the state's gas plants had added 60 million cubic feet a day of capacity. New York State's coke capacity had risen from 150,000 tons to a million tons a year in the previous three years, and Schenectady, Troy, Watertown, Buffalo, Syracuse, Rochester, and a number of other cities and towns were now using coke.[41]

The Bureau of Mines summarized the hard-coal situation in 1927: "Over a period of years there has been a gradual downward trend in the tonnage taken by certain important anthracite markets." The bureau estimated that the strike of 1925 had led consumers to replace 17 million tons of anthracite with its bituminous cousin, and "undoubtedly, some households continued to burn" soft coal. About that time, an editorial in the *New York Herald Tribune* captured the changes. "Just a generation ago coal hods were big sellers in every American city. . . . They were the symbol of anthracite. . . . Today you strain your eyes looking for [one]." In the mid-1920s, hard-coal producers and dealers struck back, trying to stem their losses, but because their major focus was on markets for space heating, their efforts will be discussed with that topic in chapter 7.[42]

Gas Goes National

In 1914, Nancy Gay's promotion of the all-gas kitchen marked the onset of national advertising by the NCGA. The association had emerged in 1904 to coordinate marketing and advertising, merging with the AGA in 1918. Its campaign kicked off with advertising in 12 national magazines with a combined circulation of 9.4 million. The campaign promoted gas as a sign of modernity, and it aimed to move gas from a summer cooking fuel to year-round use and to push gas water heating as well. Nor did imaginations stop there: an NCGA publication of 1914 featured lighting, ranges, water heaters, radiators, clothes dryers, fireplaces, flatirons, and chafing dishes. About the same time, the association also sponsored "Gas Range Week," which featured advertising in major magazines such as *Literary Digest* and the *Saturday Evening Post*.[43]

Gas-stove makers, which had typically advertised in trade publications, also began national advertising in similar magazines, as well as in periodicals aimed at women. While many stove makers also advertised equipment that would burn kerosene, promotion of coal or wood cookstoves was rare. Convenience and time saving remained the themes: in 1915, the Sentinel Automatic Cook Stove promised women "an 'afternoon off' every day," while two years later the Detroit Jewel Company urged women to "bake with ease," claiming that "every woman wants a porcelain equipped range" because "no stove polish is required." When World War I and immigration restriction dried up the supply of household help, makers of "Great Majestic," a combination gas, wood, or coal range, advertised that it "helps solve the servant problem."[44]

Utilities also continued the usual local promotions complementing the national campaigns. Rochester Gas & Electric advertised that it would install a gas stove for free and remove it later—also for free—if the customer decided not to purchase. A representative of Peoples Gas Light and Coke Company of Chicago (hereafter Peoples or Peoples Gas) claimed that his company advertised continuously on fences, bulletin boards, the sides of trucks and wagons, in catalogs, and on street barriers. He placed the greatest importance on continuous, informational newspaper advertising, although he decried its occasional placement underneath the paper's death notices. By the 1920s, dealers of hard goods were discovering that they, too, could profitably sell appliances for manufactured gas since the utilities had ended subsidies. Accordingly, department stores and appliance dealers began aggressive local marketing of stoves and gas water heaters. They sometimes coordinated with national advertisements, and they also emphasized modernity, occasionally—as did one St. Paul retailer—with a generous helping of male guilt: "You don't read by a tallow dip; you don't write with a quill pen. *Why then do you expect your wife to bake and preserve with a hot dirty coal stove?*"[45]

From their beginnings as "solicitors," women gradually expanded and formalized their roles in merchandising. By the 1920s, innovative companies such as Peoples Gas of Chicago had created service bureaus staffed by women who performed a range of sales-related work. The idea spread, and the AGA soon established a service bureau. At Peoples, under the capable leadership of Anna Peterson (1870–?), and elsewhere, women service bureau workers staffed model kitchens at company headquarters. They also performed cooking lecture-demonstrations, consulted with school home economics classes, visited homes and sometimes adjusted the gas stove, developed recipes to be included with gas bills, lectured to women's groups, gave radio talks, and staffed a "motor kitchen" that toured the suburbs.[46]

Gas Stoves Become Modern

Between about 1910 and 1930, gas cookstoves not only changed radically in their appearance; they also introduced improvements—as their sales demonstrate (fig. 5.4). By 1920, gas kitchen stoves no longer looked like smaller versions of coal ranges. Instead of cast iron and filigree, they were enameled steel, which, as advertisers noted, was easier to clean. At least some makers heeded the government's prodding to improve designs for natural gas stoves. In 1922, Roper urged dealers to sell customers its ranges featuring open tops, high burners, and low pressure that "will cut their

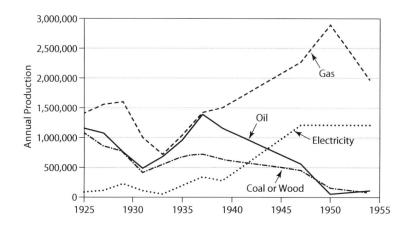

Figure 5.4. Cookstove production by type of fuel, 1925–1954. (US Bureau of the Census, *Biennial Census of Manufactures*, various years)

[natural] gas bills 50 to 75 percent." As we saw in chapter 3, in 1925, the AGA finally established a research bureau that, among other things, set safety standards for cookstoves. In 1928, Standard Gas Equipment Corporation introduced "Smoothtop" stoves with burners and oven on one level. About the same time, Roper pioneered a "color range to consummate color harmony, planned for your kitchen"—in Poudre Blue, Imperial Red, Canary Yellow, and Jade Green. By the 1930s, pilot lights for burners were becoming common, and in 1932, Magic Chef introduced the center pilot light. Other improvements included warming compartments, cabinet storage, better insulation, and broilers that did not smoke.[47]

But the most important improvement, which all but revolutionized cooking, was oven temperature control—an advantage that coal or wood could not match. In 1913, Benjamin Meacham, a superintendent at American Stove Company of Lorain, Ohio, designed an oven thermostat that he installed on his wife's gas stove. It worked from the beginning and sported the tradename "Lorain," which American Stove marketed under a variety of labels in which it installed the new device (fig. 5.5).[48]

In the Midwest, large-scale newspaper advertising began in 1918. Appealing to war-induced conservation concerns, the "'Lorain' Oven Heat Regulator" promised to conserve gas, food, time, and labor; indeed, it was the "answer to the servant problem." About 1920, American Stove began national advertisements—some of them in color after 1925—starring the Lorain "Red Wheel" (the regulator) in *Good Housekeeping, Literary Digest*, the *Saturday Evening Post, McCall's*, and other magazines. *Good Housekeeping* tested and approved the devices but, for safety reasons, only on ovens equipped with pilot lights. Early advertisements claimed that the heat regulator "ends pot watching . . . [which] gives housewives many extra hours for outside recreation," while later copy also emphasized that the ability to control oven heat precisely would greatly improve cooking. Moreover, if gas-company executives worried about American Stove's claim that the Lorain would save gas, Anna Peterson reassured them. Echoing Stanley Jevons's claim that conservation would increase energy use, she announced: "I find that women who have regulators on their stove are so much more certain of the results [that] . . . they bake twice as much as they did before."[49]

Gas Conquers the Urban Kitchen

Suddenly, coal or wood stoves seemed as dated as high-button shoes. By 1929, gas cookstoves outsold those using wood or coal by nearly a two-to-one ratio, while for hot-water heaters, the ratio was nearly four to one, and these disparities remained throughout the 1930s. While other solid fuels also encroached on anthracite markets, by the 1930s, most of these were also in decline, especially in urban kitchens, as the various gas-marketing campaigns and word of mouth increasingly made gas an attractive fuel.

In 1925, a survey of about 440,000 housewives by the General Federation of Women's Clubs found 61 percent cooking with gas. In Providence, Rhode Island, where hard coal had once ruled, by 1933, 87 percent of households cooked with gas, with only the poorest families using another fuel, although that city did not obtain natural gas until after World War II. Even in cities such as Fargo, North Dakota, where soft coal otherwise dominated, 86 percent of households cooked with manufactured gas.[50]

About this time, however, the baton also passed from manufactured to natural gas, which—because it was a far cheaper source of energy—was an even more lethal competitor of coal. Sales of manufactured gas peaked in 1930, while customers also peaked at about 12 million. In comparison, the availability of natural gas rose from 22 to 35 states between 1913 and

1940, while customers increased from about two to nine million. Census data for 1940 reveal how thoroughly gas had routed coal and wood from America's urban kitchens (table 5.2).

Nationwide, gas was the cooking fuel in 16.8 million "occupied dwellings"—nearly 46 percent of the total—and it was the fuel of choice for 73 percent of city dwellers. Even where natural gas was unavailable, as in New England, coal was in retreat: in Boston, 95 percent of households cooked with gas. In the Ohio and Indiana studies noted above, I suggested that households that shifted from coal to gas for cooking and also purchased a coal-fired water heater might have saved perhaps 2.1 tons of coal a year. Had the 16.8 million users of gas for cooking in 1940 each burned 2.1 tons of coal instead, sales of that fuel would have been about 35 million tons higher.[51]

Conclusions

In chapter 2, the long view of domestic energy use demonstrated that from 1800 to 1910, Americans' per capita domestic energy use declined, and as will be seen, but for the advent of the automobile, the decline would have continued to 1940. This chapter has provided a glimpse into how and why that decline occurred and how it shaped Americans' welfare. Anthracite led the long retreat of coal from Americans' homes, and the rout began in the kitchen because that was where coal's position was weakest. The

Table 5.2. Percentage Distribution of Fuels for Cooking, US Total and Regions: Urban, Rural Farm, and Nonfarm, 1940

	Coal*	Wood	Gas	Electricity	Kerosene	Other
US Total	11.5	23.6	45.6	5.4	9.7	0.5
Urban	8	6	73	5.1	6.9	0.4
Rural farm	14.1	69.5	2.8	2.7	9	0.8
Nonfarm	19.1	28.5	24	8.7	18.5	0.6
North total	13.2	11.7	59.3	5.1	9.7	0.7
Urban	8	1.8	79.1	4	5.2	0.5
Rural farm	24.8	54	4.7	3.8	10.9	1.7
Nonfarm	23.7	16	26.7	10.4	22.2	0.8
South total	8.7	47.3	27.1	2.9	12.5	0.1
Urban	9	17.5	54.2	5.3	12.9	0.1
Rural farm	4.7	82.9	2.4	0.9	7.9	0.1
Nonfarm	13.9	43.4	17.7	5.9	16.5	0.1
West total	10.6	22.5	51.5	10.2	2.9	0.5
Urban	6.8	10.1	70.3	10.9	1	0.2
Rural farm	19.2	57.1	7.4	5.1	6.4	0.9
Nonfarm	15.3	37.5	30	9.7	5.5	0.8

Source: Sixteenth Census of the United States, 1940: *Housing*, vol. 2, chap. 1, table 10a. There is an omitted category of "none."
*Includes coke.

comparatively high cost of gas was less of a disadvantage in cooking than in heating and was largely offset by its more efficient combustion. Coal's unattractiveness and comparative inefficiency opened a door through which gas invaded American urban kitchens.

The result was that American women used less energy to cook a meal even as they experienced an increase in the comfort and convenience that we, who never lived with the old ways, can only grasp by listening to the women who did. The reader met Ellen Murdock, who pointed out an advantage of gas that would occur only to one who had cooked with a coal stove: it saved a woman steps as she could store pantry items near the stove, which was impossible with coal owing to its dirt.

These events, which in retrospect seem inevitable, resulted from battles among the fuels for market share; the move away from hard coal reflected not only consumer responses to its rising price but also the entrepreneurial actions and creative marketing of producers of coke, fuel briquettes, and especially manufactured gas (the creators of Nancy Gay). Falling costs and smart marketing that emphasized education and stressed real differences in fuels helped spread the use of manufactured and natural gas, while the development of oven temperature controls surely consigned coal and wood cookstoves to "Old Fashionedville."

Labor-market developments and public policies shaped these events in several ways. World War I fuel shortages and the great strikes of 1922 and 1925–1926 nudged buyers away from anthracite. The Bureau of Mines performed important market-augmenting functions, providing information on briquettes, coke, and smokeless coal and helping natural-gas producers improve stoves and persuade consumers to save gas. Requirements for through-freight rates widened the area in which "smokeless" soft coal could compete, while New England's politicians made a concerted and successful effort to speed that region's transition away from hard coal.

Thus, the triumph of gas came in two forms and two stages. Manufactured gas began the rout of anthracite and other solid fuels from America's kitchens. This was a transition from coal to gas in the kitchen, even as coal was still the primary source of energy. As natural gas became increasingly available, however, it began to push both hard and soft coal from the kitchen and, until the later arrival of electricity, all other fuels as well—including its manufactured cousin.

An important implication of this story is that an exclusive focus on the energy transition from coal to natural gas misses much. Ultimately, most households did make this shift to natural gas (or, later, electricity) for cooking. But briquettes and coke, and especially manufactured gas, also contested coal. Natural gas was like the wolf that ate the rabbit *and* the coyote that was already eating the rabbit. If natural gas had not arrived when it did, manufactured gas would have pushed coal from most of America's urban kitchens, just as it did in Boston and Fargo.

6

"Cooking Shouldn't Cook the Cook"

The Kerosene Kitchen in Rural America, 1870–1940

The kitchen is a hotter place than the hay field and a "stove stroke" as bad as a "sun stroke."

—*Anon*, Michigan Farmer

If your husband can afford a riding plow, you can afford an oil stove.

—*"Critic,"* Wallaces' Farmer

In 1922, Mrs. Edith Hawley, of Agawam Massachusetts, informed the readers of *Farmer's Wife* how she used her kerosene stove to make a "'little extra money' which all farm wives feel the need of now and then." She began home-canning fruits and vegetables from her garden (which measured 131' × 232') and orchard for the local market, and she noted that "a good thing about the business of home canning . . . is that one needs little or no capital." She got the idea to go into canning from her daughter, who was in a girls' canning club, and her new stove made the work bearable. She started with a two-burner kerosene stove that she set up on the screened-in porch. The first year, 1919, she canned 1,434 jars. The next year, having added a three-burner oil stove, she put up 4,032 jars of vegetables and 1,298 jars of jelly. In 1921, she canned 6,391 jars of all kinds; her largest single order being for $332 and her total fuel bill $52.[1]

The readers of *Farmer's Wife* would not have been surprised that Mrs. Hawley cooked with kerosene (sometimes termed "oil" or "coal oil"), for many of them used that fuel as well. They were participating in one of the many energy transitions of modern times. Kerosene had once been a major source of illumination but—as discussed earlier—by the twentieth century, it was rapidly losing that market to coal—converted into gas or electric light. As Mrs. Hawley's story suggests, however, kerosene had a second act—as a fuel for cooking and space heating that was supplanting wood and coal on farms, in small towns, and among the urban poor.

The development of the coal stove, made an enormous contribution to Americans' welfare, but as we have seen, such stoves were not without their burdens. Someone—usually a woman—had to haul the coal and ashes. And they were dirty: gas was a far better fuel. But gas was not available on farms and in small towns, while kerosene was. Like gas, oil was cleaner, lighter, and required less care and maintenance than coal. Again like gas, its greatest advantage was that in summer, an oil stove did not cook the cook. When it was available, gas was always preferred to oil, but for women like Mrs. Hawley, on farms and in small towns without gas, before electricity, salvation appeared as a kerosene stove.[2]

Thus, kerosene was for rural women's standard of living what gas was for their urban sisters as it slowly drove wood and coal, corncobs, and "barnyard lignite" from family kitchens. As late as the 1920s and 1930s, nationwide sales of kerosene stoves nearly equaled—and sometimes surpassed—those of gas equipment. Surveys reveal that outside of cities, oil was a far more common fuel than was gas, and in some states, it was the dominant cooking fuel, although its expense meant that its use for space heating was less common. Especially after 1930, however, the substitution of kerosene kitchen and parlor stoves contributed to the collapse of anthracite sales and the stagnation of bituminous coal as a domestic fuel. After World War II, kerosene use gradually faded, but to ignore the rise of oil as a domestic fuel is to miss what was for many an important introduction to modern times.[3]

This chapter begins with the development of safe kerosene, for that was almost certainly a prerequisite for the wide use of oil for cooking and heating. The story reminds us that while petroleum was abundant in nineteenth century America, it had to be tamed to be exploited. Section two then discusses cooking with kerosene, traces the evolution of oil-fired stoves during the nineteenth century, and explores how women responded to the new technology and how it affected their lives. The chapter then discusses twentieth-century marketing to women and changes in stove technology. The final section explains the very rapid market penetration of oil stoves during the 1930s and 1940s, followed by an almost equally rapid collapse.

Inventing Safe Kerosene

As the reader will recall from chapter 1, the Canadian, Abraham Gesner, was a geologist and another of the many nineteenth-century scientist-entrepreneurs who propelled energy developments. Gesner first produced kerosene—"coal oil"—about 1846 from Nova Scotia asphalt deposits; a decade later he was extracting it from coal and finally oil. Kerosene qualifies as a price-induced innovation, for whale oil was rising sharply in price while available lighting substitutes (lard oil, camphene, manufactured gas, and others) were also expensive and sometimes dangerous.

Like manufactured gas, kerosene constituted a lighting miracle. Investigations done around 1860 found that even at $1.00 a gallon, kerosene lighting was a tenth the cost of tallow candles (see table 1.4). That year, one scholar has "guesstimated" that a million American families then used kerosene for illumination. With Drake's discovery of oil, prices plummeted: by 1875, it was selling at $0.35 a gallon. About 1882, kerosene lighted the homes of nearly 10 million families. Moreover, like gas lighting, kerosene lighting was improvable. A typical flat wick lamp might yield as much as nine candlepower. But by the 1890s, lamps that used a Welsbach-like incandescent mantle yielded 60 candlepower—roughly equivalent to a 60-watt incandescent bulb. Although by modern standards these provided precious little light, it must have seemed a wonder to the men and women who could remember candlelight, and it was an environmental miracle as well, for, as Gesner had predicted in 1860, "the dreams of those who advocated peace with . . . the finny monsters of the deep" were being realized.[4]

More Dangerous Than Gunpowder

Yet young technologies often bring novel risks, and kerosene proved expensive to life and property, growing quickly from its infancy into a very dangerous baby. Initially, it was anything but a standardized product. Indeed, in 1854, when Gesner applied for an American patent, he described three kerosenes: "A," "B," and "C." These differed sharply in specific gravity, boiling point, volatility and flammability with "A" and "B" too volatile and flammable to use for illumination, probably because they contained more or less naphtha and gasoline, which had a much lower flash point. Only "C" would make a good, safe illuminant, Gesner thought.[5]

Alas, market forces quickly dashed that hope. Because naphtha then had few uses, it sold at a sharp discount to kerosene, as did gasoline, thus creating strong incentives for manufacturers, wholesalers, and retailers to blend them with more expensive but less flammable kerosene, resulting in a potentially explosive mixture. Table 6.1 provides a conservative assessment of the economic payoff to adulteration in New York City, circa 1870. A retailer might buy 113°F flash test kerosene in bulk at $0.20/gallon, selling it at $0.30/gallon and realizing a profit of $0.10/gallon. But if he blended in naphtha at $0.05/gallon he could increase his profit, and a mixture of 20 percent naphtha raised his return per gallon by 30 percent—and lowered the flash point to 40°F—which authorities would

Table 6.1. The Economics of Kerosene Adulteration, 1870

Fraction naphtha	Fraction kerosene	Flash point (°F)	Wholesale price of naphtha ($)	Wholesale price of kerosene ($)	Retail price of kerosene ($)	Profit per gallon ($)	% Return on adulteration*
0	1.00	113	0.05	0.20	0.30	0.10	0
0.01	0.99	103	0.05	0.20	0.30	0.1015	1.50
0.02	0.98	92	0.05	0.20	0.30	0.103	3
0.05	0.95	85	0.05	0.20	0.30	0.1075	7.50
0.1	0.90	59	0.05	0.20	0.30	0.115	15
0.2	0.80	40	0.05	0.20	0.30	0.13	30

Source: "Report of the Committee of the Franklin Institute on the Causes of Conflagrations and the Methods of their Prevention," JFI 95 (April 1873): 261–287; Charles F. Chandler, "Report on Petroleum Oil, Its Advantages and Disadvantage," *American Chemist* 2 (June 1872): 446–448; and author's calculations.
*The extra profit expressed as a percent of the profit on unadulterated kerosene.

come to conclude was deadly. Of course, the seller might cut the retail price, encouraging consumers to purchase even more of the lethal brew. Either way, adulteration was a road to riches for some and a shortcut to tragedy for others.[6]

The price of kerosene soon became measured in lives as well as dollars, for the papers began to report an epidemic of stoves, heaters, and especially lamps that were starting fires. A report to the Franklin Institute claimed 3,500 deaths from such causes in 1871. That is a staggeringly large number: for some perspective, consider that it is equivalent to about 26 thousand deaths in the current American population, which exceeded the number of murders in 2021.

And the toll was growing by leaps and bounds. One estimate put it at 6,000 in 1877, although not all of these resulted from dangerous kerosene. Kerosene lamps were portable, and dropping them was a recipe for tragedy. Indeed, by 1866, such disasters had become so common in New York City as to merit headlines such as "Another Kerosene Explosion," when someone upset a lamp in the home of Mrs. Mary Taylor, burning her and her young son badly. The Brooklyn Fire Marshal reported on fires from kerosene in that city during 1870. "Some of the victims, poured kerosene oil on stove fires to make them burn more rapidly but have been burned to death by the experiment," he observed. In most cases, however, "the lamp has been the destructive agency," he reported. Such was the cause of the tragedy that befell the nine-year-old daughter of Thomas Peak, of Constantine Michigan, who was "burned to death . . . by the explosion of a coal oil lamp in September 1870." Lamps also made good weapons: in 1870, the *Sycamore (IL) True Republican* reported that one Mrs. Maguire "in a drunken frenzy threw a lighted kerosene lamp at her husband," killing him and burning the house down. Nor were heaters and stoves immune to disaster, and the papers routinely carried brief notices of stoves that exploded. Again in 1870, the *Chicago Western Rural* reported that Margaret Doyle, a servant girl, was burned to death when she poured kerosene on a fire. *Puck* penned an imaginary epitaph for such occasions.

> Beneath this stone sleeps Martha Briggs,
> Who was blest with more heart than brain.
> She lighted a kerosene lamp at the stove,
> And physicians were in vain.[7]

Regulations and Markets

That the use of kerosene spread as rapidly as it did despite such widely reported carnage is a testimony to how much Americans valued the new energy source. Moreover, dangerous kerosene was like a disease, for it almost immediately began to stimulate antibodies. As early as 1860, the Rhode Island Mutual Insurance Company, which insured textile mills, asked Zachariah Allen (one of its founders) to investigate the explosiveness of kerosene. Allen, whom we met in chapter 1, was one of the tinkers and thinkers whose efforts advanced the introduction of coal and contributed to safer kerosene as well. His report was one of the earliest of many scientific investigations that attempted to develop safety standards for kerosene and to discover a simple test to determine its safety. A liquid's specific gravity, Allen discovered, was irrelevant because even a quite dense liquid might contain sufficient volatile oil to render it explosive. Aided by Nathan P. Hill, professor of chemistry at Brown College, Allen described a flash test to set safety standards. Since kerosene in an ordinary lamp might heat up to perhaps 80°F, Allen thought safe kerosene ought not to flash below 100°F.[8]

Not quite a decade later, New York's Board of Health became concerned with the dangers from kerosene. The reader met the board's chemist, Charles F. Chandler, in chapter 1 in connection with its campaign to suppress the stench from gasworks. Probably at Chandler's behest, the board commissioned him to survey and assess the quality of the oil on the market. Like Allen, Chandler employed a flash test, pointing out that an alternative test—the temperature at which the liquid would catch fire—was unreliable, for liquid that would not burn at 100°F might well flash below that point. The results were eye-opening: *all* of the 78 samples of kerosene tested would flash at less than 100°, and some had been tested in the 40- and 50-degrees range. A later survey found that 615 out of 660 oils tested were unsafe. Subsequent surveys in Cleveland, Philadelphia, and Baltimore yielded similar results. Many of these were branded products with happy-sounding names. There was "Sunlight Non-Explosive Burning Fluid" and "Danforth's Non-Explosive Petroleum Fluid." Tests demonstrated that the latter product had a flash point of 44°F, and on December 28, 1870, it took the life of Mary Gibson of Poughkeepsie, New York, when it blew up in her lamp.[9]

In Chandler's report, later expanded and published in *American Chemist* (which he edited), he termed such products "murderous." Naphtha was "more dangerous than gunpowder." "No lamp is safe with dangerous oil and every lamp is safe with safe oil," Chandler claimed. He also asserted that "vapor" stoves that alleged to burn naphtha or gasoline safely were more dangerous than a keg of gunpowder. Chandler admired an 1862 English law that employed a standard test to determine an oil's flash point and described its application in detail. Others also developed tests but a review of 1880 demonstrated that many of them were highly sensitive to the use of proper protocols and often gave inconsistent results.[10]

Chandler's report was widely publicized in newspapers and periodicals, as were the many fires and explosions from dangerous kerosene. Editorials denounced the "Naphtha Fiend" and "Kerosene Murder," while the *New York Times* published lists of retailers in the city found to have dispensed dangerous kerosene. Collectively, these reports and stories began to shape public policy. A federal law of 1867 prescribed a fire test, and Maine and Massachusetts passed similar laws about the same time. A New York state regulation of 1871 employed the more appropriate flash test, and about this time, a number of cities began to regulate kerosene quality. By 1879, 21 of the 38 states had some kind of law, and a review conducted in 1914 found regulations in all but four states.[11]

Contemporaries complained that most such ordinances were poorly written and enforcement lackadaisical. Yet they must have had some impact: the Massachusetts ordinance of 1867 supported a private damage suit brought by a party whom dangerous kerosene had injured. And Michigan's 1878 law, which required inspection of all kerosene sold, found little that failed that state's 120° flash test. In any event, market forces reinforced these legal requirements.[12]

Economists have long argued that branding, linking a seller's reputation to product quality, provides a defense against shoddy goods. These were the years when Standard Oil and other large companies began to market kerosene under their own names. The Standard Oil story is well known: the company delivered kerosene in wagons labeled "Standard Oil," and it sold to retailers and sometimes directly to the public. By 1890, Standard controlled perhaps 90 percent of domestic kerosene sales. While Rockefeller was perhaps head demon in the progressive pantheon of fiends, branding kerosene provided buyers with quality assurance, for incinerating customers could not have been good for business. The company maintained quality inspectors and strove to meet differing state requirements. Standard also developed wicks and sold gasoline and naphtha stoves in order to sell those products, and it soon began to market kerosene stoves as well, which must have reinforced its commitment to product safety. Finally, changes in relative prices undermined the incentive to adulterate kerosene. Producers gradually generated a market for naphtha, so the spread between naphtha and kerosene prices at wholesale shrank to about five cents in 1873 and to less than two cents in 1880.[13]

There are no data to chart the decline of explosions of kerosene lamps and stoves, and by the 1880s, the problem had faded from the newspapers. Even more telling, the emphasis on safety gradually receded from advertisements for kerosene cookstoves and heaters. Initially, companies invariably stressed their safety. In 1877, the "Domestic" promised "absolute safety," while the Florence Stove Company advertised the same year that one of its products "cannot explode." A year later, the "Boston Gem" purported to be "perfectly safe." Such claims did not entirely disappear, perhaps because even if kerosene itself was no longer explosive, people still incinerated themselves spilling or throwing it on fires, and stove manufacturers needed to remind consumers that their products were not the source of risk. Still, by the late 1880s, safety had dwindled in importance as a selling point, and it later virtually disappeared, suggesting that the dangers from explosion also declined.[14]

The development of safer kerosene not only avoided many fires, deaths, and disfigurements; it was vital if kerosene stoves were ever to find a wide market. Before the age of electricity, kerosene had few good substitutes as an illuminant. Hence, use of kerosene for light was thriving despite its dangers. But for cooking and space heating, kerosene competed with a host of other fuels, some of which were often significantly cheaper. Kerosene that might simply explode with no warning was a mysterious, frightening risk. By contrast, reports of fires or explosions from improper use of kerosene allowed readers to imagine that they could prevent such disasters. The ability to control risks no doubt made them less scary, and the continued spread of oil for cooking and heating suggests that buyers found its benefits outweighed such risks. The story reminds us that new products, for all their wonder, bring new risks as well. That, however, is only half the story; for scientific investigation, state regulation, market forces, and marketing combined to slay the "Naphtha Fiend." Had kerosene not shed much of its reputation for blowing up without warning, oil stoves might have died at birth.[15]

Kerosene Cooking and Its Competitors

Although the predominant use of kerosene was initially for illumination (it was often simply called illuminating oil), entrepreneurs quickly saw its potential for other uses. By 1882, a census report noted that "kerosene stoves are being brought to a great degree of perfection and are found to be very useful." After worrying a bit that their use might result in "impure air," the author concluded that "they are cheap and convenient, are used by tens of thousands, and their use is increasing."[16]

Cooking Economics

Like gas, kerosene was a comparatively new fuel, and like gas, it offered decisive advantages over older cooking fuels in comfort, convenience, and cleanliness. Kerosene stoves may have been "cheap and convenient," but the fuel itself was often comparatively expensive, which profoundly influenced its market penetration. In 1885, for example, the wholesale price of kerosene in New York was $0.08 per gallon at a time when bituminous coal retailed at $4.20 a ton. If we ignore the efficiency with which the fuels burned, kerosene would be uneconomic at any price above 2.2 cents a gallon, and 194 gallons of oil would substitute for a ton of coal.[17]

But as chapter 5 explained in the case of gas, efficiency in use matters a lot. Table 6.2, contains estimates of the relative efficiency of kerosene, anthracite, and Ohio soft coal for cooking. The table indicates that it took 0.816 gallons of kerosene to cook a day's meals that required 29.4 lbs. of Ohio coal. The last row of the table indicates that because of its superior

Table 6.2. The Cost of Cooking: Kerosene vs. Coal, 1917

Daily use	Kerosene (gal.) ($.15/gal.)	Bituminous coal (lbs.) ($6.50/ton)	Anthracite coal (lbs.) ($9.57/ton)
Breakfast			
Fuel use	0.155	10.1	9.62
Btu	20,900	132,000	125,714
Costs ¢	2.5	3.45	4.84
Lunch			
Fuel use	0.338	11.5	11
Btu	45,600	138,000	131,429
Costs ¢	5.62	3.75	5.3
Dinner			
Fuel use	0.324	7.8	7.4
Btu	43,700	93,600	88,571
Costs ¢	5.4	2.5	3.5
Totals			
Fuel Use	0.816	29.4	28
Btu	110,200	363,600	345,714
Costs ¢	13.52	9.7	13.6
BCE*	55.5 gal. = ton		

Source: Department of Home Economics, Ohio State University, *Kitchen Tests of Relative Cost of Natural Gas, Soft Coal, Oil, Gasoline and Electricity for Cooking* (Columbus, 1918).
*Calculated for bituminous coal.

efficiency in cooking, only 55.5 (not 194) gallons of kerosene was the equivalent of a ton of Ohio bituminous coal. Cooking with kerosene instead of coal, that is, saved energy. Accordingly, with coal at $6.50/ton, kerosene would be cheaper to cook with at any price at or below about 11.7 cents a gallon. Put another way, kerosene at $0.15/gallon would be economic if coal cost $8.33 or more.[18]

Such a calculation implicitly values the wasted heat at zero, as would be the case in summer. In winter, however, where coal's excess heat was not wasted, it was as much as 65 percent efficient as kerosene. In the latter case, with Ohio coal at $6.50 a ton, kerosene would have had to sell for about 5.5 cents a gallon to make its energy cost equal to that of coal.[19]

Figure 6.1 contains estimated retail prices of kerosene and the highest prices at which it would be economic for winter and summer cooking compared to the nationwide retail price of soft coal, which was its major competition in midwestern and western farm states. As can be seen, kerosene was sometimes cheaper than coal as a summer fuel for cooking, but during winter, it was always uneconomic.[20]

While the conclusions from figure 6.1 probably reflect earlier conditions as well, there are simply no data for the nineteenth century. The reader should also note that these calculations ignore a number of matters. Kerosene looks more attractive when compared to anthracite. Figure 6.1 also ignores geographic difference in prices. Kerosene shipped more cheaply than coal, making it economically more attractive farther away from the

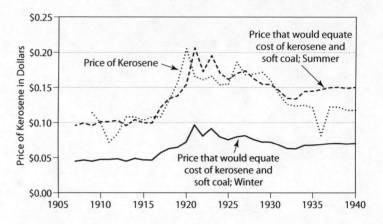

Figure 6.1. Price of kerosene and cost of energy for kerosene vs. bituminous coal, 1907–1940. Kerosene prices are based on wholesale prices. (Up to 1923, prices are from Williamson, *American Petroleum Industry*, vol. 2, tables 5.3 and 12.7; from then on, prices are from American Petroleum Institute, *Petroleum Facts and Figures*, 9th ed. [New York, 1949], 375. I have multiplied them by 1.2 as an estimate of retail prices. The equated cost employs retail coal prices from various BLS bulletins)

mines, especially in rural areas where all fuel arrived by wagon. Even more important, these comparisons ignore differences in amenity values of fuels. It wasn't just that the heat from a coal cookstove had no value in the summer: it generated genuine disutility. The ashes and dirt from all forms of coal were also undesirable, while kerosene might smoke. Despite such difficulties, what figure 6.1 depicts seems broadly correct: kerosene's strength was always in summer cooking or where use of other fuels was impossible or inconvenient. A careful study of fuel use by midwestern farm families in the early 1920s found kerosene costing $0.13 to $0.15 a gallon and soft coal $7 to $9 a ton, making the former fuel sometimes cheaper in the summer but always uneconomic in winter, thereby confirming figure 6.1.[21]

Kerosene Stoves

Such economic concerns shaped the market for kerosene-fired household appliances. Because kerosene—like manufactured gas—was usually relatively expensive, producers never tried to market it for central heating, concentrating instead on cooking and parlor stoves and water heaters and emphasizing that kerosene's amenity value—its comfort, simplicity, cleanliness, speed, and portability—offset any cost disadvantage. Gas was always the preferred cooking fuel where it was available, however. "Where gas is not available, oil stoves may be used," Mrs. S. T. Rorer advised readers of the *Ladies' Home Journal* in 1899. Similarly, in 1914, the *Efficient Kitchen* described kerosene cookstoves as a "great resource in communities where it is not possible to obtain gas."[22] For such reasons, while makers of kerosene stoves never abandoned the urban middle class, their target markets were farm families and people in rural areas and small towns not served by gas.[23]

Most early oil cookstoves were tiny; they were a natural evolution of kerosene lamps that made valuable the heat that would otherwise be a waste product—a conservation measure, one might almost say—and were

Figure 6.2. Kerosene cookstoves evolved from kerosene lights and heaters on which someone installed a top. Lamp stoves such as this one were called "babies." ("'Florence' Lamp Stove," Advertisement, *Good Housekeeping*, Sept 5, 1885, iii. Public domain)

termed lamp stoves or "babies" (fig. 6.2). These typically came in one or two burner models and looked like hotplates, although they gradually grew larger and sprouted legs as well. Others appear to have begun life as space heaters to which some entrepreneur attached a top, allowing them to heat flat irons, cookware, and portable ovens. While sometimes costly to run compared to coal, they were cheap to buy and took little space, designed— probably—to appeal to poor households with little room and inhabitants who could not afford a coal stove.

No data exist to document sales before the 1920s, but advertisements for kerosene stoves began to appear in the 1860s. Stove makers often produced a range of equipment, including gas, as well as oil stoves, and as we saw in chapter 2, they rarely advertised gas equipment except in trade publications. Advertisements for kerosene stoves appeared in these publications as well, and a decade later oil stove advertisements also began to show up in farm journals, aimed at consumers. The consumer focus of advertising for oil but not gas stoves probably resulted because there were no local utilities to promote oil equipment. There was the "Eureka," a single burner that promised to bake, broil, and boil, as well as to heat flat irons. The "Victor" and the "Domestic" made similar claims. In 1877, the Florence Company promised that its heater-cooker was "odorless and durable." In the early days, oil stoves also had to overcome that fuel's evil reputation for explosions. Not only could the Eureka bake, broil, and boil, but it "cannot explode," its maker reassured readers of *Ohio Farmer* in 1875; and "there is neither smell of oil nor smoke." Marion Harland, whose oil stove had smoked until she mastered it, concluded that such stoves were like men: "on the whole excellent . . . [but] their behavior depends largely on the care that you . . . bestow on them. Therefore, let us give both fair tests before pronouncing them useless nuisances."[24]

By the 1880s, these "babies" became common. They provided heat, cooking, and the light needed in order to eat the results, making kerosene-fueled appliances far more flexible than wood, coal, or even gas. Their small size and absence of a stovepipe made them highly portable as well. "They can be lifted about from place to place where ever it is most convenient to have them," Aunt Betsy informed readers of *National Stockman and Farmer* in 1888. Indeed, they were the first portable source of heat. Size, price, and portability also ensured a commercial market for such stoves. Florence noted that its small cookstove was designed not only for small families but for small businesses as well—"especially for milliners, dressmakers, druggists [to prepare decoctions], barbers, tailors, manufacturers [to heat glue and paste], and others." As the "Our Household" column in *National Stockman* told readers, in sewing, "a small lamp oil stove . . . is a wonderful help to the pressing arrangements."[25]

Predictably, the advertising contained a certain amount of puffery. Piggybacking on the desirability of gas, some companies advertised "oil-gas" stoves—which is to say they vaporized the kerosene—and one claiming to do this announced, as well, that it burned "90% air." Nor, as the Florence ad suggests, were companies slow in suggesting other uses for oil stoves, and they trumpeted the virtues of kerosene in space heating as well. While heating with coal was usually cheaper than kerosene, coal stoves needed a stovepipe. Kerosene space heaters, by contrast, might be unvented and could move around the house. Like cook stoves, some were even on wheels. With the majority of homes having no central heat as late as 1940, portability might reduce the number of stoves a family required—an important selling point for oil. And as I noted above, many of these might double as a cookstove as well. Florence offered a space heater for "halls, bedrooms, offices, shops, [and] conservatories" and stressed that, because it needed no flue, it was portable. Kerosene was cleaner than wood or coal and produced no ashes to dump. "Heat, where you want it—when you want it—no smoke—no smell—no trouble," makers of "Perfection" heaters told readers of *Wallaces' Farmer*. By the 1890s, some models contained internal piping that drew cold air from the floor, improving their heat circulation. Makers of "Banner" claimed that its "fire was always kindled" and offered a booklet titled "Heat without Dirt."[26]

Cooking with an Oil Stove

Women learned about kerosene stoves from advertisements and from columnists and correspondence in agricultural publications and women's magazines—and no doubt from their mothers and neighbors, too. And no claim was more common than that oil stoves reduced and lightened work—another example of households' employment of capital to save labor.[27]

Handling kerosene required less effort than handling coal. Because it was liquid and had higher energy density—a gallon weighed 6.6 pounds, whereas the energy equivalent of soft coal weighed about 10.5 pounds—and because kerosene was far more efficient for cooking, much less of it was needed. Households might buy coal either by the ton or (80-pound) bushel; for urban households, kerosene could be bought in one- or two-gallon containers from fuel dealers or peddlers, while farm families were more likely to get it from a general store. In any case, you brought in your own container, and the seller filled it from a tank. Writing in 1882, Mrs. W. A. Kellerman of Kentucky urged readers of *Ohio Farmer* to buy kerosene by the barrel or at least in 10- to 15-gallon cans, for by doing so, they might save as much as five cents a gallon, and after 1900, some farms were able to obtain bulk deliveries directly from Standard Oil.[28]

Unlike their cousins that burned manufactured or natural gas, oil stoves were by no means maintenance free, as some women learned the hard way. "Every day the burner should be washed in hot, clean suds," a columnist in *New York Evangelist* instructed readers in 1884; otherwise, it would smoke and smell. A writer in *Good Housekeeping* in 1890 also warned readers that "an improperly cut wick will . . . smoke . . . causing a disagreeable odor." Wicks needed to be turned down when not lit, or they might wick oil. Writing the next year in that publication, the columnist Helen Russell concurred and urged readers to follow instructions that came with the stove. But, she claimed, "the whole care of it [a kerosene stove] will consume perhaps ten minutes of each day." Russell contrasted that with the care required by a coal range, reminding readers of the need to get coal from the cellar, whereas a kerosene stove might be filled from a gallon can. A coal fire had to be laid and kindled, and the cook needed to learn how to regulate the drafts and dampers, while oil required only a match. Many writers stressed kerosene's comparative cleanliness, while Russell noted that with coal came the need to clean out the ashes and "the dirty task of sifting" them. "My greatest work is carrying wood and cobs to the kitchen range and removing the ashes," one woman informed the US Department of Agriculture in 1915.[29]

Women who used oil stoves also appreciated the ease of temperature regulation that improved cooking. Undoubtedly urban women who shifted to gas would have said the same thing. Coal fires seemed to die down just when the critical point came in making jam, Helen Russell recalled, while Mrs. W. A. Kellerman observed that with wood or coal, "things will burn, boil over or won't cook." Oil stoves also helped inaugurate the era of fast food: one writer of a 1923 cookbook for farm women provided receipts for quick meals that could be prepared in 30–60 minutes—perhaps on wash-day. She cautioned that they were for kerosene stoves. She also cautioned that baking with a kerosene stove differed somewhat from a coal or wood stove. While coal and wood stoves had contained a thermometer in the door since the late nineteenth century, most kerosene stoves lacked such a device until the late1920s, and the author warned that a thermometer placed in such an oven would read 50° to 70° higher than a door thermometer.[30]

Kerosene cookstoves often entered the household as a backup for summer cooking, for not only was the heat from coal stoves wasted during summers; it was the curse of cooking. A letter to the editor of *Country Gentleman* in 1875 described cooking with coal during the summer: "the cook just escapes cremation and . . . almost has to look around to find . . . [her] own ashes." That oil ranges would not turn the kitchen into such a hotbox made a powerful appeal to rural and farm housewives in a day when home canning was ubiquitous, ironing was common, and flat irons required heating on the stove. "An oil stove lessens the terrors of ironing day by half; the kitchen need not be heated hotter than Gehenna," a columnist in *Michigan Farmer* informed readers in 1883. Moreover, unlike gas, you could move an oil stove onto the porch on hot days, for they were light as well as unattached, and manufacturers began to offer some with wheels.[31]

Alternatively, women without oil stoves sometimes moved all operations save cooking to the porch, which meant many extra steps. "An oil stove and no summer kitchen is a labor saver par excellence," one writer noted in *Wallaces' Farmer*. In 1920, Mrs. James Wilson estimated that the portability of her oil stove saved her 350 steps each time she did the weekly ironing. There were fewer steps to get fuel, and the ironing board might be located closer to the (cooler) oil stove. To repeat a point made earlier in the discussions of gas lighting and gas stoves, the benefits that farm women gained in the form of cooler kitchens when they switched to kerosene are missing from measures of living standards.[32]

In 1883, Fanny Field wrote to *Ohio Farmer*: "I have used an oil stove for two summers and like it so well that I would not think of going back to wood or coal. It burns up oil like the dickens; but then it don't heat up the room, and it don't make dust and litter and ashes." Sarah Wilcox, who also used an oil stove, wrote to the *National Stockman* in 1896: "I have been canning fruit today. . . . It has been a hot sultry day when to stand over a fire . . . would have been almost intolerable." Writing in the *Cultivator and*

Country Gentleman in 1893, the columnist Katherine Johnson concluded, "An oil range for the summer is one of the most comfortable articles ever placed in a kitchen." Oil costs more to use, she noted, "but added comfort and length of days [i.e., a longer life] are of more consideration."[33]

In 1899, the perhaps pseudonymous Becky Sharp advised readers of the *National Stockman and Farmer*: "If you cannot afford a vacation get an oil stove instead." In case a husband proved unwilling to pay for a new stove, Becky had a suggestion: "Build up a roaring hardwood fire on one of the hottest days . . . and then begin to feel a little too faint [to cook]. . . . After your husband has taken one good sweat [cooking] he will begin to talk stove." A similar idea occurred to a woman who wrote the *Stockman* a poem:

If men had to do all the cooking . . .
For a week or two or three, it's all the same,
We shall find in the house by the first day of May,
A Stockman Stove with the blue flame.

Another columnist agreed: "it is not the work that prostrates one in the summer but the intense heat, many times intensified by a hot range." A later writer warned husbands: "a tired wife is a cross wife." At the turn of the century, a writer in *Good Housekeeping* concluded that summer stoves were "one of the greatest boons to the modern housekeeper." Ethel Hayward was born in rural Delaware in 1901. Years later, she remembered that her mother canned quantities of garden produce. The shed "was turned into a big summer kitchen and she had a kerosene cook stove and the big stove . . . well just wasn't used during the summer months." In 1914, a North Dakota extension agent writing in *Farmer's Wife* summarized, "The oil stove is largely taking the place of the wood or coal stove during the summer season on the farm."[34]

Not all the praise in these testimonials may have been independent. Some farm journals sold sewing machines, oil stoves (the *National Stockman*'s "Stockman"), and other appliances via mail order. Accordingly, advertising considerations may have influenced their editorials, columnists, and perhaps even the publication of letters to the editor touting such products. Despite such concerns, the close connection of these journals to their readers, and the fact that they often published letter writers' names and addresses, makes it seem unlikely that economic incentives greatly distorted the information they provided. If anything, therefore, the farm journals' merchandising of kerosene stoves functioned much like the recommendation of a trusted expert, reassuring skeptical farm women that the product really would do the job.[35]

Advertisers, of course, also trumpeted these and other advantages of oil cookstoves. As a Florence advertisement put it: "Summer Cooking Shouldn't Cook the Cook." An ad in the *National Stockman and Farmer* was even blunter: "use the 'Puritan' [oil range] and make the coroner hunt another job." Yet in the winter, the coal stove provided valuable heat in the kitchen, and families might not want the expense and space of two stoves. Hence, an early Florence cookstove could function as a space heater as well.[36]

Oil Stoves Enter the Twentieth Century

In the late 1890s the "babies" were growing up as makers of oil stoves also attempted to broaden their appeal to an increasingly affluent market, improving quality and stressing attributes other than a cooler kitchen (fig. 6.3).

An entirely new idea in
oil-burning ranges!

Figure 6.3. By the 1920s, kerosene stoves sported modern porcelain-enamel finishes and were appealing to rural and suburban middle-class women. ("An Entirely New Idea in Oil-Burning Ranges!" Advertisement, *Good Housekeeping*, May 1928, 242. Public domain)

A SWIFT-COOKING RANGE in snow-white porcelain enamel! One of 24 splendid new models, Perfection's offering for 1928. It has a new design... New burner arrangement... Speedy Giant Superfex burners... Accurate heat indicator, and built-in enamel-lined "live heat" oven.

See these 24 new models at your dealer's. All mark a big advance in oil stove manufacture, with light colors, new conveniences and new Perfectolac finish. Prices, $17.50 to $154.00. Easy terms on any of them.

PERFECTION STOVE COMPANY
Cleveland, Ohio
Sold in Canada by General Steel Wares, Limited, Toronto, Ontario.

PERFECTION
Oil Burning Ranges

Steel began to displace cast iron in oil stoves, as well as those fired by gas and coal, as the former metal fell in price. Steel was easier to machine, made for a tighter stove, and was lighter and cheaper to ship. As more families installed central heating, they no longer needed the cookstove to warm the kitchen. Accordingly, companies began to stress that oil stoves were for year-round cooking, and they offered larger four- and five-burner models that sported a broad array of amenities. In 1909, Standard Oil's "New Perfection" announced that it was the first oil stove with a cabinet top and shelves that might hold the coffee pot. An attractive advantage of oil (and gas) stoves was their ability to regulate heat quickly and precisely. Women columnists and correspondents had emphasized this feature for years, and rather belatedly, advertisers began to echo that their fuel might make women better cooks. "You can get any degree of heat you want by a single turn of the valve," proclaimed an advertisement for Standard Oil's "Wickless Blue Flame" stove in 1900. Florence advertised control of heat with "the turn of the lever." This ability to control heat application, which could not be done with wood or coal, not only made for more precise cooking; it also increased kerosene's economic competitiveness because it might reduce fuel consumption. Another Florence ad claimed that kerosene was "cheaper than coal."[37]

Makers of oil stoves benefited from the anthracite strike of 1902 as newspapers in both Chicago and New York reported that urban households were shifting from anthracite to oil for cooking and heating. The later strikes in the 1920s had similar consequences as the high prices and shortages of hard coal boosted the sale of all alternative fuels. These events probably contributed to the bite that kerosene cooking was taking from coal markets. Domestic sales of kerosene for all uses in 1920 amounted to about 33 million barrels (table A4.1), and illumination and tractor fuel still accounted for a good deal of this as electricity and gas remained uncommon outside of cities. Besides cooking and heating, kerosene had a number of other uses as well. If we suppose, however, that half of all kerosene was used for cooking, it could have replaced as much as 12 million tons of coal.[38]

Kerosene Stove Technology in 1900

By 1900, oil stove burning technology had stabilized. Cookstoves employed three different methods of combustion. The traditional short chimney burners produced concentrated heat on the utensil but, with inadequate draft, might smoke. "Blue flame" stoves with long chimneys appeared in the 1890s. These drew air from the center of the burner, and the blue flame resulted from the improved draft; their drawback was that their heat was farther below the utensil. The third technology was wickless stoves; these fed oil to a circular cup, which required that the stoves be perfectly level. Lighting an asbestos wick initially heated the oil, which then vaporized and burned without a wick. Later research by state agricultural extension services suggested that all performed about equally well, although wickless stoves required less maintenance.[39]

During these decades, makers of kerosene stoves continued to profit from favorable commentary in books on kitchens and in women's magazines. In 1918, *Good Housekeeping* published an informative review done by its institute pointing out that the new built-in ovens worked much better than had the older portable ones and that for the first time, manufacturers were beginning to deliver oil stoves that were the proper height. With kerosene, the article pointed out, "the housekeeper can have all the efficiency and convenience supplied to her city cousin" who used gas. As with gas and other stoves, porcelain enamel finishes for oil stoves arrived in the early 1920s, making cleaning easier. About that time, gas and electric stoves pioneered heavy use of insulation; for electricity, at least, that may have been because it was such an expensive energy source. Two innovations also made gas a much more formidable competitor. As we saw in chapter 5, around 1918, the American Stove Company introduced the "Lorain" automatic oven thermostat on its gas stoves equipment—a device that kerosene stoves failed to match—and several years later propane came on the market. "Yes You Can Have Gas Just Like in the City," one 1928 advertisement claimed, making gas for the first time a serious competitor in rural areas and small towns.[40]

Saving Women's Time

Much nineteenth-century advertising of oil and gas stoves had not aimed specifically at women as consumers. Beginning in the late 1890s, however, advertising began increasingly to focus on and appeal to women. In 1900, an ad for "New Process" and "Standard" cook stoves in the *American Artisan and Hardware Record* depicted a very well-dressed woman gazing rhapsodically at her stove. "To the modern housekeeper is left the responsibility of buying household necessities," it informed the reader. "The shrewd dealer . . . selects a line of stoves that is attractive." About this time, that journal began to carry regular articles on the importance of women as buyers, various ways to appeal to them, and whether women clerks could be effective. Stove ads depicting women soon became routine and moved into consumer-oriented publications as well. Some companies also realized that men might have a say in stove purchases as well. "Now is the time for every good man to come to the aid of his partner" a "Puritan" stove ad informed husbands reading *National Stockman and Farmer* in 1900. Appealing to male guilt, the ad informed farmers that "the Puritan . . . will be to your wife what the Reaper, the Binder, Thresher and Grain Drill are to you." Some men seem to have gotten the message. "I think the best thing to help the farmer's wife yet is the coal-oil stove," a Kansas farmer wrote in 1915. "Within the last four years I think one-half the homes have been supplied with them."[41]

One implication of such claims was—as gas stove makers were also rather belatedly discovering—that women's time was valuable. In 1901, Standard Oil informed readers of *The Independent* that cooking with coal or wood was "a needless *waste of time.*" The woman who uses the Wickless Blue Flame can do her work in one-quarter of the time." They were also easy to use: "no wood to chop, no coal to carry . . . it makes play of housework," the company gushed to readers of *Maine Farmer* about the same time.[42]

A USDA study of 1914 estimated that three-fifths of housework involved meal preparation, and about that time, a "Perfection" advertisement urged women to "shorten your kitchen hours," claiming that the company had already freed three million "up-to-date women . . . from coal range and wood stove drudgery." This was not entirely puffery: the study of Indiana farm women found that those who cooked with coal or wood spent an average of 22 hours and six minutes cooking a week—not counting the time spent in care and cleaning. In contrast, kerosene and gasoline stove owners spent an average of 19 hours and 25 minutes a week at the same tasks. Some of the worst drudgery farm women faced involved drawing and heating water, and about the time of World War I, Florence— perhaps in response to the "all gas kitchen"—began to advertise the "Florence Kitchen," which featured not only a kerosene cookstove but a standalone kerosene hot-water heater as well. "Water Heaters mean plenty of hot water—any time—without heating your kitchen," it announced.[43]

Saving women's time meshed easily with the national efficiency craze, which began about 1910, and it was a central message of home efficiency advocates such as Christine Frederick. In 1913, and in subsequent editions of the *New Housekeeping*, Frederick informed readers that her "efficiency kitchen" was designed with country users in mind. It contained a three-burner oil stove; she had "banished coal" because of its dirt and because hauling fuel and ashes wasted labor. Several years later, she repeated these themes in *Household Engineering*, noting, "to the rural housekeeper especially . . . kerosene offers a solution to the fuel problem." The "three burner kerosene stove and a fireless cooker are the ideal country combination," Frederick informed readers. In the early 1920s, she continued to praise oil stoves in a series of articles in *Hardware Dealers' Magazine* in which she urged dealers to push them for canning and to keep the kitchen cool. In 1923, Mrs. Edith Hawley explained that "by making a study of efficient management," she and her husband were able to put up 6,391 cans of fruits and vegetables.[44]

If, as Frederick and innumerable advertisers claimed, oil stoves improved the efficiency of farm women's work, an obvious question arose: how did they use the time that such new appliances saved? Helen Russell claimed that one of the benefits of oil stoves was "the extra half hour in bed in the winter" because they heated up much more quickly than did coal stoves. Typically, however, surveys suggest that when new appliances made housework more efficient, women simply reallocated their work time: with washing machines, they might wash more often or use the time saved to iron more; with oil stoves, they might cook more, or sew—all of which were results of conservation that Stanley Jevons would surely have understood. This does not imply, however, that purchase of an oil stove *caused* women to cook more. Rather, it might well have been the reverse: the desire to wash or cook more or better could have motivated the purchase of the new appliance. As I noted in this chapter's opening paragraph, Edith Hawley got the idea to go into canning from her daughter, who was in a girls' canning club, and her new stove made the work bearable. The oil stove did not cause more work for Mrs. Hawley; rather, she chose an efficiency-enhancing appliance in order to raise her standard of living.[45]

Housework also cut into the time that farm women could spend on farmwork, which was money making. Women helped maintain the garden, looked after bees and chickens, kept the cows, helped with the milking, and churned the butter. A 1914 study showed that farm families produced about 63 percent of the food they consumed. Of vegetables, the total was 78 percent, while 85 percent of animal products (meat, eggs, butter, and milk) were farm-produced. These activities either brought in money or avoided outlays, and it seems likely that a farm family's purchase of an oil stove was, in part at least, the result of a desire to devote more labor to such activities. The same was surely true of portable kerosene heaters, which not only might warm the parlor but also ease such chores as sterilizing dairy equipment, keeping hot frames and root cellars from freezing, protecting baby chickens, and heating honey rooms. Portable heaters lightened men's tasks as well, warming stock tanks and hog pens (the "Kozy Heated Hog House").[46]

As the payoff to such work rose sharply around the turn of the century, so did the value of women's time. The years from the 1890s to about 1920 saw large increases in the relative price of foods. For example, while all prices rose about 10 percent from 1890 to 1913, the price of milk rose 41 percent, butter 53 percent, and eggs 112 percent. Farm women's time was indeed becoming more valuable, and oil stoves that lightened the load of cooking had an increasing economic payoff. Surely this rising payoff helped lure Mrs. Edith Hawley to her canning venture in 1919, and perhaps it encouraged other farm women to invest in oil stoves as well.[47]

The value of women's time was rising in the city as well as in the country, and the pull of urban labor markets attracted young women from rural areas. From 1917 to 1919, Katherine Henry wrote several articles for *Farmer's Wife* suggesting ways to counteract the lure of the city. In

"Daughter Chooses the Farm," an oil stove was one of the magnets that Henry suggested could hold young women to the farm. A later piece subtitled "How to Tie Daughter to the Farm" made the same point.[48]

After 1900, advertisers of oil stoves began to target more affluent urban and suburban households, reflecting the rise in married women's participation in the labor force, as well as the growing trend toward suburbanization. Advertisements in farm journals began to depict stylishly dressed women. Color advertising appeared in women's magazines around World War I, and the emphasis on women's freedom continued into the 1920s. "Who has to get up to start your kitchen fire?" asked a 1924 Florence ad appearing in *Farmer's Wife*, and it went on to claim that "a stove you have to shake [to remove ashes] is as old fashioned as a car you have to crank." In a 1925 advertisement entitled "Women's New Freedom" in the *Ladies' Home Journal*, makers of "Perfection" cookstoves claimed, "Countless women in suburbs and country have learned that the secret of freedom from long kitchen hours lies in their cook stoves." Another ad by that company again took advantage of the suburbanization trend, urging women to "build your home where you like. It makes no difference where the gas mains end." Thus, with an oil stove, suburbanization and freedom would march hand in hand with no need to wait for the utility company. Stove manufacturers such as "Perfection" also imagined what women wished to do with their "new freedom." In "Makes Mother a Companion," the company claimed that an oil stove would turn "kitchen hours into play time with husband and children." These depictions reflected a rather different meaning for oil stoves than had been the case earlier. To nineteenth-century women writers and correspondents, oil stoves meant freedom from dirt and stifling summer drudgery, while advertisers depicted similar features. For twentieth-century women, companies began to suggest how they might use their newfound freedom: to move to suburbs, to raise their children.[49]

Manufacturers modestly improved oil cookstoves in a number of ways during the Depression decade. Rather belatedly, they began to advertise the importance of insulation and the advantages of the raised oven, which did not require a housewife to bend over. While only gas and electric ranges sported thermostats, Florence advertised an oven with a thermometer and "Fingertip Heat Control" that would, it claimed, help you "hold your man." Like so much else in the 1930s, oil ranges became self-consciously "modern." Stoves became streamlined and available in a variety of colors while they hid fuel containers. "See Yourself *in a* Modern Kitchen *with a Beautiful* Florence Oil Range," Florence urged women in 1936.[50]

Undoubtedly, however, the most important developments during the Depression were the mass marketing of range oil burners that could retrofit a coal or wood-fired range. Range oil was a term of art for cooking and heating fuel that included kerosene and number 1 fuel oil as well. The idea of retrofitting a coal or wood stove to burn oil was not new: an 1888 effort consisted of a can full of oil-soaked incombustible materials placed in the coal stove. Other versions employed an oil-soaked brick and in one case simply a pile of oil-soaked sand. All of these failed because they smoked. *Scientific American* depicted actual burners to retrofit a coal or wood stove in 1902, while advertisements appeared in *Popular Mechanics* in the 1920s. None of these seem to have caught on, however, and the first successful ventures were by two New England firms in 1928: Lynn Products Company and Silent Glow Oil Burner Corporation—which by then was already claiming 30,000 users. Initially, conversion units might cost $40 or more—comparable to a cheap stove—but competition and the Depression drove prices to the $5 to $8 range by the mid-1930s, and units became more technically sophisticated as well. Some sported electric ignition, while others pumped oil from a basement tank. Beginning in 1931, census data show annual production of 100,000 to 200,000 units during the 1930s, and while there are no figures for the wartime period, in 1947 sales amounted to 682,000 burners. Thus, range oil burners employed existing cookstoves to drive coal from America's kitchens, just as conversion oil burners employed existing furnaces to drive coal from America's basements.[51]

The Extent and Energy Effects of the Kerosene Kitchen

After World War I, a series of government studies of farm and rural families emerged that reflected increased interest in women's roles and in home economics. When combined with census information, the studies allow us to construct patterns of energy use in farm cooking about that time. Beginning in the 1920s, production of oil cookstoves was second only to that of gas equipment and constituted 20–30 percent of the market up to the late 1930s. Electric stoves, by contrast, were a distant third until the postwar years. As several writers have pointed out, wiring a house was expensive and many initial installations may have been inadequate to carry the load resulting from a stove.[52]

The Prevalence of Kerosene Cooking

The census of manufactures contains sales of oil stoves for alternate years beginning in 1925. If we interpolate noncensus years and sum all cookstove production from 1925, we can estimate sales of more than 14 million units. Given the long life of stoves, most of those were probably still in use in 1940, often as second, summer stoves.[53]

Indeed, a 1927 survey of Nebraska farm households found that 52 percent of them had two stoves, while a later survey found that 62 percent of the Nebraska farm women surveyed used oil in the summer, but only 8 percent did so in the winter, when they burned wood or coal. A study of

Indiana rural households about the same time found that although wood use had declined sharply—from about 12 cords per family in 1917 to fewer than six in 1926—it was still a common cooking fuel in 1930. Yet the same study demonstrated that 71 percent of families burned kerosene in the summer. Table 6.3 reveals the flexibility and complexity of rural cooking arrangements about this time.[54]

The table reveals that very few families used only one fuel year-round. Even in winter, only about half of households burned coal or wood exclusively, whereas in summers, those numbers shrank to 8.1 percent. The Indiana women in table 6.3 that cooked with kerosene used about 148 gallons a year. If they were like the Ohio women in table 6.2 who used 0.82 gallons a day, they must have used oil for about half a year.

After about 1925, the price of kerosene fell sharply, and it became increasingly economic for summer use (see fig. 6.1). Moreover, by the mid-1930s, anthracite retailed for $12.75 a ton in Boston, while kerosene prices there were $0.08 a gallon, making it the cheaper cooking fuel. Cheaper fuel spurred purchases of oil cookstoves in the 1930s and was surely what lay behind the boom in range oil burner sales in eastern states. Moreover, since families often employed range oil burners for heating and cooking, their spread had a greater impact on fuel use than sales of either cookstoves or heaters taken alone. Merna Monroe of the Maine Agricultural Experiment Station explained how families used them: "Homeowners report using 2.5 to 4 gallons of kerosene per day or 50 gallons every two weeks during the winter. They operate one burner continuously on a moderately low flame to heat the kitchen or the hot water tank and use both burners only as needed during meal preparation."[55]

Similarly, a Massachusetts commission reported in 1937 that "the use of range oil burners at the beginning and the end of the heating season tends to curtail somewhat the purchasing of solid fuel." Because of such behavior, range oil sales in Massachusetts, rose from 5.7 to nearly 10 million barrels between 1934 and 1938. New York and Massachusetts together burned nearly 16 million barrels of range oil for cooking and heating in 1938, more than the energy equivalent of manufactured and natural gas consumed in these states.[56]

The 1930s and 1940s also brought several broad surveys of household fuel use. Table 6.4 presents results from a representative survey of nearly 600,000 farm households in 1934, demonstrating that nearly a quarter of them had kerosene cookstoves, and in several regions, the numbers were above 40 percent.[57]

A series of surveys of urban, rural nonfarm, and farm households also reveal how the use of kerosene for cooking varied by location and income. Not surprisingly, it was least common among city dwellers, and kerosene there was an inferior good—that is, its use declined as incomes rose. While

Table 6.3. Cooking Fuels in Rural Indiana, 1930

| | Winter | | | | Summer | | | | Year-round | |
| | Alone or with others | | Only fuel used | | Alone or with others | | Only fuel used | | Only fuel used | |
Fuels	Number	Percent	Number	Percent	Number	Percent	Number	Percent	Number	Percent
Coal	902	64.2	423	30.1	166	11.8	32	2.2	29	2
Wood	744	53	284	20.2	398	28.3	84	5.9	66	4.7
Corncobs	17	1.2	0	0.0	24	1.7	0	0.0	0	0.0
Kerosene	288	16.2	80	5.7	991	70.6	670	47.7	78	5.5
Gasoline	65	4.6	28	1.9	179	10.5	115	8.1	26	1.5
Gas	26	1.8	24	1.7	33	2.3	31	2.2	12	0.85
Electricity	36	2.6	9	0.64	46	3.2	12	0.85	11	0.78
Acetylene	10	0.7	0	0	21	1.4	0	0.0	0	0.0

Source: Miriam Rapp, "Fuel Used for Cooking Purposes in Indiana Rural Homes," Purdue University Agricultural Extension Station *Bulletin* 339 (Lafayette, IN: Purdue University Agricultural Experiment Station, 1930), table 1.

Note: Total number of families = 1,403.

Table 6.4. Type of Cooking Stove, Farm Households: National and by Region, 1934

| Households | Percent of households using | | |
	Kerosene	Gas	Electric
United States	24.4	2.3	1.9
New England	24.0	1.9	3.6
Middle Atlantic	57.5	11.0	5.1
East North Central	43.3	45.0	2.6
West North Central	45.5	1.9	1.0
South Atlantic	9.2	1.4	0.8
East South Central	6.7	0.4	0.3
West South Central	23.0	1.7	0.6
Mountain	19.2	1.8	5.1
Pacific	15.4	9.6	15.0

Source: USDA Bureau of Home Economics, "The Farm Housing Survey," USDA *Miscellaneous Publication* 323 (Washington, 1939), table 5.
Note: Total number of households = 595,855.

40 percent of the poorest city dwellers burned kerosene, only 2.4 percent of those with incomes of $5,000 and up did so. Similar, although less dramatic, results obtained for farm and other rural households.[58]

Energy Transitions

Taken together, these data suggest that households—especially those with lower incomes, in rural areas or on farms—used kerosene as a major source of fuel for cooking and to a lesser extent for heating at least until World War II, and it came at the expense of coal and wood. Bureau of Mines data for 1940 show consumption of 45 million barrels of "range oil" (table A4.1). As I have noted, some of that went for illumination and various other uses. If we assume half of it substituted for coal in cooking, it would still have displaced 16 million tons of coal a year.[59]

Yet if kerosene was one of the fuels that was pushing the coal hod from the kitchen and the parlor, like manufactured gas, it, too, was soon the be displaced. After 1940, sales of oil cookstoves fell sharply, while range oil followed with about a decade lag. Kerosene had always been inferior to gas and electricity, but geography and poverty had sheltered its markets. Rising incomes and technological innovation brought superior fuels within reach of an increasing fraction of the population. Natural gas slowly penetrated New England and other markets in the 1950s, and nationwide, residential and commercial sales of bottled gas rose from 130 million gallons in 1940 to 2.6 billion in 1954.[60] The Rural Electrification Administration brought power to four million additional homes by 1954. Faced with competition from these technologies, the kerosene stove retreated along with wood and coal, as superior fuels gradually drove it from America's kitchens.[61]

Conclusion

The story of the kerosene kitchen reminds us that there were multiple energy paths to modern times. It also reminds us not to be so blinded by dramatic transformations, such as electrification, that we fail to see more modest changes, which collectively lightened the labors of men and women alike. And to repeat a by-now-familiar observation: the spread of kerosene cooking was not simply a reflection of American resource abundance. Its exploitation reflected the efforts of men of science, marketers and hucksters, stove manufacturers, farm and women's magazines, and Standard Oil.

For farm women, especially, oil was a blessing. It was cleaner, quicker, lighter, more portable, and, especially, cooler than wood or coal. It raised levels of comfort, accommodated changes in women's work patterns, and improved their economic status. Kerosene stoves saved women's time and strength in cooking, washing, and ironing, while portable heaters might warm drafty rooms, hatch the baby chicks, or protect the root cellar. Although we cannot measure them well, from such small improvements is a rising standard of living constructed.

7

The Battle of the Basements

Oil, Gas, and the Retreat of Coal, 1917–1940

In the end, the flivver [Model T Ford] driver will probably have a flivver [oil] burner. —*P. T. Fansler,* House Heating	Mass production calls for mass persuasion. —Fuel Oil Journal

Even as gas and oil were pushing coal and wood from America's kitchens, a similar drama was playing out in what *Printer's Ink* called the battle of the basements. In the early 1920s, a series of innovations and discoveries that promised "automatic heat" disrupted fuel markets for domestic central heating. Oil burners appeared seemingly out of nowhere, and by 1930, natural gas had slipped its regional leash and emerged as a far more formidable competitor than anyone could have imagined just a few years earlier. In fact, the competition was over all forms of heat, for stoves as well as furnaces were a battleground. Gas and oil heat were both labor- and energy-saving, and they enormously enhanced users' well-being; together, they began an assault on the supremacy of coal. On the eve of World War II, Old King Coal still fired a majority of American furnaces and stoves, but the new fuels had established a beachhead that they would expand.

In retrospect, such energy transitions seem inevitable—a result, perhaps of America's abundant endowment of oil and natural gas—and perhaps they were; yet these great changes arose out of many individual household choices, as well as entrepreneurial business ventures intended to exploit the abundance. To market oil heat, companies improved burner technology, cooperated to standardize fuel oil, developed dealer/service networks, obtained disinterested assessments from technical journals, and launched a massive advertising campaign that promised to "free the furnace slaves." Efforts to promote gas heat were similarly imaginative, and both ran circles around coal.

We begin by reviewing American heating fuel choices about the time of World War I. The next two sections analyze the rise of domestic oil heat in the 1920s and coal's efforts to counter the threat. Section 3 focuses on depression developments, one of which was a gas attack on home heating markets. The lens then shifts again to coal's efforts to retain its markets in the battle of the basements, and this is followed by a review of domestic fuel choices as of 1940.

Domestic Heating about World War I

When the nineteenth century closed, central heating with furnaces was confined to large businesses, public buildings, and the homes of the well-to-do. The systems then in use included steam heat, which was probably the most popular, hot water and hot air—moved by convection where there was no electricity—and each had a variety of subcategories.[1] No hard estimates of the extent of domestic central heating exist before the census of 1940. At that time, only 56 percent of urban households had central heat, and less than a third of rural Americans heated with a furnace. Moreover, use of central heating was strongly income dependent. A few years earlier, out of about 39,000 households surveyed in Providence, Rhode Island, fewer than 2 percent of families in the top income category heated with stoves; in Trenton, New Jersey, and Fargo, North Dakota, none of them did (see table A4.5). A statistical analysis of urban households about that time finds that increases in families' use of central heating rose a bit more than in proportion to their incomes. If we make the heroic assumption that this relationship held true historically, then with per capita GDP 76 percent higher in 1940 than in 1900, it follows that fewer than 30 percent of urban homes would have had central heat in the latter year. But whether they relied on stoves or central heat, in urban areas, in 1900, Americans overwhelmingly burned coal.[2]

The Rise of Competitive Fuels

As we have seen, by 1900, gas was beginning to nibble away at coal's markets for domestic cooking. Yet gas seemed unlikely to present significant competition for domestic heat. Natural gas was geographically confined, and nearly everyone believed that supplies would soon dwindle, reducing any apparent threat to coal. Thus, in 1920, one speaker to the Natural Gas Association predicted that "in the next ten years," there would be increased drilling, a shortage, and rising prices. As late as 1929, the Bureau of Mines published a bulletin that took for granted the imminent exhaustion of natural gas supplies and detailed the procedures for natural gas distributors to become producers of manufactured gas.[3]

As we saw in chapter 5, domestic use of manufactured gas at this time was largely confined to illumination and cooking. In 1918, the American Gas Association estimated that there were only 900,000 space heaters in use, or about one for every nine customers. This was far fewer than the gas cookstoves at the time, and many of those who cooked with gas continued to burn coal in the winter, either in a combination range or parlor stove. And because gas space heaters were expensive to run, companies marketed them as primarily useful for brief periods in fall or spring when a coal fire might be particularly inefficient.[4]

Gas for Central Heating

Just before World War I, a few companies began to market manufactured gas for whole-house heating. In 1910, the Laclede Gas Company experimented with whole-house heat in St. Louis. The company made every effort to ensure economy of use and still found that with rates that averaged $0.45–$0.50/Mft.[3] gas cost more than anthracite at $10.00/ton, limiting the market to a few well-to-do residents. Indeed, in one of its installations the homeowner had previously burned 55 tons of anthracite coal a winter.[5]

Later, in 1917, Consolidated Gas and Electric Company of Baltimore also stuck its toe into the central-heating pond, perhaps because of its access to low-cost coke-oven gas. After considerable testing and experimentation with rates and heating systems, the company offered gas at $0.35/Mft.[3] for purchases in excess of 4,000 ft.[3] With anthracite by then then selling at about $18 a ton, gas was economic.[6]

While the jump in gas prices associated with World War I set back its adoption for central heating, use of manufactured gas continued to spread slowly during the 1920s as prices again declined and those of hard coal rose. The use of special rates for heating became more common, and to avoid the shock that might arrive with January bills, companies devised ways of averaging monthly payments. Still, gas remained a pricy fuel. In 1928, when hard coal retailed at $13–$15 a ton in Massachusetts, one local gas company claimed that for it to be competitive, coal would need to cost $22 a ton. Such costs largely confined manufactured gas to the comfortable, ensuring that installations were in large houses and, accordingly, displaced a lot of coal. In Denver, in the early 1920s, gas installation went into houses that had been burning about 16 tons of coal a year. Because gas conversion burners in existing coal furnaces often proved inefficient and expensive to run, utilities had initially had been unwilling to install them—which raised installation costs and restricted markets. Gradually however, conversion burners improved, and by 1929, their sales had outstripped those of gas furnaces and boilers.[7]

The introduction of manufactured gas for domestic central heating coincided with the general rise in fuel prices during World War I that focused awareness on the benefits of house insulation. While various products had been available for domestic insulation since the 1880s, as late as 1914, the president of the Heating and Ventilating Engineers had observed that the field was "almost unexplored." Worries over wartime prices and shortages, however, soon propelled corporate, university, and government research into the complexities of heat transfer and the value of various insulating materials. As occurred in manufacturing, by the mid-1920s, an army of insulating products had invaded the domestic market, incorporating everything from sugarcane waste and diatomaceous earth to asbestos and rock wool, while research emerged to evaluate the products and calculate heat loss. Articles pointed out that in at least some cases, switching to a smaller furnace might offset insulation costs so that energy savings were effectively free, while various forms of retrofitting yielded annual returns of 20 percent or better.[8]

Insulation chipped away at coal markets in two ways. Some homeowners burning expensive anthracite insulated, taking the benefits partly in comfort and partly in reduced coal purchases. A headline in *American Builder* read "Better Insulated Homes on Smaller Coal Consumption." One engineer estimated that simply insulating the walls of a 3,000 square foot house using 11 tons of coal could shave fuel consumption by a quarter. In addition, manufactured gas and insulation were complementary products. As a high-priced fuel, manufactured gas also yielded a high payoff to insulation, while the insulation, in turn, made the use of gas more economic. Gas companies such as Peoples of Chicago kept records of the payoff to insulation; it strongly encouraged the use of better house insulation and offered gas conversion packages that included insulation and weather stripping. The AGA featured "Blue Star" all-gas insulated homes, and Denver's gas company advertised that with insulation, you could have all the benefits of gas at the cost you would pay for coal.[9]

Yet even when packaged with insulation, manufactured gas hardly threatened coal at this time, for its home-heating market was small (comprising only about 76,000 houses as late as 1929) because demand remained limited to upper-income households. Looking back on the 1920s, one writer claimed that the market for both gas and oil during these years was limited to homes worth at least $7,500—slightly less than a quarter of the population in 1930. Accordingly, use spread slowly; in 1940, of 9.3 million customers burning manufactured gas, only 261,000 heated with it, and they burned 68 billion cubic feet—equivalent to about 2.1 million tons of hard coal. Any serious threat from this manufactured gas to the dominance of coal for domestic heating would require some combination of rising incomes and a decline in the relative price of gas; it was likely, therefore, to be a long time coming. In short, in the early 1920s, coal producers seemingly had little occasion to worry that they might lose the house-heating market.[10]

Straws in the Wind

Yet technological change combined with rising incomes would soon bring coal producers an unpleasant surprise: rising incomes not only encouraged the shift to central heat; they also lessened families' tolerance for the dirt and inconvenience that came with coal. "Grimly it [the furnace] demands a feast of coal six times a day"—while cleanup moved one woman writing in *Good Housekeeping* to long for "The Ashless Isle":

> For this trouble with the ashes
> Wears the heart and wastes the body
> Makes us long with bitter longing
> For a residence in some region
> Where the "natural gas" flows freely—
> For some island of the blessed
> Far removed from dust and ashes.

Combined with increasing affluence, such complaints made coal vulnerable to cleaner fuel—a conclusion borne out by statistics: in Providence, Rhode Island, 60 percent of the wealthiest families heated with oil in 1933; in Cleveland, the figure was 44 percent and in Fargo, North Dakota, nearly 87 percent (table A4.5).[11]

The Arrival of Oil

By 1920, oil, in the form of kerosene, had long been a source of cooking—and more rarely heating—fuel in rural households, while heavy oil was widely used in transportation as well. By the early 1920s, large commercial buildings in eastern cities were also shifting to oil. In Boston, the Copley Plaza hotel switched from anthracite to fuel oil and saved $30,000 a year, about 10 percent of which was from reduced labor cost. Oil was also cleaner and saved valuable space. As one newspaper summarized in 1923, "Installations of oil burners in commercial, industrial and public buildings have been extensive in New England, especially at Providence and Boston." In New York by this time, the Ritz-Carlton and Park Avenue Hotels, the Singer Building, the Metropolitan Life Building, Macy's, and many other large buildings were rapidly turning to oil as well. But in the early 1920s, oil for domestic heating remained rare; it was expensive, and early oil burners suffered teething problems.[12]

In July 1922, *National Petroleum News* concluded that "domestic heating with fuel oil . . . is still an untouched field . . . east of the Rocky Mountains." The problem, it opined, was the lack of an adequate oil burner. Solutions, however, were already at hand. About this time, technical journals such as *Fuel Oil Journal, Heating and Ventilating,* and *Domestic Engineering* began to feature articles evaluating the technologies and economics of domestic oil heat, as did construction industry publications such as *Building Age* and *National Builder.*[13]

Sales took off (fig. 7.1; table A4.6). According to the *Fuel Oil Journal,* there were only about 7,500 domestic oil burners in the entire country in 1920. Within a decade, however, the numbers had risen to half a million, and on the eve of World War II, nearly two million were in operation.

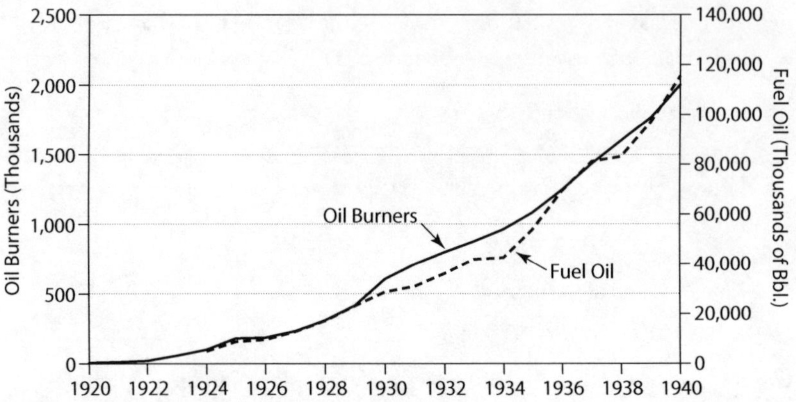

Figure 7.1. Domestic oil burners in use and fuel-oil consumption, 1920–1940. (Chart data drawn from Bishop, *Retail Marketing*; and "Replacement Market," FOJ [July 1939]: 15)

Heating-oil sales also exploded, rising from a few million barrels in 1923 to 115 million barrels in 1940, which equaled roughly 21 percent of gasoline consumption that year. In Massachusetts alone, fuel-oil consumption rose from about one million barrels in 1927–1928 to roughly 19.6 million barrels in 1938–1939—the equivalent of at least 4.5 million tons of coal. Hard-coal consumption in that state, by contrast, fell about 2.8 million tons over this period.[14]

Technology and Economics

Burner technologies differed in a variety of ways. Older models operated on a vaporization principle, whereas a newer approach was to atomize the oil. The early vaporizing burners were usually gravity-fed and employed a pilot light to vaporize the oil, which could then be ignited. These were typically manually controlled and worked best with a fairly light (and expensive) kerosene-like oil. They were therefore expensive to run and no more automatic than coal. The future lay with newer models that atomized the fuel—a technique that originated in oil fields and on railroads, where companies employed high-pressure steam to atomize oil. Reliance on steam was obviously unsuitable for domestic purposes, and further experimentation resulted in burners that atomized fuel mechanically or by pressurizing it and supplied an electric spark or pilot light for ignition. In either case, of course, oil heat required that the house have electricity. By about 1920, companies were marketing such equipment complete with a thermostat that could automatically control household temperatures to within one or two degrees and when attached to a clock could automatically adjust for day and night temperatures. Such a system began to be termed "automatic heat," and it promised a level of comfort, cleanliness, and ease of operation that suddenly made heating with coal seem as old-fashioned as long skirts (fig. 7.2).[15]

The economics of burning oil were similarly complicated, for they depended on capital costs (fuel prices varied across time and space), on the

Figure 7.2. This 1902 cartoon appeared twenty years too soon. Oil at that time was only a modest worry to coal producers, but in the 1920s, domestic oil heat began to threaten Old King Coal's crown. ("Old King Coal's Crown in Danger," *Puck*, Sept. 28, 1902, cover. Public domain)

heating values of fuels that were not homogeneous, and on combustion efficiency, which varied sharply among differing fuels. Table 7.1 indicates that oil and gas were energy-saving fuels, burning at 70 percent and 80 percent efficiency, respectively, versus soft coal's 50 percent. Thus, burning coal took 40 to 60 percent more energy than burning oil or gas (70% or 80%/50%) to generate the same amount of useful heat.

Writing in 1927, P. E. Fansler (1881–1937), associate editor of *Heating and Ventilating* magazine, suggested that installing a conversion oil burner

Table 7.1. Relative Costs of Fuels for Domestic Heat, 1925

Fuel	Price ($)	Btu	Combustion efficiency (%)	Btu per dollar	Dollars/ Mm Btu
Bituminous	9.24/ton	26.2 Mm/ton	50	1,417,749	0.71
Coke	6.98/ton	26.2 Mm/ton	70	2,627,507	0.38
Anthracite	15.45/ton	25.4 Mm/ton	65	1,068,648	0.94
Fuel oil	0.09/gal.	138,500/gal.	70	1,106,000	0.90
Kerosene	0.13/gal.	135,000/gal.	73	758,076	1.32
Natural gas	0.56/(Mft.^{3})	1,075/ft.^{3}	80	1,535,714	0.65
Mfg. gas	1.23/(Mft.^{3})	550/ft.^{3}	80	364.228	2.75

Source: Adapted from P. E. Fansler, *House Heating with Oil Fuel*, 2nd ed. (New York, 1925), 6.

and tank was likely to cost nearly $1,000—at a time when Ford was sell-ing Model T Runabouts ("Flivvers") for $260. But while the annual cost of oil heat was far above that of soft coal (table 7.1), it was competitive with anthracite. Still, it seems unlikely that many early adopters of domestic oil heat could have done so based on cost alone. Indeed, as noted above, data on fuel use by income level in the early 1930s (table A4.5) indicate that the well-to-do were the first purchasers of oil heat, implying that coal by then was viewed as an inferior fuel.[16]

The new technology attracted entrepreneurs by the hundreds. Writing in 1925, Fansler observed, "The market is flooded with burners of various types," and he guessed that there might be 50 to 150 serious manufactur-ers. A year later, *Forbes* magazine put the number at about 200 and noted that some thought the number might be as high as 520 for oil burners of all types. In 1927, a private survey listed 26 manufacturers of oil burners that it claimed held 89 percent of the market. Many of these were small, new entrants.[17]

A Californian, M. A. Fessler, had invented a mechanical atomizing burner as early as 1902, which he attempted to sell through the Fess System Company. Fessler moved east and installed his first "Fess" burner in Boston in 1915. Fess sold out to the Petroleum Heat and Power Company (later "Petro") in 1920, and that company began to sell fuel oil as well. Petro later bought up "Nokol," a similar system that Amalgamated Machinery had developed and marketed in Chicago beginning in 1920. The Williams Oil-O-Matic was the brainchild of Charles Williams and his son of Blooming-ton, Indiana. Williams was a car dealer, and the coal shortages associated with World War I apparently motivated his development of an oil burner that would burn used motor oil. The company marketed about 200 burners in 1920, 9,000 in 1924, and—*Forbes* claimed—was planning on sales of 50,000 in 1926. By the late 1920s, there were large companies as well, in-cluding Timken-Detroit, Bethlehem Shipbuilding, Sundstrand Adding Machine, and Socony (bought by Timken in 1925). Somewhat later, Esso Marketers, GM-Delco and General Electric also entered the industry.[18]

Firms, Institutions, Networks

In 1922, manufacturers formed the American Association of Oil Burner Manufacturers, and in early 1923, they met with representatives of the American Petroleum Institute (API), as well as architects, lawyers, combustion engineers, and furnace makers, to discuss common interests, a primary one of which was how to market oil heat. Alfred Chandler has pointed out that first movers in industries producing complex products face special difficulties. Oil-burner manufacturers needed adequate (and evenhanded) fire codes, standardized fuels, and dealer and repair networks, and they needed to protect themselves from shoddy products, dishonest practices, and advertising that would tarnish the industry. *Heating and Ventilating* told of manufacturers of poor-quality equipment that would set up a dealer, who might have money but little skill, and then force him to take a number of burners. When they failed to work well, the dealer might go broke while the manufacturer walked away. False claims abounded in early advertising. One burner guaranteed perfect combustion, while another alleged to be more than 100 percent efficient. Obscure, meaningless jargon such as "aero-atomization" was also common, and in 1926, at the behest of *Heating and Ventilating*, the industry developed a code of ethics for advertising. In 1927 manufacturers renamed the American Association of Oil Burner Manufacturers the Oil Heating Institute. Collectively, the association and then the institute published numerous technical/advertising bulletins on oil burners and heating. By 1925, Underwriters Laboratories had begun to test burners for safety, and companies quickly employed its findings in advertisements.[19]

The Oil Burner Manufacturers also countered a perceived threat of government regulation. The idea that petroleum should be saved for essential uses dated back at least to World War I. The possibility that heating houses might be a "nonessential" use of fuel oil was a lively topic when the Federal Oil Conservation Board held hearings with oil industry representatives in 1926. Walter Teagle, president of Standard Oil of New Jersey, explained how relative prices determined whether fuel oil would be cracked. Leon Becker of the Oil Burner Manufacturers drove the point home: the "question of superior uses . . . is primarily the consumer's problem," he asserted. Coal also could be substituted for gasoline, he pointed out, but people would rather pay 10 to 20 cents a mile to travel by car when they could go "26 miles on the subway for a nickel." The notion that a wise government could discriminate among petroleum uses for the public good would return later in the National Resources Committee's *Energy Resources and National Policy*, but as policy, it remained a dead letter before World War II.[20]

The Oil Heating Institute also pressed to standardize fuel oil. Oil burners needed oil with a standard pour point, gravity, and flash point, and in the early 1920s, fuel oil was anything but standardized. Some of it was described by color ("straw"); writers often used the term "furnace oil" without defining it. The *Oil Trade Journal* divided products by API gravity, but that was all, and others sometimes employed Baumé gravity. In 1924, Underwriters Laboratories pressed the interested parties to agree on standards, and in 1927, the editor of *Heating and Ventilating* described fuel oil specifications as "one of the unsolved problems of both the petroleum and oil-burner industries." Finally, in 1928, the Oil Heating Institute, working with the API, developed fuel-oil standards that, with some modification, are those used today, and at its behest, the United States Bureau of Standards adopted them in 1930.[21]

As makers of complex products, oil-burner manufacturers needed to establish dealer networks to service, as well as sell, their products. While an automobile that would not start was inconvenient, a broken oil burner might mean frozen pipes. One writer noted the "vital importance" of proper installation and the need "to intelligently service the burners that have been sold . . . [owing to the] bitter experience with burners that would not stay sold." He went on to note that owners were either "boosters or knockers," depending on how well the product worked, so dealers had to develop a "permanent relationship" with their customers, as they were really selling not just oil burners but in fact a "heating service." Successful dealers sometimes refused to install burners that were too distant for proper service. About the same time, a USDA pamphlet advised purchasers of oil burners "not to seek the 'best' burner but rather one that is handled by a reliable sales organization."[22]

Major manufacturers such as Socony, Williams, Delco, and Petro-Nokol advertised for dealers in trade journals such as *Domestic Engineering* and the *Fuel Oil Journal*, and all established chains of dealers ("factory representatives"). Manufacturers also marketed through local appliance, hardware, plumbing, or electrical-supply companies and even at times furniture and automobile dealerships. In the 1930s, companies also began to sell through large department stores, and here, too, they offered service as well. In 1934, *Printer's Ink* also stressed that appliance dealers could be particularly effective at selling complex machinery by emphasizing its value rather than its mechanics to potential buyers. A contrast with the coal industry is instructive. While oil-burner manufacturers and oil companies developed close relationships with dealers out of necessity, coal—probably because it faced little competition—had not done so. Its belated efforts to catch up will be noted below.[23]

Attack of the Oil Burners

About 1925–1926, major oil-burner manufacturers launched a massive sales campaign. They picked a good time, for strikes and shortages had soured many households on anthracite. Thus, in late 1925, one oil-burner maker advertised to Massachusetts households that switching to oil was "how to avoid fuel uncertainties." Moreover, fuel-oil prices were declining (table A4.3), while real per capita GDP increased by 25 percent from 1920 to 1929, making the market receptive to luxury goods such as oil burners.[24]

The campaign kicked off with producers placing large—sometimes full-page—advertisements in important eastern newspapers and national magazines, spending in 1926 about $1.5 million. Williams's Oil-O-Matic was one of the top advertisers. It had spent $2,700 on advertising in 1921; in 1926, it spent more than $411,000—nearly 10 percent of sales—most of it in magazines and newspapers. These were immense sums for the time. The advertisements appeared in big-city newspapers and major popular journals. Industry advertising by the Oil Heating Institute began in 1927, and it, too, appeared in major newspapers and national magazines. As the institute put it, "mass production requires mass persuasion." By the mid-1920s, market research by the institute and industry trade journals was guiding these campaigns, identifying the Northeast as the most promising market. Major northern cities also featured oil-burner shows, while newspapers often ran rhapsodic "news" stories of oil heat that sometimes amounted to infomercials. Manufacturers also sponsored radio shows. Newspaper advertisements usually listed local dealers, who often supplemented manufacturers' ads with offerings of their own. Indeed, one writer claimed that in 1925, Williams got its dealers to run 10 lines of advertisement for each one of its own. Sometimes dealers of all brands of burner combined with oil retailers in what were, in effect, industry ads.[25]

The goal of these efforts was not to sell oil furnaces—although that would come—but to persuade owners of homes and apartment buildings to convert existing coal furnaces to oil. This was a vital distinction, for it sharply reduced households' costs and risks. Advertisers, therefore, had a twofold persuasion job: coax buyers to shift to oil and then persuade them to buy that company's product. As I have noted, early domestic oil burners were expensive to buy and to run. Like gas heat, oil's one cost advantage was that in spring and fall, it made part-time heating much easier and cheaper than was the case with coal.

Again, like gas heat, what sold oil burners was convenience, cleanliness, and comfort: they were one of the many labor-saving devices sweeping through both home and factory in the 1920s. "Automatic, . . . Built to Give Service, Not to Require It" was how Wayne summarized and sharpened the contrast between its oil burners and hand-fired coal equipment. "Automatic" soon became shorthand for all gas and oil heat. Occasionally, advertisers also stressed environmental benefits; in 1926, Chicago Oil Burner Association billboards promised "cleaner homes and a cleaner city." Coal was indeed dirty; part of the cellar was devoted to the coal bin, and even with anthracite, coal dust permeated everything. Oil's cleanliness and the demise of the coal bin might make the basement habitable; it was like adding a free room to your house, advertisers claimed. Oil also promised comfort. Coal furnaces, in contrast, required attention day and night; they needed stoking, and the ashes had to be removed and then hauled away; and regulating the temperature with a coal fire was difficult, for there was a considerable lag between adding or reducing fuel and changes in temperature.[26]

Convenience, cleanliness, and comfort even at some cost appealed to the increasing numbers of upper- and upper-middle-class families. The content and placement of oil-burner advertisements indicates that this group was their intended domestic market. In 1929, one trade journal claimed that the ideal customer was a family building a $20,000 new house—which then included only 3.4 percent of all families. *Heating and Ventilating* magazine estimated that buyers of oil burners were willing to pay a 100 percent price premium over coal. Oil also continued to spread among owners of commercial and apartment buildings, for with oil, they no longer needed to hire anyone to stoke the furnace and remove ashes. They appreciated, as well, that oil did not go on strike as anthracite had so recently done.[27]

All of this was well known to the men who wrote the advertising copy for oil burners, and they knew, as well, from personal experience that keeping the coal fires burning was both women's and men's work. Men, apparently, brought the fire up early in the morning, but when the husband left the house, tending the fire and removing the ashes was women's work. Women, who did the house cleaning, also appreciated the lack of dirt that came with oil, while both men and women might enjoy the finer regulation of temperature that came with oil and the extra usable basement space as they waved goodbye to the coal bin.[28]

Early advertisements included a barrage of selling points; they stressed the manufacturer's dealer network and experience, as well as that the product was noiseless. There was the "Quiet May" and the "Silent Nokol," while Timken's product was "quiet as a drifting cloud." Although companies initially advertised the price, by 1928, about 15 percent of sales were on the installment plan as surveys revealed that without such procedures, the oil-burner market would remain confined to the well-to-do. As some gas companies had done with stoves, Nokol even offered a trial rental for the skeptical.[29]

Because coal furnaces demanded both female and male labor, oil-burner manufacturers pitched advertisements to both women and men; and because the target audience was middle- and upper-class households, advertisements portrayed tending a coal furnace as inappropriate work for women. Williams was the largest advertiser and the leading producer in the late 1920s. It favored full-page ads, the first of which appeared in the *Chicago Tribune* on May 18, 1924. In 1926, the company spent $185,000 on magazine ads alone. Its most famous slogan was "No Coal Shovel Was Ever Made to Fit a Woman's Hand." Another depicted a woeful-looking wife and daughter waving goodbye to the husband/father heading off to work, the image titled "Don't Leave Them in the Shadow of the Coal Shovel." Another producer, May, showed women dressed in high heels wearing furs down in the cellar forlornly looking at a cold furnace. Norge sold oil burners in Boston by using a comedy skit entitled "Freedom of the Shes." Such ads were aimed as much at husbands as wives, even when the advertisements appeared in women's magazines such as *Better Homes and Gardens*. A Williams ad in that magazine announced that "Men May Buy the Coal but Women Pay for It" in the form of a dirty house, suggesting that male guilt—and wallets—were the targets (fig. 7.3). Since starting the fire in the morning was typically the man's job, however, the company

Men may *buy* the coal *but women Pay for it!*

LOOK at your walls back of pictures, if you think your coal heated home is clean!

Start washing the woodwork and note the contrast. Give a party and count the days spent before in washing, scrubbing and dusting!

Then ask yourself if there is any special reason why your neighbor's wife who has an Oil-O-Matic should be relieved of this drudgery.

If you believe your family is comfortable with coal-heat, hide the shovel from your wife while you are downtown. Or start the furnace too early or too late in the fall. Then ask any Oil-O-Matic owner to tell you what comfort really means.

If you think coal isn't a nuisance, try leaving the house for a few days. Your wife can't leave it for five hours!

But with Oil-O-Matic heat you may stay away for as long as you wish.

Spasmodic coal heat is unhealthful. How many winters does your family go through without colds due to fluctuating temperature and drafty rooms? Compare your doctor bills with those who have Oil-O-Matic heat.

Yes, you may buy the coal, but your wife and family pay for it.

Selection of an oil burner for your house, doesn't require that you know anything

about oil burning. Most owners of Oil-O-Matic couldn't tell you how it works. But they will hasten to tell you that it has never failed to work.

It is a cautious man indeed who would question the judgment of some of the greatest engineers. These men know the basic soundness of the four natural laws of oil combustion. For no other reason could Oil-O-Matic have become the world leader.

Your local oilomatician can tell you how much a guaranteed installation will cost in your heating plant. Whether it's steam, hot water, or a warm air furnace. Let him explain his deferred payment plan that spreads the initial cost over a full year.

For the complete story, send the coupon below for "Heating Homes with Oil" and basement plan for ideal arrangement of space. It's sent free and postpaid.

Williams Oil-O-Matic Heating Corp. Bloomington, Illinois

Without obligation, please send me "Heating Homes with Oil," by return mail. BHG-86

Name

Street

City   State

No part inside the firebox

WILLIAMS OIL-O-MATIC HEATING
World's Largest Producer of Automatic Oil Burners
Printed in U. S. A.

Figure 7.3. In the 1920s, oil-burner manufacturers advertised in magazines aimed at middle- and upper-class women, emphasizing the comparative ease of control and cleanliness of oil heat. ("Men May Buy the Coal but Women Pay for It," Advertisement, *Better Homes and Gardens*, June 1926, 53. Public domain)

warned that "The shadow of the coal shovel is darkest . . . just before the dawn." Such thoughts seem to have moved this basement bard to convert to oil:

> Last year I rose at half past six
> My darned old furnace fire to fix
> And as my feet would hit the floor
> 'Twould feel like Greenland's Icy shore.[30]

Appealing to the compassionate husband, Timken asked men at the office, "equipped with modern business . . . conveniences . . . What about your wife? . . . Is it fair to waste her talent . . . [by adding] janitor's work to her other housewifely duties?" Timken also asked the husband to consider "the embarrassment she frequently suffers when discussing home equipment with the girls of her group." Another Timken presentation showed a woman lugging ashes outside on a winter day. It asked, "Does she remember how careful and considerate you used to be?" And it included romantic pictures suggesting how an oil burner might pay off for him as well.[31]

Oil, advertisers also stressed, was "modern." One ad for Gulf burners asked, "Still Living in the Coal Age?" Companies appealed to status envy as well, which was evident in ads that ran in magazines such as *Vogue* and *House Beautiful*, publications aimed at upper-middle-class readers. Furnace tending was clearly just as inappropriate for middle- and upper-class men as it was for their wives. Williams's copy sometimes depicted clearly upper-class households, whose members, one suspects, would not know a coal shovel from a codfish. One featured a man with an unhappy expression on his face, shoveling coal while wearing a dressing gown over his tuxedo; the title read, "Every Modern Convenience Except One OIL HEAT." Others featured testimonials by very wealthy individuals. Initially, at least, these ads were highly productive. One writer in 1932 noted that Williams's initial splurge in the *Chicago Tribune* in 1926 had netted 1,460 inquiries. The same writer also claimed that early magazine ads might cost as little as $1.85 per response while by 1932, as responses dwindled, the figure had risen to $50.[32]

Critics of advertising have often complained that consumers faced a vacuum of independent advice, but that was far from true of the oil-burner market. Presumably the demand for advice depends on its value, which might be small for products bought repetitively. But oil burners were a one-time purchase, with no trade-in value, so mistakes were expensive. For homeowners in the 1920s, many independent publications contained valuable information. The pamphlets by P. E. Fansler in 1925 and 1927 are examples, as is the 1927 USDA circular and the evaluations provided by Underwriters' Laboratories. Magazines also supplied objective information. From 1925 through 1928, *Popular Science* ran a series of informative articles on oil burners. And like modern *Consumer Reports*, the magazine surveyed and reported on buyers' experiences with oil burners. In 1928, Wanamakers held an oil-heating exposition in New York City that featured a speaker from the Popular Science Institute. Similarly, the Good Housekeeping Institute offered a booklet on oil burners, and in 1925 the magazine carried several long, technical articles on oil burners by P. E. Fansler. He provided similar information in *Garden & Home Builder* the next year. In 1925, *Better Homes & Gardens* carried an article by the chairman of the Oil Heating Institute that promised "Pointers on Types of Burners Which Will Help You in Selecting Yours."[33]

Builders of new homes could obtain information from technical journals such as *American Builder* and *Building Age*, which featured articles on oil heat (some, but not all, of which were penned by employees of the Oil Heating Institute). These stressed a point also emphasized by advertisements in popular magazines and newspapers: oil did away with the coal bin, thereby essentially adding a room to the house. By 1929, *American Builder* was proclaiming "Oil Burners Help Sales," while builders were stressing oil heat in their advertisements.[34]

Selling oil burners required companies to develop sophisticated marketing procedures to supplement their advertising. Manufacturers assigned their dealers sales quotas based on market demographics, and they sometimes ran dealer sales incentive campaigns. Williams sent its dealers' sales personnel through a course created by the Business Training Corporation that developed selling, as well as technical, skills. A survey by the *Fuel Oil Journal* revealed that 39 percent of leads resulting in new sales resulted from word of mouth from satisfied customers. One West Coast Williams dealer won the incentive campaign for 1927 through door-to-door canvassing and production of its own literature. Williams published *Oil-O-Matic News* (later *News-O-Matic*), a colorful monthly for its dealers full of merchandising ideas, along with *The Home of Today*, a monthly that it sent to buyers. It contained photographs, articles on matters such as how to use the old coal room, and cards the recipient might use to alert the company to new sales prospects. One Williams dealer in New Jersey assigned salesmen to a territory, providing a preliminary canvass to provide leads and developing information on housing size, heat source, and appliance ownership (having a refrigerator, for example, suggested a likely prospect). Companies also employed tax lists to generate prospects. The result was live leads and high sales productivity.[35]

Coal Strikes Back

Because the market for oil burners was comparatively wealthy families with large homes, sales displaced a lot of coal—especially because many houses were in cold climates and uninsulated. In 1930, engineers estimated the coal requirements of a modest (2,250 square foot) uninsulated house in Saint Paul, Minnesota, to be 13 tons of anthracite a year or about 55 bbl.

of fuel oil. A much larger, probably uninsulated, house in Springfield, Massachusetts, that converted to oil had burned 104 bbl. of oil instead of 25 tons of hard coal.[36]

Yet even as oil was biting off large chunks of the coal market, coal producers long ignored the threat. Not until the mid-1920s did anthracite producers deign to notice the inroads in sales. Bituminous producers were even more tardy (their efforts will be discussed in a later section). In 1922, *Coal Age* reported on a large anthracite producer that still seemed indifferent to the concerns of its customers. Two years later, S. D. Warriner, president of Lehigh Coal and Navigation, one of the largest anthracite producers, informed readers of the *Mining Congress Journal* that the industry's major problems were its relations with labor and government. In 1925, just as oil-burner producers were beginning their massive advertising blitz, *Coal Age* decried the lack of hard-coal merchandizing and reported that "from producer to retailer [the industry] has been nothing but an assemblage of order takers." Several years later, it recalled the "dead level of complacent self-sufficiency" that characterized most producers right after the war. Dealer relations were often poor. The industry's trade association—the Anthracite Operators Association—largely focused on labor relations and did no research. In contrast with the oil-burner makers, who quickly allied with oil producers, wooed architects, and established dealer networks, there was little coordination among coal producers, equipment makers, and dealers.[37]

Some of the complacency reflected the belief that oil and gas competition were merely short-term problems that would soon go away. In 1925, New England anthracite dealers concluded that bituminous coal and its derivatives (coke, electricity, and manufactured gas) were "more menacing" than oil. A year later, Edward Parker of the Anthracite Information Bureau (and late of the USGS) informed the American Mining Congress that oil competition would "grow less menacing." Such hopes reflected the pronouncements of a chorus of experts, such as George Otis Smith, who opined that the United States would soon run out of oil. Indeed, as late as 1932, two years after Dad Joiner's well struck oil in the enormous East Texas field, the Federal Oil Conservation Board was still warning of a "paradox of a present oversupply in the face of ultimate shortage."[38]

Quality and Service Improvements

Anthracite producers at last woke up to the threat of oil about 1925, and they finally adopted industry-wide quality standards. Companies had rejected such standards when the U.S. Coal Commission recommended them in 1923 but apparently reconsidered, as poor-quality control resulted in a chorus of complaints from dealers. "The present lack of uniformity . . . [in] sizing and preparing coal . . . gives rise to much of the criticism, complaint, and ill-will on the part of consumers," the National Retail Coal

Merchants Association reported. Moreover, the industry was then trying to interest consumers in smaller sizes of coal, and if these contained too much ash, they simply would not burn in domestic furnaces. The first standards governed size and percentage of impurities, and the industry tightened them in 1927. A new inspection service was created to ensure compliance. Because they reduced the quantity of marketable output per ton of coal mined, the standards had the effect of reducing productivity and raising costs, but they did improve quality: a 1935 survey found that impurities in Buckwheat #1 dropped from 18 to about 11 percent.[39]

In 1927, a subgroup of the Operators Conference began the Anthracite Coal Service, to improve dealer relations with an eye to better customer service. Dealers and producers had long been wary allies, at best. In 1925, New England dealers complained that producers sometimes sold to large customers at the same price they gave to dealers, while many dealers returned the favor by carrying coke, bituminous coal, and fuel oil. The focus of the coal service was on heating, suggesting that by this time, producers had largely conceded the loss of cooking-fuel markets. A large-scale investigation into consumer complaints in 1928 revealed continuing problems of coordination. The investigation demonstrated that most difficulties stemmed from furnace problems rather than coal quality, yet coal dealers rarely offered service. In response, the Coal Service established regional offices as far west as Minneapolis that offered dealer training in combustion so that they could provide furnace services to customers. By 1929, it had supplied instruction to employees of nearly 1,800 retailers in 111 cities. The Coal Service began *Anthracite Salesman*, distributing about 14,000 copies a year to dealers. Many consumer complaints, dealers discovered, reflected either improper firing techniques by consumers or faulty equipment. Soot buildup on the boiler might interfere with heat transfer, resulting in skyrocketing coal consumption to maintain comfortable temperatures.[40]

Also in 1927, industry representatives assembled at what the Bureau of Mines described as the "Mount Carmel Conference" to provide a "united effort on the part of operators, miners, distributors, consumers, and all others interested in the economic welfare of the anthracite region" to address the industry's loss of markets. The conference resulted in an Anthracite Cooperative Association (later, the Anthracite Institute); it focused on public relations, taxes, and freight rates. In 1929, producers also established a credit bureau to aid company sales.[41]

Stokers for Domestic Furnaces

But coal's greatest problem was that burning it was labor-intensive at a time when rising incomes were leading consumers to demand convenience even as oil salesmen were trumpeting automatic heat. Here, anthracite producers were late to the party. Although domestic stokers had been advertised in major newspapers since the early 1920s, coal producers had ignored them. *Coal Age* caught the significance of their tardiness: "The time to meet the competition of rival fuels . . . is at the inception of their development. . . . The oil burner, once installed, will not quickly be abandoned." Later, the journal pointed out that the coal industry had been ten years behind oil in providing automatic heat. Finally, in 1928, hard-coal producers began "assisting in the development" of mechanical stokers that would feed smaller, cheaper sizes of anthracite. In 1929, producers formalized an Anthracite Equipment Corporation to encourage technical improvements in furnaces.[42]

Mechanical stokers fed coal from a hopper and might be thermostat-controlled, requiring much less tending. Stokers seemed to be the key to meeting gas and oil competition, and soon, a number of coal companies began to manufacture their own; coal dealers were encouraged to carry them as well. Stokers reemphasized the need for clean coal, for—as the author knows from personal experience—stone might clog the mechanism, potentially breaking a shear pin or causing motor overload. While companies trumpeted stokers as "automatic" heat, in fact, the hopper had to be loaded about once a day; moreover, a householder still had to shovel the same amount of coal as with hand firing—while ashes also needed handling. Later, bin stokers arrived; these drew coal right from the bin, and some came with automatic ash removal and really were nearly automatic. But they were expensive.

Yet stokers were a mixed blessing for anthracite because they used smaller, less-profitable sizes. As *Coal Age* pointed out in 1925, inducing consumers to purchase (say) Buckwheat #1 could *reduce*, not increase, profitability and might prove suicidal if its sales came at the expense of domestic sizes. Thus, for stokers to improve hard coal's profit position, prices of smaller sizes would have to rise, yet their inexpensiveness was a major selling point of stokers. Moreover, bituminous producers also awakened to the need for better marketing, and they, too, began to develop stokers.[43]

Marketing and Branding Coal

Anthracite began a marketing campaign about 1927. The Philadelphia & Reading Company inaugurated newspaper advertising and a few magazine advertisements. Rather belatedly, it touted the smoke-control benefits of anthracite with ads urging consumers to "let a little sunshine in." A number of other producers and dealers combined to advertise "cert-i-fied" anthracite, stressing quality. There were trademarking efforts as well: Reading advertised that it had "Fyrewell" coal, a small amount of which would allegedly work wonders when added to the coal pile; and in 1929, Glen Alden and the Delaware Lackawanna & Western mines began to dye their product and advertise "Blue Coal." Nor were coal dealers averse to an occasional smear campaign suggesting that oil burners were likely to blow up—a claim hotly disputed by the Oil Heating Institute. Yet anthracite producers had little political power outside Pennsylvania, so oil-heat interests never experienced a concerted political attack from anthracite interests. These early advertising campaigns compared unfavorably with those of oil-burner producers. They rarely employed humor, and visual images were often uninteresting. The campaign soon petered out.[44]

Depression Developments

The Great Depression, inaugurated by the stock market crash of 1929, proved catastrophic for coal. While the decade after 1918 had been one of stagnation for bituminous coal and decline for anthracite, sales for both collapsed during the great contraction. This collapse occurred even as longer-term changes in living patterns were reducing domestic fuel demand as well. Increasing urbanization was shifting people from single-family dwellings to apartments that were more fuel-efficient. And while they might burn hard or soft coal, if it was anthracite, they burned the smaller, less-profitable sizes. A New York City fuel survey of 1936 pointed out that single-family dwelling had fallen from nearly 32 percent of the total in 1921 to 20 percent in 1936, with the remainder being, of course, multiple family dwellings. Moreover, the largest apartment buildings used about 24 percent less coal than the smallest buildings to heat a given volume of space. Similar trends were occurring nationwide. These events, the survey concluded, were "most unfavorable to anthracite," and they provided no more encouragement for the domestic use of soft coal.[45]

Oil Heat: From a Class to a Mass Market

The Great Depression hastened changes in the oil-heat market that had begun in the late 1920s and would worsen coal's domestic difficulties. In 1929, as the market of high-income buyers became saturated, companies began to develop smaller, lower-cost burners, and they tried to standardize and cheapen installation procedures. Market saturation in the Midwest also induced burner manufacturers to move east. There had long been a trend for oil and oil-burner dealers to join forces, and the decline in household incomes and rise in unemployment in the 1930s encouraged such activities. In 1932, when Standard Oil entered the Washington, D.C., market to sell fuel oil, it offered "free" burner service, thereby squeezing local oil-burner dealers. Two years later, Esso Marketers also offered Washington residents low-priced oil burners. By the end of a decade, nearly a fifth of Washington homes with central heat burned oil, while another 10 percent used gas.[46]

Local oil dealers also pushed prices down. Many found that to sell oil, they needed to sell burners as well, and this reasoning precipitated a price war in Fall River, Massachusetts, in 1933–34 that drove down the price of installed burners to $150. By 1940, oil had grabbed a quarter of the market

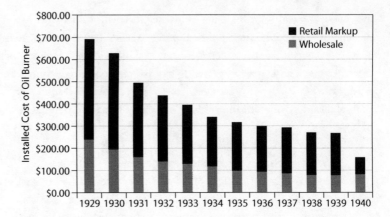

Figure 7.4. Oil burners evolve from a class to a mass market, 1929–1940. (Chart data drawn from "Prices in 1940," FOJ 18 [Jan. 1941]: 14)

for central heating in that state. In the mid-1930s, Williams and Sears were both advertising oil burners installed with tank for $250. As finance companies became familiar with the oil-burner market and willing to supply credit to dealers, purchases on time rose to about 45 percent in 1934. Down payments also fell, and they disappeared almost entirely after the Federal Housing Act of 1934 reduced sellers' risks by allowing federal insurance of such household improvements as oil burners. In 1929, the average dealer price of a conversion oil burner stood at $240 and the retail installed price at $692; depression and competition reduced these sums to $82 and $158 by 1940 (fig. 7.4). Schumpeter had argued that the great virtue of capitalism was to make silk stockings available to factory girls, and in the 1930s market, pressures began to make oil heat affordable so that "flivver [Model T Ford] drivers" could have "flivver burners."[47]

Competitive pressures and technological change also improved burner technology and reduced operating costs. Newer equipment was significantly quieter ("hushed heat"), and beginning in the late 1920s, improvements in their electrical operations reduced radio interference. In 1933, Williams stressed that the Oil-O-Matic could burn number 4 fuel oil at $0.055 a gallon, making it cheaper to run than other oil burners (or anthracite). About 1934, oil dealers also began automatic deliveries based on degree-days. This had the effect of raising the average number of gallons delivered and reducing truck-miles and delivery costs—in some cases as much as 35 percent.[48]

The 1930s brought both continuity and change in advertising. Companies continued to stress comfort and cleanliness. In 1930, Petro reported that Petropolis "is a state of comfort where everyone sleeps later in the morning." Timken urged purchasers to "say goodbye to ashes—forever." In 1934, it claimed that its products would banish dirt and dust forever. In a timely spread of March 1934, Timken pointed out that while Lincoln freed the slaves, it was "freeing the *furnace slave* mother." Not to be outdone, two years later, Williams showed pictures of women and men refusing to shovel coal and reported a "Mutiny in the Basement as "thousands are rebelling against furnace slavery."[49]

As companies took aim at a middle- and working-class market, advertisements depicting the mansions of the super-rich disappeared. In 1939, a Timken ad depicted two apparently working-class men as hunters with shotguns, leaning over a pickup truck discussing their oil burners. Women were less likely to appear in stylish dress and high heels tending the coal furnace than as housewives wearing aprons, lugging out the ashes. And they, too, were thoroughly practical: "I am buying heating results," a woman informed readers of a Timken ad in 1939. Advertisers continued to stress modernity. In 1930, Timken suggested that the coal shovel belonged in a museum, and a 1937 Williams add claimed, "Your coal shovel is as out-of-date as the spinning wheel." That same year, in a nod to recent labor developments, another Williams ad announced that wives were instituting a sit-down strike against furnace tending.[50]

Although burner manufacturers liked to stress that oil heat might make the basement habitable, such claims were undercut as long as oil burners looked like machinery and furnaces looked like furnaces. By then, burner manufacturers that had concentrated on selling conversion units for coal furnaces were beginning to market oil furnaces, sales of which rose from about 8,000 in 1933 to 50,000 in 1940. To capitalize on their claim that with oil heat, the cellar (without the coal bin) became usable space, manufacturers invented the furnace beautiful. Petro seems to have begun the parade in 1931 with a burner and boiler "in one beautiful, automatic self-contained unit" that came in "two tones of soft velvety tangerine, or blue or green or cardinal." In 1934, the oil burner also became a streamlined work of art, as May added a sheet metal covering to its "Quiet May" and advertised the "Burner Beautiful." In 1935, May offered a furnace with a "beautiful cabinet of chrome finished ebony and gray." American Radiator's oil furnace of 1936 came in "sea green enamel, chromium trimmed." *Business Week* claimed it "looked more like a refrigerator" than a furnace. General Electric advertisements claimed its new oil furnace was "beautiful in appearance" and "as far ahead of 'oil burners' as a streamlined train is ahead of the first locomotive." The 1930s also saw a host of improvements in furnaces. They shrank in size, were optimized for oil, featured rock-wool insulation and improved draft, and sported hot-water heaters. Furnace manufacturers also began to offer "air conditioned" (humidified) heat as well.[51]

Competition spilled over from basements to parlors, as the 1930s saw a boom in oil space heaters (fig. 7.5). These were the old kerosene stoves wearing a new dress of sheet metal just like the new oil furnaces. While marketing was surely part of the motive for the new styles, they also heated by convection rather than radiation, yielding more even temperatures. With oil burners and furnaces appealing only to those with central heat—just 42 percent of the population as late as 1940—the potential

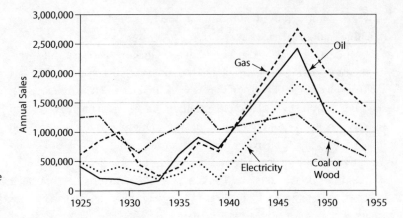

Figure 7.5. Space-heater sales by type of fuel, 1925–1954. (Chart data drawn from US Bureau of the Census, *Biennial Census of Manufactures*, various years)

market for such heaters was immense. The new term was "cabinet heaters." Manufacturers included makers of furnace oil burners and range oil burners, and oil-burner and furnace dealers carried them as well. Oil space heaters benefited from the same decline in fuel prices that boosted oil heat generally (table A4.3). It also helped that department stores began to market them. In Boston, Summerfield's and Kane's, both large furniture dealers, and in Chicago, General Furniture, which had twelve stores in the region and claimed a million customers in 1935, took the lead. In Providence, Rhode Island, in 1933, nearly a quarter of the lowest-income groups heated with kerosene (table A4.5). Sales of oil space heaters rose from about 100,000 in 1931 to more than 900,000 in 1937.[52]

Gas Attack

We saw in chapter 3 that by 1930, natural gas had evolved from a regional fuel, with what appeared to be a dim future, into an increasingly national competitor as expanding production and transportation pushed it into new markets. There had been about 3.5 million domestic and commercial natural gas customers in 1925; fifteen years later there were almost 10 million (table A1.6).

The forces driving the spread of oil heat were primarily the individual burner manufacturers, many of whom sold only oil equipment. The structure of the gas industry led to a different approach to marketing gas heat. With enthusiastic support from the AGA, local utilities—sometimes in groups supported by local hardware dealers—took the lead, just as they had with gas cookstoves. Accordingly, most sales campaigns were local, and after 1930, the arrival of natural gas touched off basement battles in Atlanta, Chicago, Detroit, Indianapolis, Minneapolis, St. Louis, and many smaller cities and towns.

Utilities with access to natural gas were far better positioned to enter the heating market than either coal companies or oil-burner or fuel-oil producers. Many had been marketing manufactured gas for heating, and nearly all had decades of experience selling and servicing cookstoves and gas space and water heaters. Most already had sales and service departments; they had experience working with local contractors and equipment dealers, and they knew their customers. For smaller companies or those new to heating, the AGA proceedings contained a mine of information that covered all aspects of market research, service, and selling.

While some companies marketed straight natural gas, others blended it with the manufactured product, resulting in a gas that was significantly higher in energy content and cheaper than manufactured gas per Btu. Either way, overnight, natural gas became a new competitor in many home-heating markets. The arrival of natural gas also propelled the utilities to charge by energy content rather than by the cubic foot. Great Britain had begun to sell gas by the therm—100,000 Btus—about World War I, and there was desultory discussion of the idea in the United States in the early 1920s but little action. The therm arrived in 1930, when natural gas came to Chicago. Because natural gas contained more energy than its manufactured cousin, selling it by the foot at the same price as manufactured gas would result in a sharp drop in volume purchased—and revenue. Peoples Gas petitioned the regulators to sell by the therm and other utilities slowly followed.[53]

Still, even natural gas was usually expensive compared to soft coal, so, as with oil, early installations were typically in the large houses of the well-to-do, ensuring that it replaced a good deal of coal. Selling the new product for house heating thus required the proper rate structure, as was belatedly brought home by a study undertaken by the AGA in 1936; the study demonstrated how price-sensitive heating sales were. In addition, a proper economic package required attractive installation costs and financing—all of which required advertising. Moreover, like oil, gas heat sold by word of mouth and a few unsatisfied customers might sink the entire campaign. Accordingly, companies chose customers with care; they were careful not to oversell the product, urged use of insulation and weather-stripping, reported degree-days along with gas bills and shared outcomes at meetings of the AGA, so each could learn from the experience of others.

One illustrative campaign, termed the "Master Stroke Heating Sales Plan," included six gas utilities and began in 1932, when natural gas arrived in parts of Iowa, Nebraska, and Minnesota. The plan was designed to make gas economically competitive and—perhaps because it began in the worst year of the worst depression in American history—risk-free to customers. What were termed "promotional" rates (i.e., declining blocks) typically began at $0.52 to $0.65/1000 ft.^{3} of 1000 Btu gas and declined with volume, making it competitive with soft coal at $10/ton. In addition, for customers passing a credit check, the company would install a conversion burner for $135—significantly less than an oil burner—with no down payment. There was a monthly "rental" plan for the cash-short, but in either case, the company agreed to remove the burner at any time at no cost. Independent dealers might participate, receiving a bonus for each installation, and a campaign of billboards, direct mail, and newspaper advertising trumpeted the comfort of gas and the desirability of the companies' offer. The companies worked closely with purchasers, explaining bills based on degree-days, adjusting furnaces, and suggesting cost-effective insulation. In three months, the program sold 3,024 burners, furnaces, and heaters to 43,000 gas customers. By 1940, in Lincoln, Nebraska, alone, gas had signed up about 26 percent of households with central heat, while another 11 percent burned oil. Similar campaigns erupted all over the Midwest as cheaper gas arrived: as one representative told the AGA in 1936, "every gas company in the middle west is actively interested in home heating."[54]

Although the emphasis of all these efforts was on central heating, companies knew that the majority of their customers used space heaters. Indeed, in 1934, one speaker informed the AGA that space heating was the largest single source of domestic demand for natural gas since only about half of all his customers had central heat. The market for new heaters, he thought, comprised not only those customers without central heat but also those that burned coal who might want a space heater to avoid using the furnace in fall or spring. Before natural gas arrived, the heaters—like gas furnaces—had been a hard sell. But as with oil, lower rates widened the market. Typically, space heaters had been sold mostly by dealers, but in Pittsburgh, as gas sales plummeted during the Depression, Equitable Gas Company entered the fray. Allied with local dealers, it sold 3,370 heaters in March 1933—the very bottom of the Depression. Beginning in 1925, the census published sales data on such space heaters in alternate years, and until 1939, these totaled more than five million, while adding in noncensus years might double that figure, helping to drive coal from parlors, as well as basements.[55]

In Minneapolis, competition from electricity was squeezing the gas company's sales for cooking, leading it to move into central heating after 1932, when natural gas arrived from the Hugoton, Kansas, fields. The company offered installation of conversion burners for $10 down and gas at $.62/1000 ft.^{3}. Oil, which had once commanded the entire market for automatic heating installations, saw its share fall to 52 percent as the usual battle of the basements erupted, with gas grabbing 35 percent and coal stokers bringing up the rear. The arrival of natural gas allowed the St. Louis Missouri County Gas Company to increase the energy content of its product from 570 to 800 Btus even as it cut rates by 20 percent. The company developed an elaborate sales campaign—the "ONE YEAR PLAN"—that included engineering studies of customers' houses and promised to remove equipment if first-year costs exceeded their estimate by more than 10 percent. There, cheap coal retarded progress, however, for only about 2.5 percent of houses had switched to gas by the end of the decade, while another 3.3 percent burned oil. When natural gas invaded Grand Rapids, Michigan, about 1936, the gas company offered conversion burners for $99.50 with four dollars down and five years to pay. Customers bought 1,000 Btu gas and paid $0.50 per thousand for all gas purchased in excess of 3,000 ft.[3] By the end of the decade, 14 percent of central heating there used gas.[56]

The Battle of the Basements

Coal producers and dealers continued to fight these incursions by improving their service and products and with advertising campaigns that stressed these improvements and targeted the new fuels. Following the lead of the Anthracite Coal Service, by the early 1930s, some Massachusetts and New York dealers began to explain that they were selling heat, not coal, and they installed and serviced stokers and removed ashes, all for one flat fee per ton.[57]

Gradually, bituminous coal producers also woke up from their century-long slumber. In late 1929, the National Coal Association heard from retailers who were thoroughly dissatisfied with the marketing help they got from producers. Regional marketing ventures began to appear about this time that relied on newspaper advertisements. Blackwood Coal & Coke began to trademark its fuel as "Red-Bar" coal. By the late 1920s, soft-coal producers began to offer coal in sizes, as well as run of the mine, and coal washing also spread. Somewhat later, hard-coal dealers stole the idea of degree-day delivery from the oil companies' playbook. In early 1931, in a rare show of unity, some hard- and soft-coal producers and equipment manufacturers combined in the "Committee of Ten—Coal and Heating Industries" to push stokers. By early 1932, some bituminous producers realized they, too, were selling heat, not just coal, and began offering to install and service stokers and remove ashes. Soft-coal companies also improved cleaning, and they began to sell "dustless" coal (that had been treated with oil) as well. About this time, bituminous producers started to offer bagged coal for stoves while "MinePakt" anthracite also arrived.[58]

To trumpet coal's new virtues, in July 1936, hard-coal producers formed "Anthracite Industries" to launch a new advertising offensive aimed at the home-heating market. In 1937, with an annual budget of $750,000, it inaugurated a three-year campaign with advertisements in 91 newspapers in 59 East Coast cities and supplied copy to dealers as well. The advertisements were far more visually appealing than in the previous campaign and compared favorably with offerings from oil-burner makers. They, too, depicted women and children and emphasized safety and comfort. "Automatic anthracite . . . saves on curtain and rug cleaning," one 1937 ad announced, while "transform[ing] the basement."[59]

They also focused on anthracite rather than individual companies' products. There was "special emphasis" on automatic stokers with slogans like "Anthracite's 7 Star Features Make a Champion Fuel" and "Solid Fuel for Solid Comfort." Anthracite industries also featured exhibits at National Home Show Week exhibitions in many eastern cities and funded a display at the 1939 World's Fair. Stoker manufacturers such as Iron Fireman also advertised independently, not only in eastern newspapers but in national magazines as well. Bituminous producers also counterattacked. In 1937, the National Coal Association instituted a similar campaign for bituminous coal interests that coordinated with dealers and architects. Stoker manufacturers joined in as well, with advertisements in major newspapers and national women's magazines. In 1939, producers and dealers in Kansas, Missouri, and Oklahoma formed the Bituminous Coal Utilization Committee. It pledged $15,000 for an advertising campaign and allied with the Kansas City Power & Light Company to promote (electrically powered) stokers just as gas companies marketed conversion burners.[60]

Thus, by the mid-1930s, heating with coal was far different from what it had been only a decade earlier. The fuel itself—whether anthracite or bituminous—was far cleaner. Buyers could purchase soft coal sized for stokers, and hopper stokers lengthened the chain that tied homeowners to the furnace, while bin stokers and thermostats were almost automatic. Coal dealers provided furnace service, delivered coal via degree-days, and would remove the ashes for a price. Yet the statistics were not heartening. Stoker sales never made up the decade they had lost and during the 1930s were typically less than half those of conversion oil burners (table A4.6). In 1940, their best year to date, companies marketed about 148,000 stokers; but oil-burner sales totaled 265,000 that year, and gas equipment manufacturers recorded sales of 117,000 units of central heating equipment. Worse, not only did stoker sales typically lag behind those of oil burners, but as of 1939, only about 14 percent went into new houses. *Coal Age* caught the significance of these facts: "lack of installations in new homes will stamp the use of coal as a fading one in modern house heating."[61]

Part of the problem was that stokers were not really automatic, and they were just as dirty as hand firing. Bin stokers with ash removal were automatic but, to make them cleaner, required sealed coal and ash storage bins, which were expensive to install and space-intensive even as oil and gas were advertising that they added another room to your house. While an oil tank might take 15 square feet of a basement, in one installation of a bin stoker, the coal and ash storage alone took up 90 square feet in a 240 square foot basement.[62]

Coal prices were another part of the problem. Hard coal had always been an expensive fuel, but average retail prices for hard coal nearly doubled between 1915 and 1930, while soft-coal prices rose 54 percent (table A4.3). Because transport costs along with wholesale and retail dealer handling constituted such a large portion of retail prices, the ability of producers to meet competition was limited. In 1940, for example, when the mine price of hard coal was $3.99, it retailed for more than $11.45 a ton in Buffalo and $13.00 in Boston and other important cities. Anthracite would have been expensive if producers had given it away. Similar price relations held for soft coal as well.[63]

These various promotions resulted in what *Printers' Ink* termed a "battle of the basements" among coal, oil, and gas heat. Although the reference was to the battle in Chicago, it applied equally well to events in St. Louis, Minneapolis, Detroit, and other cities. In Chicago, oil-burner makers such as Williams had been active for some years when, in mid-1933, Peoples Gas Light and Coke Company, along with several other regional gas companies, introduced therm rates and launched a house-heating campaign. Peoples had long been an aggressive marketer of manufactured gas, but with the arrival of natural gas from Texas, it now had a much more attractive product.

Relying on the 1934 census and its own surveys, Peoples divided the city into mile-square blocks and focused on the 143,000 single-family dwellings with central heat. Within that group, its target market was the 17 percent of households owning homes worth $7,500 and up. The company cut the cost of gas in half to $0.07/therm and developed a "standard package" of a conversion burner for $149.50 with no down payment marketed as "Installed in your home at our expense. . . . Removed at our expense if you don't like it." It focused salesmen and brochures on those city blocks with a high proportion of target households, but it also blanketed Chicago with newspaper and radio advertising. The advertising appropriation amounted to $1.5 million; compare that to the $15,000 pledged by the Bituminous Coal Utilization Committee noted above. Gas company advertising featured testimonials from satisfied customers. "No More Shoveling Coal or Hauling Out Ashes for Me," was Joseph Cepek's happy assessment. These advertisements, which gave the name and address of the families, were powerful indeed, and got under the skin of the Chicago Coal Merchants Association, which responded with reverse testimonials. "I couldn't keep warm with gas heat," complained one dissatisfied customer. Oil-burner makers fought back as well: a Williams ad had one happy woman saying, "Take a Long Rest, Coal Shovel." About this time the Chicago Oil Heat Committee that represented oil companies, as well as burner dealers, piled on, and the battle continued. By 1940, gas and oil had clawed about 7 percent of the market away from solid fuels.[64]

Nationwide, in 1940, gas and oil were the fuels in nearly 20 percent of all houses with central heating and about 22 percent of the houses using stoves (table 7.2). These changes in heating patterns were less dramatic than those for cooking and water heating. Coal furnaces and stoves were long-lived, while the investment in oil or gas for central heat, even if one bought a conversion burner, considerably exceeded that required for an oil or gas stove, limiting the size of the market. The figures varied sharply

Table 7.2. Number and Percent of Dwellings Using Specified Fuels for Heating, with and without Central Heating, Nationwide, 1940

Dwellings	Reporting	Coal or coke	Wood	Gas	Fuel oil/kerosene	Other
With central heating						
Number	14,152,024	10,909,163	373,322	1,109,587	1,687,737	78,215
Percent	100	77	2.6	7.8	11.9	0.6
Without central heating						
Number	19,732,355	7,622,427	7,362,156	2,728,381	1,706,722	50,022
Percent	100	38.6	37.3	13.8	8.7	0.3

Source: Sixteenth Census of the United States, 1940, *Housing*, vol. 2, part 1, tables 12a and 12b.

by income and region, but coal heat was embattled everywhere. The amount of coal that oil and gas displaced cannot be known with great accuracy; however, some plausible estimates are possible. We know that those with central heat who switched from coal or wood to oil or gas were the comparatively affluent and would have burned perhaps 10 to 12 tons of coal a year. The 2.8 million houses with central heat that burned oil or gas in 1940 must have meant a loss to coal, therefore, of between 28 and 34 million tons.

Another 4.7 million dwellings without central heat used gas or oil or kerosene. A study of consumption patterns in the late 1930s found that low-income families in villages who mostly heated with stoves and burned coal consumed about 6.9 tons a year. Accordingly, the 4.7 million families using gas or oil stoves might have burned 32 million tons had they employed coal instead. While these figures are hardly exact, they suggest that by 1940, the battle of the basements—and parlors, as well—had already cost coal some 60 million tons of sales a year.[65]

Nor did the future look bright. *Coal Age* commissioned two surveys of consumer attitudes in 1939 and 1940. They found that 62 percent of families burning coal would prefer oil or gas, and when asked what fuel they could choose in a new home, only 28 percent chose coal. In a second survey, they found that a third of those using stokers now wished they had oil.[66]

It is tempting to speculate how this battle might have played out if coal companies had made the various improvements I have outlined above a decade or so earlier. Had oil and gas confronted coal furnaces fed clean coal by stokers, supported by producers with an aggressive dealer service network and promoted by smart advertising, the new fuels would surely have progressed more slowly. But such speculations reveal a misunderstanding of the way in which markets drive economic change, for one cannot imagine coal's attempted renaissance without the threats from the new fuels.

Domestic Energy Use in 1940

As this and previous chapters have shown, by about 1940, American households were well into the energy transitions that would continue after World War II. The major changes were the absolute decline in the use of wood and hard coal even as the population rose rapidly and—unsurprisingly—the rise in automobile gasoline consumption, which was entirely responsible for the increase in per capita energy consumption from 2.07 tons BCE in 1910 to 2.71 tons in 1940 (table A4.7). Indeed, gasoline amounted to a third of all domestic energy consumption on the latter date.[67]

Per capita fuel use for purposes other than transport was less in 1940 than in 1910. This long-term decline in energy use for heating and cooking probably reflected the continued retreat of the fireplace in favor of more efficient stoves and furnaces and the switch from wood and coal to more efficient fuels. It also was a result of increasing house insulation. Finally, increasing urbanization from about 46 percent of the population in 1910 to 57 percent in 1940, reduced fuel use per person, for apartment buildings could heat more efficiently than detached houses.

One underappreciated result of these developments was that they must have slowed the growth in air pollution from households. On an energy-equivalent basis, oil emits only 3 percent of the particulates of coal, while such pollution from gas amounts to about 0.003 percent of coal, and, of course, both oil and gas burned more efficiently. Accordingly, pollution would have been substantially worse had households not shifted to oil and gas and instead burned an additional 60 million tons of soft coal in 1940. Within another decade, largely voluntary use of the new fuels would become important enough to have dramatic impact on the urban smoke problem.[68]

Municipal smoke concerns seem to have had little impact on domestic energy choices, and, clearly, that represented a missed opportunity. Only St. Louis was willing to mandate that households reduce smoke either by using less-polluting fuels such as low-volatile coal, gas, or oil or by installing stokers. I calculate that as of 1940, in major cities of the North Central states (e.g., Chicago), the costs of mandating the cheapest cleaner fuel—low-volatile coal—would have been modest: about $2.69 per ton of coal. With middle-class consumers burning perhaps 10–12 tons of coal a year, the annual cost would have been perhaps $29.59, which amounted to roughly 2 percent of median incomes in cities such as Minneapolis and Cleveland. Poorer stove users would probably have burned about seven tons, costing about $18.60 a year. Such a policy would have sped the transition to cleaner fuels; the failure of politicians to mandate cleaner fuels suggests that they believed the cost exceeded citizens' willingness to pay for cleaner air.[69]

Conclusion

There was much talk in the late 1930s of how coal might recapture lost domestic sales, but even enthusiasts must have had their doubts. Stokers might modernize a few homes, but more people bought oil or gas conversion burners, and most new construction chose gas or oil. Moreover, the battle was hardly joined. Oil heat was only 15 years old, while gas was just starting to shed its regional skin. After booming during World War II, by 1955, retail coal deliveries were down to 54 million tons; by 1960, they had declined to 30 million tons, and anthracite was all but extinct.

The spread of oil and gas heat was not the inevitable result of their abundance: neither oil nor gas leaped from wells to furnaces. Rather, their increasing use was an outcome of the sort of capitalist development that Schumpeter lionized. Even crops grown from the best seed require a good deal of tending, and the arrival of gas and oil in America's basements and parlors required innovations in oil burners and pipe and pipeline construction, along with creative marketing campaigns to close the deal. Initially, the market was the upper crust; eventually, it would be nearly all of us. For soft coal, loss of domestic markets was a serious blow; for hard coal, it destroyed an industry and a way of life.

Heating and cooking with gas and oil were energy-saving innovations, for both fuels burned with considerably greater efficiency than did wood or coal, reducing domestic energy use per person and air pollution. And they were welfare-enhancing innovations as well. In one of the Peoples Gas advertisements, Joseph Cepek happily contemplated a future without coal and ashes to shovel. There is no reason to think this portrayal was not genuine, while the depictions of women who enjoyed the cleanliness of oil heat were surely accurate. Here again, as with gas and kerosene cookstoves, the comfort, convenience, and cleanliness of oil and gas heat surely increased households' welfare in ways not captured by statistics on the standard of living. Such are the small building blocks that form the foundations of modern affluence.

Part IV

COUNTERATTACK

8

Coal Fights Back

Machines, Markets, and Research, 1880–1945

Stripping methods [of mining coal] furnish complete conservation.

—*Eli Conner, "Anthracite and Bituminous Mining, 1911"*

A review of the world-wide technical developments ... shows little tangible progress in providing new markets for coal.

—*Arno Fieldner, "Recent Developments in Coal Utilization," 1933*

As coal markets crumbled during the interwar years under the combined assault of conservation and superior fuels, the industry and its allies battled back in a number of ways. As we saw in the previous chapter, some producers responded by upgrading their marketing and promoting stokers for domestic markets. There were political efforts to cartelize coal, in an effort to raise profitability, but little came of them. This chapter focuses on two other fronts on which coal fought back. The first is mechanization as manifested in the rise of strip mining. I ignore underground mechanization because the emphasis here is on coal's competitive responses to market losses, and even when mechanized, underground mining was less competitive than was stripping.

The second topic of this chapter is the pursuit of research to improve coal's competitive position. Mechanization and research represented two very different solutions to coal's market problems. Mechanization, whether underground or via stripping, was a cost-reducing strategy; it aimed to make coal more price-competitive in existing markets. Ultimately, mechanization, and especially stripping, would transform mining, but it could not transform coal. The aim of much coal research was also simply to make it more competitive, but this chapter emphasizes efforts to develop new markets for coal by transforming it into a source for liquid fuels. Previous chapters have explored the increasing role that scientists have played in midwifing energy transitions from wood to coal to oil and gas. Efforts by scientists to develop economic liquid fuels from coal would have reversed these transitions, but here science failed. The new liquid fuels failed to pass a market test, remaining but a glint in the eye of policy makers. Coal was like a pig that scientists could make run faster, but they could not transform it into a race horse.

The Rise of Strip Mining

While American strip mining of coal employing power equipment began around 1870, the technique only became important about the time of World War I. There were and are several common types of strip mining. The most widely employed during these years was area mining, which worked well in comparatively flat parts of the Midwest. Circa 1914, a company using this approach might blast the overburden—with either dynamite or black powder—if necessary. Typically, a large power shovel or dragline (or both) would then proceed across the coal, stripping the overburden and dumping it in a ridge in a previously worked area directly or employing a conveyer to do so. Workers would then blast the exposed coal and load it into railroad cars that would haul it to market without cleaning or sizing, although early on, small steam shovels were beginning to replace hand loading, When the end of the seam or property had been reached, the process worked backward, depositing the spoil in the previously mined area.[1]

Comparative Advantages of Stripping

Strip mining possessed certain inherent advantages over underground work, which contributed to its relative profitability. Most obviously, there was no need to support a roof, and thus no need for the vast amounts of timber consumed by underground operations. Strip mining, that is, conserved forests. Stripping also required no ventilation. Machinery, including mine cars and locomotives, might be much larger as well. Companies could open strip mines comparatively quickly, and because much of the capital investment was mobile, it had a higher salvage value than that of underground mines. The net effect of these and other advantages was that labor productivity in stripping was higher than at underground mines. In bituminous mining in 1915, output per worker-day in underground work was fewer than four tons, while in stripping, it was six tons (fig. 8.1)—a

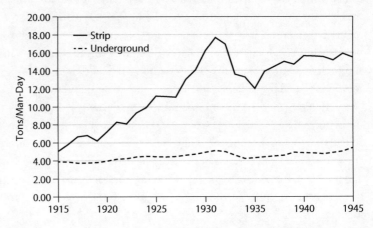

Figure 8.1. Output per man-day in strip and underground bituminous-coal mining, 1915–1945. (Chart data drawn from USBLS, "Technological Change and Productivity in the Bituminous Coal Industry, 1920–1960," *Bulletin* 1305 [Washington, 1961], table 8A; and USGS/USBM, MR and MY, various years.)

50 percent advantage—and thereafter, its advantage widened. Stripping operations could undersell coal mined below ground, still turn a profit, and expand.[2]

Strip Mining as Coal Conservation

Modern writers emphasize the environmental damages of stripping, and while Progressives of Theodore Roosevelt's day shared such concerns, they worried more over the waste and exhaustion of coal reserves. Stripping needs to be understood against this backdrop, for it provided an important market response to such worries. For some contemporaries, stripping was a form of resource conservation because it sacrificed low-quality farmland for higher-value coal, which seemed an acceptable trade-off between different resources. In the phrase of the time, it was "wise use," not an affront to the environment.[3]

In his address to the Governors' Conference in 1908, Joseph Holmes had characterized "necessary waste" as coal that was lost because it was uneconomic to mine under current conditions. He stressed that the loss of life in coal mining was even more serious than the waste of resources. Yet even as he spoke, strip mining was making economic some of what had once been "necessary waste," and it was beginning to save lives as well. Stripping thus provides another example of the way that America's resource abundance was socially created, for by making available coal that could not be mined by other methods, in an economic sense, stripping "created" resources.[4]

Eli Conner (1864–1938) was a mining engineer who claimed that his father had introduced steam-shovel mining for anthracite in 1881. In 1911, the son rehearsed the by-then standard indictment of America's wasteful mining techniques but pointed out: "stripping methods furnish complete conservation." Stripping conserved coal in two ways. One was because the steam shovel allowed mining of coal too shallow for underground methods. On the eve of World War I, *Coal Age* reported on stripping operations in Pennsylvania, Ohio, and Kansas, where "the extraction of coal . . . could not be secured by any other means." A writer in 1920 reported stripping where "the cover is so shallow and the roof 'so rotten'" that underground mining was impossible.[5]

A second conservation advantage was that strippers could rework fields previously mined by underground methods that had left coal either because it was too shallow or because it was needed for pillars or roof. In 1916, *Coal Age* reported a large underground anthracite mine near Hazleton, Pennsylvania, that had been abandoned as worked-out but that when reworked by strippers was expected to yield "many thousands of tons of coal." In the 1920s, some anthracite strippers were mining the same bed for the third time; there had been an underground working and then a previous stripping, but with better equipment, a second stripping extracted even more coal. A 1929 headline in *Coal Age* read, "Stripping Completes Recovery of Deep and Already-Mined Beds." In the late 1930s, Illinois stripping companies reported working much land in the southern part of the state, where underground mining had long since been abandoned as unsafe and unprofitable. "As a natural resource conservation measure," the strippers crowed, "coal stripping is the ideal."[6]

The net effect of these advantages was that stripping generated much higher recovery rates than did underground mining. In Illinois, in the late 1920s, stripping obtained 90 to 98 percent of the coal; by contrast, at about the same time, the recovery rate for underground work was about 50 percent. Such conservation differences shaped views on stripping throughout these years. As late as 1947, the Bureau of Mines was claiming that "strip mining . . . recovers more coal . . . than underground mining and therefore is an important factor in the conservation of this valuable natural resource." To repeat: when resources are viewed in economic terms, stripping "created" coal.[7]

Yet if stripping made more coal available, in the process, it "absolutely destroys the land for farming purposes," a writer in *Coal Age* bluntly asserted in 1916. There were occasional dissenters, such as E. C. Drum, who informed *Coal Age* readers in 1917 that "disturbance of the surface improves the land for agricultural purposes." A later writer noted that when stripping turned over the soil near Frontenac, Kansas, it replaced clay with shale, thereby improving its fertility. If such improvements were probably rare, neither was there much criticism of the effects of stripping on the landscape during these years. A 1916 story in the *Kansas City Star* did report on the destruction of farmland occasioned by stripping in that state, noting that "twelve hundred acres of prairie and farm land has been turned upside down." There was criticism of stripping, the paper pointed out, but it came from unions worried about job loss from more productive, lower-cost competitors and from real estate dealers, bankers, merchants, and others who worried about property values. In the 1930s, some Illinois farmers worried that land that had been stripped would yield the community less in taxes, while underground mineworkers supported land reclamation as a way to raise the costs of stripping and reduce its competitive advantage. Thus, the critique of stripping was mostly economic, not environmental, but the *Star* also pointed out that the farmers in question had sold the land that was otherwise worth $20/acre for $500 to $1,200 an acre. A search of Illinois and Indiana farm papers yields very little criticism of stripping before World War II. In the 1920s and 1930s, with much agriculture in depression and farmland depreciating, farmers with coal underlying their lands could not afford the broader concerns that have motivated more well-to-do, urban environmentalists in the postwar years.

For many, strip mining offered an acceptable trade-off. Mining engineers pointed out that most stripping worked low-value land and that the coal recovered was of far greater value than the farm production lost. Market forces were, as they saw matters, generating an efficient result. As one writer put it, this, too, was "real conservation," for "the world needs to be warmed as well as fed." All the same, many contemporaries wished to minimize the environmental costs of stripping. To mining companies, leveling the land was unthinkable because that would have essentially doubled excavation costs, but as the *Kansas City Star* noted, some other form of economic reclamation was being talked about. In the 1920s and 1930s, a number of states saw efforts to regulate stripping and legislate reclamation, though little seems to have come of them.[8]

Beginning in the late 1920s, however, coal companies, worried over such potential legislation, began reforestation and other efforts to make strip-mined land again economically productive. By 1942, Illinois strippers had formed an association to take care of reclamation, and working with state agencies, had reforested 7,250 of the roughly 16,000 acres that had been stripped, planting about seven million trees. In Indiana, a strippers' association planted about 3.4 million trees on reclaimed land from 1934 through 1938 and similar efforts were under way in Ohio. In Pennsylvania, market forces were moving *underground* coal companies to begin reforestation because they had stripped local forests bare, using the wood for roof support.[9]

In the late 1930s, the Bureau of Mines studied reclamation in Illinois, Indiana, Kansas, and Missouri and was cautiously optimistic. It found stripped land that had lain idle for several decades where natural vegetation had come back strongly and, in one instance, was growing trees 12–24 inches in diameter. Some of the land was being leased for hunting and (because the last cut of a strip usually filled with water), fishing; some had become a state park. The bureau was skeptical that reclamation for farming would pay, but it reported successful reforestation efforts in all the states it studied. The early postwar years generated much research into these early efforts that found considerable success in reclamation for ranching, as well as forestry.[10]

Economics and Technology of Stripping

As I have noted, power strip-mining for coal became important on the eve of the First World War. The keys to its expansion lay in the interaction of geology, technology, and economics. Technology and the price of coal determine the maximum amount of overburden that a company can remove per ton of coal and still turn a profit:

$$\frac{Overburden(yd.)}{Coal(ton.)} = \frac{\dfrac{Price}{ton} - \dfrac{Profit}{ton} - \dfrac{Mining\ cost}{ton}}{\dfrac{Excavation\ cost}{yd.}}.$$

The size of this maximum will, in turn increase with the price of coal and the productivity of stripping equipment. For example, using the formula above, suppose that the mine price of coal is $2.00/ton. Mining costs are $0.75/ton, the target profit is $0.25/ton, and the cost of removing a yard of overburden is $0.20. The maximum amount of overburden per ton of coal that it will pay to remove is five yards, which would increase if excavation costs were to fall or the price of coal rise.[11]

Nineteenth-century stripping equipment was simply not sufficiently large or efficient to make very much stripping economic. That is, with excavation costs per yard high (and coal cheap), the ratio of overburden to coal was necessarily low. When stripping began in the 1880s the economic ratio of overburden to coal was about 1:1; hence, little coal could be stripped. Early power equipment included dredges, drags, and, most often, shovels. These latter were small by later standards: those first employed in anthracite stripping had a 1.75-yard bucket. They were made at least partly of wood and often not sturdy enough for stripping. All early equipment was steam powered—which meant that it required a coal and water supply and a fireman. Early railroad shovels ran on tracks and could not revolve a full 360°; both characteristics slowed down production and raised costs. Such equipment could only mine shallow coal, and beds close to the surface are likely to be of poorer quality, weathered, and contain reduced energy content. Heavier equipment allowed deeper coal, and, initially at least, deeper coal meant better coal.[12]

In the early twentieth century, stripping technology improved rapidly and dramatically. The first fully satisfactory stripping shovel did not arrive until 1911, when a mining company approached the Marion Steam Shovel Company with a proposal for a heavier, self-propelled, revolving shovel with longer range. Marion responded with its Model 250 with a 65' boom, a 40' dipper stick, and a 3.5-yard bucket. It was fully revolving and had a hydraulic leveling system. Bucyrus followed the next year with similar equipment, and the two companies' rivalry and close association with buyers would spur rapid technological change in stripping equipment from then on. A Marion catalog explained how closely it worked with buyers. "The builder is seldom called upon to exactly duplicate a former machine," the company noted, and it stressed that "our engineering department is at all times available for the adaptation of our machinery to special requirements." Just as later innovations in pipeline technology would revolutionize natural gas transport, so makers of mining and construction equipment midwifed modern strip mining.[13]

The USGS first took official note of what it termed "steam shovel mining" in 1914, doing a partial survey. In 1915, it found such operations digging soft coal in Alabama, Illinois, Indiana, Kansas, Missouri, Ohio, and Oklahoma, where a total of 87 shovels and 2,300 workers produced 2.8 million tons of coal—about 0.6 percent of that year's total. In 1914, the USGS reported that the average ratio of overburden to coal was about six, but by 1930, it had risen to 7.6 and above in some states where coal prices were higher. *Coal Age* thought these estimates were conservative, claiming in 1929 that better technology had raised the economic stripping ratio to 15:1.[14]

Like most new technologies, early stripping had a number of teething problems; whereas underground mines operated 238 days in 1914, the strippers averaged only 187 days—a result of equipment failures and bad weather, for strip pits had a propensity to flood. Yet economic and technical pressures encouraged improvements in drainage and led to other changes that reduced the impact of bad weather on operations. By the late 1920s, strip pits operated about as many days as did underground mines. Initially, dirty coal was also a problem. Some strip mines had no tipple, so uncleaned coal was simply loaded into freight cars to be sold. In 1920 *Coal Age* noted that poor quality was giving stripping a bad name. Companies responded; after removing top cover, they began to use tractors and power sweeps to remove dirt that remained, and by the late 1920s, most companies shipped coal to a tipple for cleaning and sizing. Because large shovels increased the amount of coal that could be economically mined from a pit, they reduced the annual fixed costs of a tipple, thus making cleaning and sizing more economic. By the late 1920s, coal washing was beginning to spread, and the ability to clean small coal increased its value, thereby widening its market and encouraging recovery.[15]

Table 8.1 and figure 8.2 provide an overview of the explosive expansion of bituminous stripping during these years. From beginnings in six midwestern states with 87 shovels at work in 1915, by 1945, stripping had moved into 25 states with more than 3,400 shovels. During these years, production spread from the Midwest, east into Pennsylvania, Kentucky, and West Virginia and west to Montana, South Dakota, and other states. Productivity rose from not quite six tons per worker-day in 1915 to more than 15 by 1945. Output over that period rose 27 times—an annual rate of not quite 16 percent per year.

Changes in the costs of labor, materials, and land, along with improvements in product quality, also aided strip mining during the 1920s. Timber costs were twice as high as in prewar years, while, as I have noted, miners'

Table 8.1. Bituminous Coal Stripping, 1915–1945

	1915	1925	1935	1945
No. of states	6	18	19	25
Pits	—	227	368	1,370
Shovels	87	144	497	3,439
Coal (M tons)	2,832	19,871	23,647	109,987
Percent of US total	0.6	3.2	6.4	18.9
Employment	—	8,609	10,484	33,569

Source: USGS/USBM, MR and MY, various years and author's calculations.

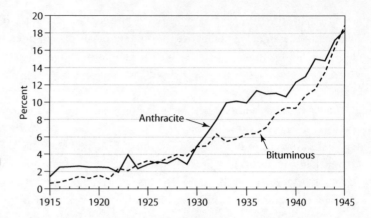

Figure 8.2. Percent of coal produced by strip mining, 1915–1945. (Chart data drawn from USGS/USBM, MR and MY, various years)

wages rose relative to prices. Both developments handicapped underground work, which was both timber and labor intensive. Meanwhile, the agricultural depression that settled over the Midwest depressed farmland prices, cheapening the costs of land acquisition to coal-stripping companies.[16]

Stripping thus evolved from a technique that might mine coal unavailable with underground methods to an approach that was economically preferable under an ever-widening range of circumstances. Accordingly, the market share of stripping for both hard and soft coal reached 18 percent by the end of the war and continued to expand, reaching 55 percent for bituminous mining by 1975. By the twenty-first century, two-thirds of all coal came from surface mining. And while much of stripping's gain came at the expense of underground coal, as will be seen, it almost certainly saved coal markets that would otherwise have gone to oil.

World War I kicked off this expansion, for it generated an immense demand for coal. As the USGS observed, the spread of stripping reflected "the dispatch with which a property suitable for this method of mining can be opened up and equipped for large scale production." At a time of extreme labor shortage, stripping's conservation of labor also made it attractive. The trend toward giantism in shovels and draglines continued. Stripping shovel size and efficiency grew steadily; by the early 1920s, five-to-eight-yard buckets with 85-to-90-foot booms were common. By 1940, there were 35-yard stripping shovels, and some mines employed 7-to-10-yard loading shovels. Very large draglines also appeared—by World War II some had 215-foot booms and 14-yard buckets—sometimes employed in concert with power shovels and sometimes as substitutes. Draglines could move spoil longer distances and thus allow economic mining of deeper coal. By 1945, draglines were mining deeply pitching seams of anthracite down to 400 feet.[17]

Digging equipment became more sophisticated as well. In the 1920s, revolving shovels supplanted older types. Caterpillar traction first appeared on smaller coal loaders in the early 1920s, and by the end of the decade, it had become common on the largest stripping shovels. As commercial electric power spread in the 1920s, electric operation gradually replaced steam, accounting for 35 percent of all shovels by 1931. Electric shovels could swing faster than steam and experienced far less expensive downtime; one company found them operational 90 percent of the time versus 68 percent for steam. In addition, electric operation required three operators, while steam required five. Use of aluminum and alloy steel in buckets and booms arrived in the mid-1930s and since their lighter weight allowed bigger loads, companies switched out 10-yard for 12-to-15-yard buckets. In the 1940s, bucket-wheel excavators appeared that could rapidly remove cover that was not too compacted. Around 1940, shovels with knee-action booms that increased range also appeared. Photoelectric leveling of equipment arrived in the 1940s, as did more sophisticated controls that sped up movement.[18]

The late 1920s saw the beginning of a revolution in haulage as well, as trucks began to supplant railroads, and companies switching to truck haulage reported cutting costs by 50 percent or more. For example, in 1938, one Indiana mine reported trucking reduced its costs of hauling coal from 16 cents to 7 cents a ton. Trucks were more flexible as well; they could climb steeper grades and navigate sharper curves. In haulage, as in digging, equipment size steadily rose. In 1932, one company began haulage with six-ton trucks; by 1939 their size had risen to 80 tons.[19]

The increasing size of equipment reshaped stripping in complex ways. By the 1930s, stripping was far more capital intensive than underground work, with $380 of machinery and equipment and 26 horsepower per worker versus $63 and 8.5 horsepower for deep mining. Stripping was, therefore a capital-intensive technique that saved both labor and coal. This heavy equipment allowed economic stripping of increasingly heavy cover, and by the late 1930s some pits removed 17–20 yards of spoil per ton of coal. Companies learned to employ double stripping of two separate seams, thereby increasing recovery. The expense of the giant shovels led to their operation two or three shifts a day. Companies employed time and motion studies to increase swing efficiency. One company estimated that with coal selling for $2.50/ton, every swing the shovel failed to make cost it $2.50. Such calculations help explain why stripping spread at the expense of underground mining, even as it, too, experienced rapid mechanization.[20]

Although most stripping operations employed ever-larger equipment, some small coal beds could not be economically mined with the giant shovels. In the late 1930s, construction companies with equipment idled by the Depression moved into stripping. While such operations might be comparatively high-cost, lighter cover, better coal, or market access might provide an offset. The rise in such smaller operations probably accounts for the slowdown in productivity growth during the 1930s depicted in figure 8.1. The wartime boom after 1940 resulted in an explosion of small operations as it shut off civilian construction and idled many contractors. These developments also moved stripping into eastern Ohio, western Pennsylvania, Kentucky, and West Virginia. Between 1940 and 1945, the number of strip pits in Pennsylvania increased from 107 to 542.[21]

Stripping continued to make available coal that would otherwise have been uneconomic to mine—reducing "necessary waste" in Holmes's language. This conclusion remained valid at least through the war years. Smaller contractors stripped coal close to the surface, while reworking older beds remained common. In 1944, *Coal Age* reported a large strip mine where 73 percent of the coal could not be recovered by underground methods. Of the coal that might have been mined with underground techniques, stripping generated a 30 percent higher yield.[22]

Although underground mines began to respond to these competitive threats in the 1920s by mechanizing loading, stripping retained an enormous productivity advantage (see fig. 8.1)—an edge that the labor shortages of World War II sharpened. As the USGS put it: "a smaller number of men is required for a given output in this method of mining."[23]

Figure 8.3 presents the labor productivity in stripping relative to underground mines along with the ratio of mine prices for stripping and underground mining. As can be seen, by about 1925, stripping's productivity advantage allowed it to underprice the underground competition. For such reasons, underground mining lost markets to strippers. Between 1914—the first year such data are available—and 1940, stripping output rose about 37 million tons. Over the same period, total coal production rose only 40 million tons. In 1930, the Washburn lignite mine in North Dakota that had been worked underground for decades converted to stripping, probably because they were able to cut the workforce from more than 270 to 70 employees. But where such conversion was uneconomic, many underground mines simply closed.[24]

Stripping protected at least some coal markets that would otherwise have been lost to oil or gas. That is, had stripped coal not replaced

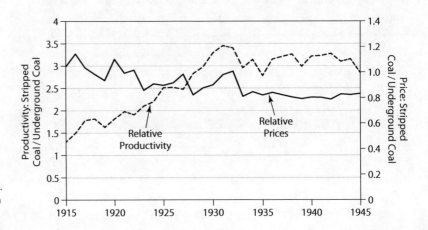

Figure 8.3. Relative labor productivity and relative prices of strip- and underground-mined coal, 1915–1945. (Chart data drawn from USBM, MY, 1946)

underground-mined coal, the latter might well have lost out to gas or oil. For example, when stripping of Texas lignite began in the mid-1920s, coal prices collapsed, and power plants that had previously switched to oil promptly switched back to lignite. Similarly, when the Northern Pacific Railroad discovered vast amounts of lignite on its North Dakota properties in 1923, it immediately began stripping and converting locomotives from oil to coal. In the Southwest, the spread of natural gas pipelines led companies to begin stripping as a defensive response.[25] Thus, one of the results of stripping was to limit the transition away from coal, much as fracking has limited the transition from oil and gas in more modern times. Writers on energy transitions have observed that it is often difficult to dislodge an existing system because, when pressed, it becomes more efficient, and strip mining of coal provides a historically important case study.

Stripping Coal, Saving Lives

When Progressives of Teddy Roosevelt's time spoke of the need to conserve resources, they meant not only coal and oil but human life as well. At the state level, this impulse propelled workers' compensation legislation, and it also encouraged the formation of the US Bureau of Mines. The bureau's greatest success was in reducing coal-mine explosions, but otherwise, underground mine safety showed little improvement in the interwar decades. Perhaps the brightest spot in the mine safety picture resulted not from federal regulation but from the market. Stripping saved more than labor-hours and coal: it saved lives as well, compared to underground mining.[26]

Early mining literature contains many references to the safety of strip mining in comparison to its underground cousin. Contemporaries believed that stripping was inherently safer than underground mining. There was no roof to collapse—a danger that typically accounted for half of all fatalities in underground mining. Gas and dust explosions were unheard of, and without the problems of confined spaces, electricity and haulage could be made safer, while there was less risk of silicosis and black lung disease. "For speed and safety . . . the stripping method is ideal," a writer in *Coal Age* informed readers in 1916. The next year, H. H. Stoek (1866–1923), professor of mining engineering at the University of Illinois, made the same point: "from the standpoint of safety . . . strip mining is always to be preferred," he noted. In 1926, it seemed self-evident to Frank Kneeland (dates unknown), editor of *Coal Age*, that stripping was the "safest form" of mining, and in 1928, the Bureau of Mines simply observed that "accidents and fatalities are low."[27]

The first data on strip mining fatalities appear in 1930, and they confirm that stripping was a far safer way to mine bituminous coal (table 8.2). From 1930 to 1945, workers' risks in bituminous coal mining averaged about 0.84 per million employee hours in stripping versus 1.45 in

Table 8.2. The Human Costs of Coal, 1930–1945

	Fatalities			
	Per million employee-hours		Per million tons of coal	
Year/Type of coal	Underground	Strip mining	Underground	Strip mining
1930				
Bituminous	1.89	0.98	4.15	0.50
Anthracite	1.76	1.66	6.52	3.15
1935				
Bituminous	1.43	1.22	2.74	0.59
Anthracite	1.79	1.60	5.58	1.42
1940				
Bituminous	1.65	0.99	2.82	0.52
Anthracite	1.54	0.82	3.97	0.81
1945				
Bituminous	1.16	0.58	1.90	0.35
Anthracite	1.08	0.11	3.16	0.10
Average 1930–1945				
Bituminous	1.45	0.84	2.63	0.44
Anthracite	1.64	0.92	4.92	0.92

Source: USBM, *Coal Mine Accidents in the United States,* various years and titles.

underground work, making the latter about 73 percent more dangerous. Similarly, in hard-coal mining, underground work was about 78 percent more dangerous than in stripping.[28]

Productivity and Safety

While fatalities per worker-hour measure the risks of death to employees, fatalities relative to coal production capture some of the human costs of producing that form of energy. Using this measure, as table 8.2 also reveals, stripping's advantage in mining bituminous coal was even more impressive. Again, over the whole period, there were about 2.63 fatalities per million tons of underground soft coal, while in stripping the rate was 0.44—about one-sixth as high. There were two reasons for such striking disparities in fatalities per ton: risks per worker were lower in stripping, and stripping required far fewer workers to mine a ton of coal.[29]

We can illustrate the safety payoff from stripping by constructing the following counterfactual. Suppose that the coal stripped each year had come instead from underground mines, with their associated higher fatality rates per ton. This represents the fatalities *avoided* because the coal came instead from stripping. This calculation (not shown) demonstrates that if stripped coal had instead come from underground mines at existing fatality rates per ton, there would have been about 1,400 more miners killed from 1930 to 1945. In short, compared to underground work, stripping traded land for lives. But coal and petroleum or natural gas are good substitutes, and oil-field work was even safer than strip mining. Thus, a long view suggests that as markets shift to new ways to mine coal and to newer forms of energy, and as labor's rising productivity requires fewer workers in such dangerous jobs, the net result is to improve overall work safety.

Research: The Science of Salvation

As we have seen in previous chapters, the 1920s and 1930s witnessed the blossoming of corporate research laboratories, along with rapid technical progress and a flowering of new products and processes such as radio, petroleum cracking, oil burners, and nylon. Although coal producers funded comparatively little research, studies by users of coal, along with investigations by the industry's allies in supplier industries and governments, accomplished much. By the early 1920s, research had already yielded coal stokers and pulverization, which improved combustion efficiency, protecting markets from rival fuels. Accordingly, coal's advocates increasingly saw salvation in the laboratory. "Research Not Rhetoric," *Coal Age* urged in 1927. Salvation, they hoped, might come in two forms. New technologies might protect and expand demand in existing markets, as had the development of pulverized coal. Advocates also hoped that research might yield dramatic, transformative innovations that would create entirely new markets, allowing coal to turn the tables on oil and gas.[30]

Policy makers and others less closely connected to coal mining also began to urge coal research, for the loss of markets to oil and gas worried them as well. World War I alerted officials to the military importance of oil to fuel trucks, tanks, airplanes, and naval vessels, thereby creating a public interest in oil conservation and shifting the focus of experts' resource pessimism from coal to the newer fuels, which they believed would run out far more quickly. They saw the rise of natural gas and oil and the decline of coal as a short-term aberration that would soon be reversed rather than the beginnings of a long-term energy transition, and they feared that the nation might soon need to turn again to high-quality coal. Such worries had led President Coolidge to create the Federal Oil Conservation Board, which appeared in 1924.[31]

As I noted in chapter 3, the board provided a gloomy assessment of oil's future. The board investigated the prospect of obtaining oil from coal or shale ("synfuels") and raised the possibility of restrictions on demand that would confine petroleum to motor fuels and lubricants. The board lingered on until 1932; its importance was not for what it accomplished, which was little, but as a reflection of policy makers' changing ideas on resource conservation. Throughout this book, I have examined the ways that market incentives encouraged companies to avoid being locked into a particular fuel. Thus, if there was any danger of being locked into coal in the 1930s, it stemmed not from technology or markets but from policy makers.[32]

By the mid-1920s, coal research had a broad constituency that included not only those associated with the industry but other energy experts, as well. For the industry, propping up coal markets was an end in itself; for government officials, however, the purpose of coal research was to expand its use, thereby preserving oil. The research during these years broadly divides into investigations that might help protect or expand already existing markets and investigations that held the promise of promoting entirely new markets for coal or its byproducts. Although researchers pursued both goals throughout the interwar decades, in the interests of clarity the following sections discuss them sequentially.

Expanding Traditional Markets

Both types of research were in evidence at the (First) International Conference on Bituminous Coal held in Pittsburgh in 1926 under the auspices of the Carnegie Institute of Technology. Two subsequent conferences followed in 1928 and 1931, and all were the brainchild of Thomas S. Baker (1871–1939), president of Carnegie. Baker seemed an unlikely advocate of coal—his advanced training was in German language and literature—yet the conferences would attract worldwide attention, and he would transform Carnegie into a major center of coal research, inaugurating a coal research laboratory in 1930.[33]

Coal Age devoted nearly an entire issue to the conference, describing it as a "meeting of far more than ordinary importance." Baker gave the opening address, hymning the virtues of research. The conference, he forecast, "will illustrate in a most striking manner what can be accomplished when the principles of pure science are applied to practical questions." Baker also noted the "highly unsatisfactory condition" of the coal industry and urged mine owners to seek advice from the scientists.[34]

Several speakers linked coal's short-run problems with experts' longer-run concerns over conservation, with research being the connecting thread. A paper by Marius Campbell of the USGS took the long-run perspective. Campbell was reporting on the first major upgrade of estimates of coal reserves undertaken by that agency in two decades. For the first time, they were able to break down reserve totals by quality, and he concluded gloomily that "exhaustion [of our best coal] may occur in the not very distant future." Several years later, he would be more specific, claiming that "perhaps within 50 years, much of the high-rank coal will be exhausted." John Hays Hammond, late of the US Coal Commission, and Frederick Tryon (1892–1940) of the Bureau of Mines argued that fuels were becoming increasingly interchangeable, which accounted for the languishing of coal, but they noted that the "supply of oil is . . . exceedingly limited compared with coal." Coal research was the key to both problems, they thought, for it might make coal an increasingly valuable substitute for oil. Conservation of coal would not only economize on that resource; it would make coal more competitive, thereby protecting coal markets and thus saving gas and oil as well.[35]

The conference saw many papers on the potential for expanding coal's traditional markets—which would, of course, also save oil—by making coal smokeless or improving its pulverization. There were several papers on recovering byproducts and expanding use of chemicals derived from coal such as nitrogen for fertilizer. Another weapon to defend coal markets was the improvement of coal quality by washing and sizing. While both processes dated from the nineteenth century, their large-scale employment for soft coal in the United States was comparatively recent. Market competition was propelling producers to clean and size coal. Utilities and other large companies were becoming increasingly sophisticated buyers, as research and engineering experimentation discovered the economic costs of impurities and improper sizing, while households that bought stokers also wanted cleaned and properly sized coal.

Coal cleaning and sizing were technically and economically complicated, requiring a good deal of experimentation and engineering investigation. Initially, there were only water-based processes; pneumatic cleaning arrived in the early 1920s. Each process came in a number of variations, and many—such as float and sink methods that made use of the different specific gravities of coal and impurities—had been developed by metal mines for ore processing. To clean and size coal, companies needed to know its physical and chemical properties and its intended use, as well as the costs and payoff to the process. Coals differed markedly in chemical composition, and they varied as to washability, friability, and other physical characteristics. Small sizes, for example, were more expensive to clean. Similarly, buyers' demands for cleanliness, chemistry, and sizing differed. Buyers of steam coal wanted exact size mixes, while in coke making, some impurities were more injurious than others. Impurities cost buyers more the farther they had to be shipped; accordingly, longer shipments required greater cleaning, or to put the matter another way, greater cleaning widened companies' geographic markets. Such complexities ensured that among large producers, experimentation and testing became routine.[36]

Finally, efficient cleaning and sizing required the selection of proper techniques for any given coal; wet methods were preferable for smaller sizes. As one expert described the best process of coal cleaning as "the one particularly suited to the kind of coal to be washed, the nature of the impurities . . . the final distribution of the fuel and the cost of . . . cleaning."[37]

In the late 1920s, the share of soft coal that was cleaned rose dramatically—from about 5 percent in 1927 to 22 percent in 1940. Undoubtedly, it slowed coal's loss of markets and perhaps won some back. Yet developments in oil markets were undermining coal's efforts to protect or enhance already existing markets, for after 1926, oil prices fell sharply, the result of large-scale new finds that expanded supplies. This did nothing to erode the experts' pessimism over long-run supplies: in 1930, the

Federal Oil Conservation Board was still warning that "even the most generous estimates place the date at which our oil reserves will be practically depleted . . . in the comparatively near future." But to the coal industry, which needed immediate markets, it dashed any hopes that incremental improvements could save the day.[38]

Creating New Markets for Coal

The investigations that seemed to show the most potential for developing new uses for coal involved liquefaction. In the first conference, the German chemist Friedrich Bergius (1884–1949) described his investigations into hydrogenation of coal to produce oil—a process for which he would receive the Nobel Prize in 1931. Franz Fischer (1877–1947) also presented a paper on synthesizing oil from coal, foreshadowing what became the Fischer-Tropsch process. Yet at the time of the first conference, both processes were in the early stages of development.

Most of the interest in obtaining liquids from coal at the 1926 conference, however, centered on low-temperature carbonization. In 1906–1907, research on low-temperature carbonization in England by Thomas Parker (1843–1915) and in the United States by Samuel Parr (1857–1931) had originally focused on the possibility that it might produce a more easily combustible coke ("artificial anthracite"), although it also yielded more tar and liquids than did the high-temperature process. In response to the booming demand for byproducts induced by World War I, combined with the worries over oil and gas shortages, researchers dusted off this earlier research but with a new focus on the liquids.[39]

Should low-temperature carbonization become commercially viable, it would thoroughly disrupt energy markets. Gasoline was, then, the most rapidly expanding market for energy; the ability to extract liquids from coal cheaply might become the industry's salvation. And to conservationists, it seemed to open the door to a paradise of efficiency. Since the process could use low-ranked coal, they dreamed of a day when virtually all coal would be carbonized with the coke burned, rather than coal, and when liquids would be refined into a host of fuels, dyes, explosives, medicines, and so on, thereby reducing pollution, rejuvenating languishing coal markets, and saving precious petroleum. Among these happy results, liquid fuels stood out as the pot of gold at the end of the rainbow. Europeans found low temperature carbonization especially attractive, for war had highlighted the national security vulnerabilities of nations without oil. The result was a worldwide blizzard of research extending well into the late 1920s.

The Bureau of Mines kept abreast of these developments, and in 1924, it sent Arno Fieldner (1881–1966) to survey the state of low temperature carbonization in Europe. Fieldner had graduated from Ohio State in 1906 with a bachelor's degree in chemical engineering. He joined the USGS and then the Bureau of Mines as a chemist shortly thereafter, where he had a long and distinguished career, receiving many awards, accepting two honorary doctorates, and becoming president of the ASTM in 1937. He was an expert on coal utilization and conservation. In Europe, he visited British and German plants and returned in 1926 only to pour cold water on enthusiasts' schemes. Fieldner thought commercialization was unlikely. Nor would the process save much petroleum: he estimated that a ton of coal might yield six gallons of fuel; carbonizing 100 million tons of coal would generate the equivalent of only 8 percent of that year's motor fuel consumption.[40]

Fieldner's report fell on deaf ears; researchers ignored his warning, and the conference revealed work on many different approaches to low-temperature carbonization in England, Germany, and the United States. European research involved governments, as well as private companies, and was protected by tariffs on imported petroleum products. In the United States, the ventures were private, undertaken by a few large American coal companies, including Consolidated Coal and Old Ben Coal, as well as various independent researchers, but none of the American ventures were yet commercial.

In 1928, Baker tried again with a second international conference on bituminous coal. In his welcoming address, Baker was still hymning the virtues of conservation: "we all recognize the propriety and wisdom of every effort that is made to prolong the life of our stock of raw material," he observed. Yet he acknowledged that because of "the advance in our knowledge of coal, the mining industry is in a languishing condition." He again urged a partnership between scientists and the industry because the conference "would illustrate . . . what can be accomplished when the principles of pure science are applied to practical questions." Further research, he urged, "may find new uses for . . . coal, [because] the same sort of ability that has narrowed . . . markets can discover the means of enlarging them."[41]

Again *Coal Age* prominently featured the conference with an upbeat title borrowed from Baker: "The Ability That Narrowed the Coal Market Will Expand It." The journal editorialized: "there is no escaping the conclusion that coal as a source of raw material for other industries must grow in importance." It suggested that the industry might integrate into the production of products derived from coal similar to the way oil refiners had become petrochemical companies.[42]

Frederick Tryon and H. O. Rogers (dates unknown) of the Bureau of Mines buttressed Baker's claim that conservation could save the coal industry, quoting Stanley Jevons that "it is a confusion of ideas to suppose that the economical use of fuel is equivalent to a diminished consumption." This prophecy had held true for 50 years after Jevons uttered it in 1865. Tryon and Rogers asked, "Is it not conceivable that it will be found true again?" The conference revealed considerable research devoted to enhancing coal's competitive position in traditional uses. There were papers on pulverization and use of stokers and on deriving nitrogen for fertilizer as a byproduct of coke manufacture—although for this to expand coal demand would have required that the byproduct become the product. A paper by Arthur D. Little (1863–1935) and R. V. Kleinschmidt (dates unknown) provided a skeptical assessment of such conventional efforts to save coal's fortunes. They agreed that conservation might expand demand for coal as Jevons claimed, but to do so it would need to create "new uses for . . . [the industry's] product."[43]

By this time, the industry was beginning to show interest in research. A few large coal companies maintained testing laboratories, mostly to accommodate the needs of coal washing. Hard-coal producers formed the Anthracite Institute in 1927 to coordinate responses to competitive threats. In late 1928 and 1929, it held a research conference and soon contracted with Frost Research Laboratory, which undertook investigations aimed at discovering new markets for anthracite, but their findings showed little promise.[44]

The leading candidate to create new markets for coal still seemed to be some version of low-temperature carbonization, which was the subject of more papers than any other topic at the 1928 conference. There were almost an infinite number of ways to modify the process, involving temperature, pressure, type of coal, and its physical state, among other considerations. Again, British and German research continued, while the Japanese government had begun investigations in 1921. The American Gas and Electric Company had entered the field with the "Carbocite" process, and the International Coal Carbonization Company was building a commercial plant in New Brunswick, New Jersey. The International Bituminoil Corporation also seemed well on its way to commercialization. Putting on his rose-tinted glasses, one writer optimistically summed up these developments: "low temperature distillation . . . offers much promise."[45]

Fading Hopes

When the third—and last—coal conference convened in 1931, attendees were still waiting for research to come to the rescue as the Great Depression further eroded coal demand, even as it caused an utter collapse in the price of oil. Again there were papers on coal washing and sizing, on pulverization, on small stokers, and on ways to prevent smoke, none of which showed much promise to expand coal markets. The conference also saw papers on market stabilization (cartelization). The industry, of course, saw cartelization as a way to restore profits, while conservationists imagined that it would result in higher recovery rates—as they had since the time of Joseph Holmes. The idea was that higher prices resulting from market restrictions would encourage production of coal currently too expensive to mine, yet this makes little sense given that higher prices would also restrict sales, so companies would still mine cheaper coal first.[46]

By the time of the third conference in 1931, none of the American ventures in low temperature carbonization had panned out, and Americans had largely abandoned the field. The problem was simple: as kerosene manufacturers had discovered about 1860, the products that could be derived from the process could also be created more cheaply using oil or some other feedstock or process. As Fieldner had concluded in 1926, "the time when such a process can be profitably worked will be determined by the exhaustion of our present abundant supply of petroleum."[47] For the coal industry, needing immediate help, low-temperature carbonization was a dead end. Although that process continued to attract attention in Europe and Japan, only a few companies in the United States pursued it, and they focused on production of coke rather than liquids. One American researcher reviewed the many investigations, concluding that Fanny Brice had sung their epitaph in "Rose of Washington Square": "I ain't got no future, but oh, what a past."[48]

Coal research continued throughout the 1930s, and it remained largely dominated by the Bureau of Mines and various state agencies and universities. In the early 1930s the Anthracite Institute created its own research division, and it cooperated closely with scientists at Penn State. About the same time, the National Coal Association formed a research committee, but that seems to have accomplished little. In 1931, *Coal Age* published the first of several annual reviews of coal research, listing 81 projects at 29 institutions, virtually all of which were government or educational establishments. Some utilities and railroads also did coal research, as did a few private universities in coal-mining states. Some of this was parochial, as states tried to develop markets for the particular coals found within their boundaries. In late 1934, the National Coal Association finally formed Bituminous Coal Research Inc. The next year, the organization contracted with the Battelle Institute for research on domestic stokers and with Penn State University to study hydrogenation—programs that ended in 1938.[49]

In 1937, a number of coal companies and users signed a four-year $350,000 research contract with Carnegie Coal Research Laboratory. There, and elsewhere, there were studies of air pollution, briquetting, combustion, coal properties and preparation, mine safety and ventilation, uses for fly ash, and the value of coal as a fertilizer. Much of this work was valuable; finding new uses for coal ash might not expand coal markets, but it could make a waste product marketable. Studies of combustion and coal properties and preparation all improved the efficiency of coal use by helping match particular coals to buyers' needs. Yet none of these topics had the potential to salvage coal's fortunes. Even research that yielded new coal-based products held out little hope. In 1940, *Coal Age* noted that coal had been the feedstock for the recent development of nylon and Lucite by DuPont, along with Goodrich's Koroseal. Yet these and other synthetics would soon be made more cheaply from petroleum. By about this time, production of coal-based organic chemicals amounted to but 300 million pounds—a mere lump in the coal scuttle. By contrast, from 1921 to 1939, organics from other sources, mostly oil and natural gas, had risen from 21 million to three billion pounds. Coal gave birth to many of the new synthetics, as it had given birth to kerosene 80 years earlier, but like kerosene, they grew to maturity on a diet of petroleum.[50]

Only one strand of inquiry continued to hold out promise that it could transform coal's fortunes. At the first bituminous coal conference in 1926, Friedrich Bergius had described his work on the liquefaction of coal by hydrogenation. Bergius had begun his efforts in 1913. He found that under pressure and at high temperature in the presence of a catalyst, coal could be made to take up hydrogen and could be converted to liquids and gases. Using similar processes, Bergius also found that heavy oils might be cracked, yielding much lighter products. By about 1927, Standard Oil was able to employ Bergius's methods to crack petroleum to increase gasoline yields. A second approach to liquefaction of coal derived from the work of Franz Fischer and Hans Tropsch. By the mid-1920s, relying on prewar work at BASF Industries, they were able to derive gasoline and similar petroleum products using water gas as a feedstock with a catalyst at high temperatures and pressures.[51]

Worried about potential loss of oil supplies during war, Britain, Germany, and Japan each developed a domestic "synfuel" industry based on the work of Bergius and Fischer-Tropsch. The Third Bituminous Coal Conference in 1931 included six papers on coal liquefaction—all by Europeans. German work was the most advanced; by 1938, Germany had the capacity to produce about 17 million barrels of gasoline a year by synthetic methods, while the figures for Britain amounted to 1.3 million barrels. Japan had small plants on line but was planning to become self-sufficient by 1945.[52]

One can see why synfuels were attractive to coal producers and resource pessimists. At that time, Fieldner estimated that a ton of coal would produce about a fifth of a ton of gasoline (roughly 66 gallons). In 1938, with gasoline consumption of about 21.6 billion gallons, the potential market was therefore as much as 327 million tons of coal. Alas, estimates placed the cost of European gasoline by these methods at 17–19 cents a gallon—far more than petroleum-based imports—and European governments sheltered all of these ventures from competition through some form of tariff or subsidy. While differences in the prices for coal and labor might make American synfuel costs somewhat higher or lower, with gasoline then wholesaling at less than five cents a gallon in the United States, neither process for obtaining that fuel from coal seemed remotely economic.

In the absence of government protection, aside from the National Coal Association's brief foray into hydrogenation noted above, American companies showed little interest in either method.[53]

Yet synfuels remained a glint in the eye of policy makers during the 1930s, and with enthusiastic support from the coal industry, government officials would resurrect them in the postwar years. As I have noted, the Federal Oil Conservation Board had investigated synfuels, and the Bureau of Mines followed foreign research closely. To keep abreast of European developments, the bureau had begun laboratory work on coal liquefaction via the Fischer-Tropsch process in 1927, which Depression-era budget cuts ended in 1930. In 1936, the bureau recommended synfuel work, when it built a small unit that could hydrogenate coal at its Pittsburgh experiment station, which it continued during World War II and beyond.[54]

The most important expression of experts' thinking about energy appeared in the 1939 volume *Energy Resources and National Policy*, by the National Resources Committee. While the impact of the book on current policy was, as one historian has put it, "undetectable," many of its ideas would echo throughout the post–World War II decades. The central messages of the book reflected the extent to which resource pessimism and market skepticism had come to dominate policy makers' thinking.[55]

"Oil must be regarded as a distinctly limited resource," the committee summarized, while its assessment of natural gas reserves was equally gloomy. As I have noted, the rise of these fuels and coal's decline were not, in this view, evidence of an energy transition but, rather, a short-term aberration that would soon reverse. Yet such pessimism was oddly out of step with the studies of petroleum reserves presented in the same document and elsewhere. Petroleum reserves had been rising relative to production, and in the future, when increased scarcity eventually would set in, the studies explained that the ensuing rise in price would induce both conservation and a hunt for more oil.[56]

Ignoring this inconvenient evidence, the committee dismissed the notion that markets might play a major role in conservation: "the free play of undirected competition cannot be relied upon," it intoned. Instead conservation required "continuous planning" along with "fundamental and applied research . . . stimulated and supplied by the federal government." Sounding somewhat like Thomas Baker a decade earlier, it enthused that research "can extend the reserves of petroleum . . . develop methods for making gasoline . . . from coal; . . . develop methods of recovering most of the coal from thin and thick beds," and in many other ways "retard the inevitable depletion of the fuels . . . considered to be most valuable."[57]

Carried over into the postwar years, these ideas would help support federal funding for synfuels and similar programs, all of which were expensive and none of which panned out. But despite the committee's optimism, the one thing research could not do was rejuvenate Old King Coal.

In a detailed review of coal research in 1933 that remained true for the remainder of the interwar period and beyond, Arno Fieldner found "little tangible progress in providing new markets for coal."[58]

Conclusion

In the interwar decades, strip mining was coal's most effective way of defending its markets from marauding substitutes. It also contributed to important social goals, for two underappreciated progressive concerns were conserving coal and conserving lives. Most progressives and subsequent historians have failed to see that market-driven stripping supported these goals, allowing exploitation of otherwise uneconomic or unavailable resources while reducing their human costs. Yet at best, stripping only slowed the decline of coal, for its expansion came in good part at the expense of underground mining. Accordingly, the unemployment and bankruptcies that characterized underground coal mining after 1918 were in considerable measure self-induced, and they would have come about, although perhaps not as rapidly, even if oil and gas had never existed. Stripping was creative, in Schumpeter's terms, and it contributed to the destruction of underground mining as well.

Stripping and underground mechanization could sharpen coal's ability to compete on price, but coal remained a dirty, hard-to-handle solid. The holy grail of the industry and outside experts was that research could change coal's form to a liquid or gas to compete with petroleum. This possibility attracted allies for coal among scientists and policy experts who were resource pessimists. They worried over oil and gas reserves, imagining that liquids from coal would soon be needed—a belief that lingered on well into the post–World War II years. Ultimately, coal liquefaction proved to be an enormously expensive dry hole. Once again, the Red Queen's insight seems appropriate. Sometimes "it takes all the running you can do to keep in the same place," but during the interwar decades, neither strip mining nor research could make Old King Coal run fast enough.[59]

Conclusion

American Energy Transitions in the Age of Markets

In 1800, Americans had few choices for energy—horses, wood, wind, and water. Coal at that time was nearly nonexistent, and surely no one could have imagined that a bit over a century later, in 1918, production would be nearing 580 million tons, while horses and wood had been shouldered offstage. Nor in that year could anyone have thought that natural gas, oil, and waterpower would soon begin to push coal from center stage. Energy futures are hard to predict.

Yet by 1940, the broad outlines, if not the details, of coal's future in America had become clear, for coal languished during the interwar decades as energy markets unleashed a host of conservation measures along with waves of new competition from oil, gas, and water. Since that time, coal's career has not been particularly surprising, for despite considerable federal efforts, these same market forces continued to chip away at coal markets. Coal's domain has shrunk to where its great virtue—cost—protected it, and that is mostly in electricity generation. When you flip the light switch, you burn coal, albeit less so now than you did a decade ago as natural gas and renewables—with generous assistance from government—are again biting chunks from coal's last major American market.

The marshaling of America's resources, rather than just their sheer abundance, has driven American energy history. Anthracite coal did not simply hop from the hills to the hearth, nor were there signs saying "Drill Here" in the East Texas oil fields. The United States was born on a mountain of natural resources, and perhaps the energy transitions depicted in these chapters were inevitable. But there was nothing inevitable about the pace, pattern, and extent of the new fuels' exploitation. That history was contingent upon the work of tinkers and thinkers and entrepreneurs, as well as everyday women and men responding to market signals, trying to better their lives. Mrs. Edith Hawley was one of the entrepreneurs whom the reader met earlier. She was a farmwoman, and in 1919, when she canned 1,434 jars of preserves with her kerosene stove, her motive was "to make a little money." Mrs. Hawley was choosing a new form of energy to market a new product to boost her standard of living. That, in capsule form, has been America's energy history.

It is useful to take a long view of these matters; between about 1800 and World War I, the United States experienced three energy transitions, each of which varied across regions and industries and was shaped by patterns of settlement, geography, geology, and technological change. The keys to these energy transitions, as I have argued, were entrepreneurs responding to market signals and supported by an increasingly complex technical network that included not only the tinkers but scientists and engineers as well. They were heirs to the growing body of knowledge about geology, mining engineering, fuel chemistry and combustion, heat transfer, and much else.

The first of these transitions involved the shift from wood and horses and waterwheels as sources of heat and power to anthracite coal. In 1800, the American population was perched largely on the Eastern Seaboard. As population grew and the economy expanded, wood became scarcer and more costly, encouraging the search for substitutes. Entrepreneurs abounded, some of whom possessed extraordinary technical skills and curiosity but had little or no formal scientific training. And because most Americans were literate, these tinkers could read and borrow from each other and from Europe. Geography placed anthracite near to eastern markets and potential availability through river transport, while most soft coal resided farther south and west. By the 1820s, entrepreneurs and tinkers were developing ways to burn anthracite and transport it to eastern markets, beginning America's first coal age.

Anthracite was simply a better fuel than wood; it was cleaner and easier to transport. Hard coal was thus a price-induced innovation; yet it raised standards of living, and after much experimentation on how to burn it, anthracite, by the time of the Civil War, had driven wood from the stoves and fireplaces of eastern city dwellers and from the iron industry. Hard coal was also driving wood from steamboat and locomotive fireboxes, while steamboats and locomotives in turn mostly drove horses from long-distance transport. Where wood departed, however, it did not go without a fight; even with coal in the offing, railroads improved the efficiency with which they could burn wood. Old technologies facing new competition rarely fade away entirely. Still, without the transition to coal, wood would have become increasingly dear, raising households' fuel bills and the costs of transporting all goods. Industrial development would have been retarded and standards of living reduced.

The coal that would win the day, of course, was bituminous, for geology made hard coal expensive to mine, and geography confined it largely to eastern Pennsylvania. Soft coal was more widely diffused and cheaper to extract; it had mostly been a western fuel until the late 1850s, when railroads began hauling it to eastern markets. The combination of wide availability, an expanding transportation network, and gradual technical progress in mining allowed an enormous growth of output at roughly constant prices. Moreover, by the Civil War men and women with scientific training from newly founded engineering schools were increasingly joining the tinkers and entrepreneurs, to work on energy problems, performing chemical analyses of coal and oil and founding technical societies and journals. Accordingly, a second energy transition began about the time of the Civil War, as nearly everywhere soft coal began to replace wood and anthracite.

Up to about World War I, this rising tide of market-driven innovation expanded the kingdom of coal, for until the early twentieth century, bituminous coal faced little competition from other primary fuels. As we have seen, this reflected geography and settlement patterns. Good coal was available in the East and throughout the Midwest, while early discoveries of natural gas and oil during these years were small and—thanks to the rule of capture—quickly depleted, and gas could not then be shipped long distances. Only on western railroads and as fuel for lighting and rural cooking did coal face serious competition from oil. Thus, on the eve of World War I, the dominance of soft coal reflected a combination of geography, the timing of resource discoveries and the progress of technology, but its crown would soon begin to slip. In the twentieth century, the combination of large discoveries of oil and gas, along with new ways to ship gas long distances, began a third energy transition that would bring an end to coal's reign.

Yet when President Roosevelt called the National Conference of Governors in 1908, these developments lay in the future, and the centrality of low-cost coal to America's industrial prominence seemed obvious. The limited quantity of coal reserves also seemed obvious, even as Americans were squandering it at a breakneck pace. Clearly, it appeared to those who attended the conference, conservation of such a vital national resource should not be left to the market, so I have termed them and their descendants "resource pessimists" and "market skeptics."

As the president and other critics then and since have emphasized, America's energy history did indeed reveal much waste. Sometimes waste resulted from perverse incentives due to the rule of capture. Yet the waste of coal, oil, and gas was not always wicked, and sometimes it was self-correcting. For waste also resulted from rapid development along with the inevitable need to learn about new fuels and technologies and to develop markets. Early farmland clearing resulted in excess wood, which soon found a market on steamboats and in cities. Similarly, innovations such as Coxe's stoker and strip mining led to reclamation of what was once waste—while coal stripping fulfilled another progressive goal as it saved lives, as well as coal. Finally, some of what critics called waste resulted from a misunderstanding of market economies. It is not waste to leave in the ground thin seams of coal worth one dollar if mining them would cost two dollars. Implicitly, such critics were calling for Americans to remain poorer longer in order to save coal for modern Americans who do not wish to burn it.

The broader point is that precisely because markets reward conservation, they have routinely discovered new ways to save energy. The energy scares associated with World War I, combined with the increasing role of scientists and engineers in industry, resulted in dramatic gains in energy efficiency during the interwar years. Technological convergence ensured that fuel-saving innovations in steam generation led by utilities and shared through technical journals contributed to about a 44 percent increase in manufacturing energy efficiency between 1910 and 1940. Households gained in two ways from such efficiency gains: the profusion of consumer goods then arriving became cheaper to buy and cheaper to run.

The demands of World War I for petroleum to fuel trucks, tanks, and ships shifted the focus of energy experts from coal to oil. This reinforced their skepticism and pessimism because it seemed clear that market decisions would not adequately weigh national defense needs and because they believed that oil and gas reserves were far more limited than coal. Thus, the languishing of coal after World War I appeared to be a temporary market aberration that would soon end as oil and gas reserves diminished, rather than as the beginning of a long-term energy transition away from solid fuels. Accordingly, many energy experts during the interwar years wished to impede the transition away from coal as shortages of oil and gas loomed before their eyes.

Their concern, that is, shifted from running out of coal to running out of coal markets because they hoped to return Old King Coal to his throne. The invisible hand, they felt, needed guidance from experts with superior foresight, which might take the form of a ban on using oil to raise steam or heat houses. None of these ideas went anywhere, but the expensive and detailed efforts to slow the transition away from coal that characterized government interventions in energy markets in the years after World War II originated in the resource pessimism and market skepticism that reached back to Teddy Roosevelt's Conference of Governors in 1908.

American policy makers no longer worry about running out of anthracite or soft coal or oil or gas. They worry, instead, about the need to transition from such fuels to protect the global climate, but they are just as skeptical of energy markets as were their progressive ancestors. Yet historical evidence suggests that such skepticism is unwarranted, for until World War II, America's energy history abounds with examples of market-directed innovations that drove energy transitions. Relative fuel prices and attributes such as the convenience and cleanliness of the fuel have been central, as markets focused inquiry into prospectively high payoff areas. The many efforts to burn anthracite provide early examples of this, as does the later spread of techniques to burn pulverized coal.

Government research supported these market developments. Early studies by the New York Board of Health along with state regulations backed up market forces to ensure that Mrs. Hawley's kerosene would not explode. The widely published and publicized coal researches by the USGS and, later, the Bureau of Mines emphasized the value of coal chemistry to buyers, while the bureau's work on oil and gas conservation demonstrated the payoff to such activities.

Research was also one of the ways that coal battled its loss of markets. Most coal research derived from government agencies, fuel users, and equipment makers, and the results were considerable. On the eve of World War II, coal was more mechanized, cleaner, better sized, and contained fewer impurities than in 1910, and its loss of markets would have been far greater without washing, sizing, stokers, pulverization, and the arrival of strip mining. Yet as fast as coal ran, it lost markets; its great liability was its solidity. The focus on deriving liquid fuels from coal in the 1920s reflected the hope by coal interests and by energy experts worried over oil supplies that the process might turn the tables on coal's competitors. Its failure to become economic largely sealed coal's fate.

If research and innovation have propelled America's energy transitions, from the dawn of the coal age, marketing also midwifed the new ways, shaping the speed and extent of their market penetration. Not only did entrepreneurs have to assist anthracite in its hop from hill to hearth; they also had to persuade buyers that the balky fuel was flammable. To promote the new fuel, Josiah White and other anthracite entrepreneurs, whom the reader met in chapter 1, decided to "make a noise in Philadelphia," which they did through pamphlets, promotions, and door-to-door solicitations. A century later, gas companies employed Nancy Gay to tout the all-gas kitchen, and makers of kerosene stoves stressed that cooking shouldn't cook the cook. Oil companies and oil-burner manufacturers' massive advertising campaign of the 1920s sped the transition away from coal heat. Nor was marketing limited to consumer goods, as a perusal of technical energy journals reveals. The AGA combined research and marketing to lure producers of ceramics, porcelain, and nonferrous metals from coal to gas.

Formal and informal market structures shaped energy transitions as well. The structure of gas utilities made them especially formidable marketers of fuel for cooking and heating, and they received strong support from the AGA. Oil-burner manufacturers and oil companies quickly standardized fuel oil and developed strong dealer and service networks backstopped by the Oil Heating Institute. Coal, in contrast, was slow to coordinate marketing and service.

Gas utilities, of course, had been selling light since the 1830s, and gas's gradual evolution was simply a lighting miracle. It reveals that toting up Americans' total energy use in Btus or BCEs is of very limited use for capturing its impact on human welfare. This point is a general one, for other examples abound. The reader met Ellen Murdock in chapter 5, where she pointed out that her gas stove saved her many steps; Mrs. James Wilson estimated that her kerosene stove saved her 350 steps a week doing the ironing. "No More Shoveling Coal or Hauling Out Ashes for Me," was Joseph Cepek's explanation when he turned to gas heat. In short, cooking or heating with gas or oil immensely improved women's and men's lives. Yet the national product statistics never captured these gains in welfare. We are richer than our parents in ways we cannot measure, indeed, can barely comprehend, for the benefits we have gotten from energy have risen much more rapidly than has energy use itself.

These welfare gains resulted because energy innovations during these years were market-driven. New fuel uses emerged from free choices like those of Edith Hawley and Joseph Cepek, and, accordingly, there were many energy transitions that proceeded at different rates in different circumstances and were usually incomplete. Incomes, geography, personal tastes, the price and availability of substitutes, and season of the year all shaped individual families' choice of heating and cooking. The reader may recall Indiana farm women's complex seasonal use of cooking fuels that ranged from electricity to corncobs. Similarly, companies burned soft coal in pulverized form or employed stokers, or chose gas or anthracite or oil; they bought electricity or generated their own, all depending on shifting patterns of fuel costs, characteristics, and availability. Collectively, these changing individual and company choices generated a kaleidoscope of shifting regional and historical patterns of energy consumption. Hard coal had been a luxury fuel in 1880; it was an inferior good by 1930. Kerosene was mostly used as a rural cooking fuel and then mostly in the summer. Until the late 1930s, only well-to-do households heated with oil; until the diesel, only western railroads burned oil and then only if the price was right. Just as natural selection tailors a wondrous variety of evolving species, so energy markets reflect the complexities of humans and their changing circumstances.

Much of this book has stressed the enormous benefits Americans have gained as markets began to transition away from coal, but it is useful to remember just how much the transition from wood to coal had improved Americans' welfare and how important coal once was—in, say, 1910. It is hard to imagine—indeed, it is impossible to imagine in any detail—what the America of that day would have looked like had it not evolved during nearly a century of very cheap, coal-fired energy. Yet glimpses are possible. It seems unlikely that oil and gas could have ridden to the rescue much earlier, awaiting as they did the progress of discovery and technology, so much higher energy costs would have prevailed. Over decades, growth would have been slower, stunting standards of living, reshaping patterns of industrialization and urbanization, and perhaps reducing the rate and changing the course of technological change. Economic development is cumulative; without the age of coal, the Americans of 1910 would have been poorer, and so, in turn, would we be poorer today. To paraphrase Isaac Newton, we are rich in part because we stand on the shoulders of those men and women of 1910, who stood, in turn, on the shoulders of Old King Coal.

Appendixes

Appendix 1
Basic Data

This appendix presents and explains the basic data on primary energy that provide the under-pinnings for much of this book. Most of the data are presented in five-year intervals; readers want-ing information for other years should consult the sources. The forms of energy include wood, animals (horses, mules, oxen, and whales), coal, natural gas, petroleum, waterpower, and wind power. Electricity is absent because it is derived from fuels and waterpower and thus is not a pri-mary energy source. For various reasons, I exclude human energy.

Energy from Wood

The data in table A1.1 on the quantity of wood from 1850 onward derive ultimately from the USDA, and they are of unknown accuracy. For these and all other BCE calculations, see Measurement and Conversion Factors.[1]

Animals: Horses, Mules, Oxen, Whales

The data in Table A1.2 are largely from the census. BCE figures represent the amount of energy it took to feed a horse, converted to its coal equivalent. I estimate that farm work animals consumed about 22,710K calories a day or 90,103.5 Btu/day. This amounts to 32,887,783 Btu/year, which is 1.26 BCE. I have found no data that would allow such a calculation for oxen.[2]

Table A1.1. Fuel Wood, 1800–1940

Thousands of cords or tons					
Year	Wood	BCE	Year	Wood	BCE
1800	23,000	18,400	1875	137,000	109,600
1805	27,000	21,600	1880	136,000	108,800
1810	31,000	24,800	1885	128,000	102,400
1815	36,000	28,800	1890	120,000	96,000
1820	42,000	33,600	1895	110,000	88,000
1825	48,000	38,400	1900	100,000	76,900
1830	57,000	45,600	1905	95,000	70,395
1835	65,000	52,000	1910	91,000	67,431
1840	77,000	61,600	1915	87,000	64,467
1845	88,000	70,400	1920	83,000	61,503
1850	102,000	81,600	1925	79,000	58,539
1855	114,000	91,200	1930	75,000	55,575
1860	126,000	100,800	1935	72,000	53,352
1865	132,000	105,600	1940	70,000	51,870
1870	138,000	110,400			

Table A1.2. Horses, Mules, and Oxen on Farms, 1800–1945

Year	Thousands				Millions of horsepower-hours	Oxen (thousands)
	Horses	Mules	Total	BCE		
1800	1,342	—	1,342	1,690	990	470
1840	4,337	—	4,337	5,465	3,199	1,518
1850	4,338	560	4,898	6,171	3,612	1,701
1860	6,249	1,150	7,399	9,323	5,459	2,255
1870	7,633	1,245	8,878	11,186	6,550	1,319
1875	9,333	1,548	10,881	13,710	8,028	—
1880	10,903	1,878	12,781	16,104	9,430	994
1885	12,700	2,102	14,802	18,651	10,921	—
1890	15,732	2,322	18,054	22,748	13,320	1,117
1895	17,849	2,708	20,557	25,902	15,167	—
1900	17,856	3,139	20,995	26,454	15,490	—
1905	18,491	3,586	22,077	27,817	16,288	—
1910	19,972	4,239	24,211	30,506	17,863	—
1915	21,431	5,262	26,693	33,633	19,694	—
1920	20,091	5,651	25,742	32,435	18,992	—
1925	16,651	5,918	22,569	28,437	16,651	—
1930	13,792	5,382	19,174	24,159	14,147	—
1935	11,861	4,822	16,683	21,021	12,309	—
1940	10,444	4,034	14,478	18,242	10,682	—
1945	8,715	3,235	11,950	15,057	8,817	—

I also estimate the number of horsepower-hours that might have derived from horses and mules but found no evidence to allow such a calculation for oxen. The census of 1910 reveals that about 85 percent of all horses on farms were more than two years old and, accordingly, were work stock. The horsepower-hour calculations assume that the animals themselves did not change much throughout the nineteenth century and that they worked about the same number of hours during those decades as they did in the early twentieth. In fact, horses almost certainly grew more powerful, and they probably worked more hours before the advent of tractors. While these biases tend to offset each other, they need not cancel.[3]

I found thirteen USDA and state agricultural department studies that estimated work hours for horses between 1909 and 1930; collectively, they included not quite 21,000 horse-years. Weighting the annual hours worked in these studies by horse-years yields an average of 868 hrs./year—substantially more than some previous researchers have found. Horsepower-hours per year are therefore horses and mules multiplied by the fraction that were work stock (0.85) and then by 868.[4] These calculations reveal that a farm horsepower-hour required about 3.42 pounds of BCE—probably better than most steam engines until the twentieth century. In addition, farm horses and mules supplied more power than did steam engines in manufacturing in 1880 (table A2.6). They probably underestimate the importance of animal power because they ignore oxen.

Table A1.3 presents decadal estimates of nonfarm work animals that ultimately derive from the USDA. I calculated the energy needed to supply them the same way as for farm work animals. I assumed that all animals were working stock, and I employed a conservative estimate that they averaged four hours of work per day during a six-day week. While such calculations are anything but precise, they remind us of the importance of animal power well into the industrial age.[5]

Whales were important not because of the vast quantities of energy they supplied but because they were a source of high-quality light before the advent of gas and kerosene. Based on the work of

Table A1.3. Nonfarm Work Animals, 1850–1940

Year	Animals (M)	BCE (M)	Hph (Mm)
1850	800	1,008	998
1860	1,200	1,512	1,498
1870	1,600	2,016	1,997
1880	2,100	2,646	2,621
1890	2,600	3,276	3,245
1900	3,100	3,906	3,869
1910	3,500	4,410	4,368
1920	2,100	2,646	2,621
1930	300	378	374
1940	300	378	374

Lance Davis and others, I have estimated quantities of whale oil and their BCE for selected years. For example, I estimate whale oil for 1800 as one-fifth of Davis's 1800–1804 totals converted to (42-gallon) barrels. Following Nordhaus, I assumed these had half the energy content of crude oil, while 4.5 bbl. of crude is equivalent to a ton of bituminous coal. Accordingly, in 1800, American whale oil consumption amounted to about 2,671 bbl. and had the energy equivalent of about 297 tons of coal. Similar calculations for 1860 yield 11,257 bbl., equivalent to 1,251 tons of soft coal.[6]

Coal

The data in table A1.4 on anthracite and bituminous coal production come from two sources. Up until 1850, they are from Eavenson; thereafter, they are from the USGS/USBM, MR, and MY. I take consumption to be production + net imports and ignore changes in stocks. The figures in parentheses are net exports. The prices in table A1.5 are per ton average receipts at the mine, and they, too, are from MR and MY.[7]

Natural Gas

The natural gas data in table A1.6 are marketed production and are all from MR and MY, except the production figures from 1882 to 1905, which are from Schurr et al. The data do not separate domestic from commercial use before 1930. After this date, however, roughly three-quarters of both customers and gas use are domestic. Here, in the category "industrial," I have excluded field use of gas and its use to make carbon black.[8]

Crude Petroleum

The data in table A1.7 on crude oil production and apparent consumption are from Schurr et al. The production data are straightforward and available elsewhere as well. The apparent consumption figures are *not* simply production minus net exports of crude oil. Rather, because the United States exported considerable quantities of refined products, as well as crude, the apparent consumption data also account for the crude oil equivalent of such net exports. The results provide a better estimate of Americans' actual consumption. The BCE of apparent consumption is my calculation.[9]

Waterpower

Kilowatt-hours of waterpower comprise those from waterwheels and water turbines. In table A1.8, for the years down to 1920, I converted these to equivalent tons of bituminous coal by employing

Table A1.4. US Coal Production, Net Imports, and Consumption, 1800–1940 (all figures tons)

| Year | Production | | | Net Imports* | Consumption |
	Anthracite	Bituminous	Total		
1800	250	108,100	108,350	9,229	117,579
1805	400	146,100	146,500	17,245	163,745
1810	2,200	176,300	178,500	14,035	192,535
1815	2,400	252,650	255,050	3,921	258,971
1820	4,065	329,700	333,765	—	333,765
1825	43,119	437,270	480,389	28,724	509,113
1830	234,790	646,109	880,899	65,113	946,012
1835	759,939	1,059,216	1,819,155	66,881	1,886,036
1840	1,129,206	1,345,113	2,474,319	176,292	2,650,611
1845	2,625,757	2,097,175	4,722,932	83,341	4,806,273
1850	4,326,969	4,028,770	8,355,739	151,444	8,507,183
1855	8,606,687	7,542,990	16,149,677	196,017	16,345,694
1860	10,983,972	9,056,897	20,040,869	58,371	20,099,240
1865	12,076,996	12,349,348	24,426,344	95,946	24,522,290
1870	19,958,064	20,471,006	40,429,070	215,890	40,644,960
1875	23,120,730	32,656,911	55,777,641	(87,082)	55,690,559
1880	28,649,812	50,757,119	79,406,931	(154,531)	79,252,400
1885	38,335,974	71,770,408	110,106,382	(508,930)	109,597,452
1890	46,468,641	111,302,322	157,770,963	(1,066,408)	156,704,555
1895	57,999,331	135,118,193	193,117,524	(2,718,036)	190,399,488
1900	57,367,915	212,316,112	269,684,027	(6,490,853)	263,193,174
1905	77,659,850	315,062,785	392,722,635	(8,730,180)	383,992,455
1910	84,485,236	417,111,142	501,596,378	(13,984,397)	487,611,981
1915	88,995,061	442,624,426	531,619,487	(19,240,588)	512,378,899
1916	87,578,493	502,519,682	590,098,175	(24,575,571)	565,522,604
1917	99,611,811	551,790,563	651,402,374	(29,776,106)	621,626,268
1918	99,611,811	579,385,820	678,997,631	(27,483,126)	651,514,505
1919	88,092,201	465,860,058	553,952,259	(24,098,567)	529,853,692
1920	89,598,249	568,666,683	658,264,932	(43,537,168)	614,727,764
1925	61,817,149	520,052,741	581,869,890	(20,408,168)	561,461,722
1930	69,384,837	467,526,299	536,911,136	(18,455,644)	518,455,492
1935	52,158,783	372,373,122	424,531,905	(10,872,768)	413,659,137
1940	51,484,640	460,771,500	512,256,140	(13,527,210)	498,728,930

*Figures in parentheses are net exports.

Table A1.5. Mine Prices of Coal, 1880–1940 (dollars per ton and ratios)

Year	Bituminous ($)	Anthracite ($)	Pa/Pb*
1880	1.25	1.47	1.18
1885	1.13	2.00	1.76
1890	0.99	1.43	1.44
1895	0.86	1.41	1.64
1900	1.04	1.49	1.43
1905	1.06	1.83	1.73
1910	1.12	1.90	1.70
1915	1.13	2.07	1.80
1920	3.75	4.85	1.30
1925	2.04	5.30	2.60
1930	1.7	5.11	3.00
1935	1.89	4.03	2.13
1940	1.91	3.99	2.10

*Price of anthracite divided by the price of bituminous.

Table A1.6. Natural Gas Production, Use, Customers, and Prices, 1882–1940

| Year | Production and Use (Mmft.^{3}) | | | Customers | Prices (¢ per Mft.^{3})* | | |
	Total	Domestic and Commercial	Industrial	Domestic and Commercial	Wellhead	Domestic and Commercial	Industrial
1882	3,400	—	—	—	—	—	—
1885	76,000	—	—	—	—	—	—
1890	239,000	—	—	—	—	—	—
1895	137,000	—	—	370,130	—	—	—
1900	236,000	—	—	706,309	—	—	—
1905	351,000	—	—	779,639	—	—	—
1910	509,155	169,823	—	1,327,722	—	24.4	—
1915	628,579	217,201	—	2,195,081	—	28.3	—
1919	745,946	255,743	270,490	2,501,462	—	35.0	—
1920	798,210	286,001	269,293	2,615,043	—	38.2	—
1923	1,008,135	277,218	277,745	3,233,800	10.0	51.0	—
1925	1,188,429	272,146	352,404	3,536,000	9.4	56.0	—
1929	1,917,693	359,853	590,810	5,097,860	8.0	62.0	19.7
1930	1,943,421	376,407	575,210	5,448,000	7.6	63.5	19.3
1935	1,916,505	413,685	673,977	8,004,000	5.8	68.5	16.9
1940	2,660,222	578,290	995,337	9,986,000	4.5	66.7	16.7

*Except for wellhead, prices are at point of use.

Table A1.7. Crude Petroleum Production and Apparent Consumption, 1860–1940

| Year | M bbl. | | M tons BCE consumption |
	Production	Domestic consumption	
1860	200	500	111
1865	2,498	1,700	378
1870	5,261	2,011	447
1875	8,788	2,002	445
1880	26,286	17,203	3,823
1885	21,859	7,172	1,594
1890	45,824	27,652	6,145
1895	52,892	29,726	6,606
1900	63,621	39,564	8,792
1905	134,717	105,119	23,360
1910	209,557	173,559	38,569
1915	281,104	243,230	54,051
1920	442,929	454,242	100,943
1925	763,743	716,096	159,132
1930	898,011	970,762	215,725
1935	996,596	945,857	210,190
1940	1,353,214	1,284,954	285,545

Table A1.8. Waterpower, 1850–1940

Year	Waterwheels	Water turbines	Total	Coal/kwh (lbs.)	M tons (BCE)
	Mm kwh				
1850	1,033	—	1,033	15.32	7,911
1860	1,436	—	1,436	13.16	9,448
1870	1,830	—	1,830	11.47	10,495
1880	2,025	—	2,025	10.28	10,406
1890	1,961	250	2,211	8.43	9,318
1900	1,950	2,786	4,736	6.85	16,215
1910	2,220	8,626	10,846	4.77	25,870
1920	3,150	18,779	21,929	3.04	33,332
1930	1,100	35,878	36,978	1.62	29,952
1940	650	50,131	50,781	1.35	34,277

Table A1.9. Wind Power on Land and Water, 1850–1940

Year	On land		On water		Coal/hph (lbs.)	BCE (tons)	
	Windmills (M)	Hph (Mm)	Gross tons (M)	Hph (Mm)		Land	Water
1850	70	14	3,010	1,400	11.43	80,010	8,001,000
1860	100	20	4,486	2,100	9.82	98,200	10,311,000
1870	150	30	2,363	1,100	8.56	128,340	4,705,800
1880	200	40	2,366	1,100	7.67	153,400	4,218,500
1890	400	80	2,109	1,000	6.29	251,600	3,145,000
1900	600	120	1,885	900	5.11	306,600	2,299,500
1910	900	180	1,655	800	3.48	313,605	1,393,800
1920	1,000	200	1,272	600	3.56	356,000	1,068,000
1930	1,000	200	757	400	2.13	213,000	426,000
1940	650	130	200	100	2.02	131,300	101,000

the coal per horsepower-hour data from Dewhurst et al. and then converting horsepower to kilowatts—that is, coal/kwh = (kwh/hph) × (coal/hph). For 1920 onward, the coal-per-kilowatt-hour data are from my table A3.3.[10]

Wind Power

The estimates for wind power on land in table A1.9 derive from two sources. The number of windmills is from Dewhurst et al. I calculate kilowatt-hours using the estimate that windmills averaged ½ horsepower each and worked an average of 400 hours per year. I used those estimates for all years. Thus, for example, in 1870, 70,000 windmills generated (.5 × 400 × 70,000) = 14 million horsepower-hours. Both tonnage of shipping and horsepower-hours again come from Dewhurst et al., as do the estimates of coal per brake horsepower.[11]

Appendix 2
Chapters 1 and 2

This appendix provides and explains the materials that form the basis of the tables and figures in chapters 1 and 2 under the following headings: Energy Overview; Transportation; Industrial Fuel and Power; Manufactured Gas, 1890–1920; and Domestic and Retail Fuel and Light.

Energy Overview

The data in tables A2.1 and A2.2 are the source of table 1.1 and are all derived from the tables in appendix 1. They are production, not consumption, data. I have excluded whale oil because as a resource, it is tiny, but it is discussed below under Domestic Fuel Use. I also ignore a small amount of natural gas liquids. The most serious omission is probably the absence of data on wind and waterpower before 1850.

Transportation

Data for wood use by antebellum western steamboats in table A2.3 are derived as follows. Steamboats and the percent by weight class are from Louis Hunter. I have taken the midpoint of his weight classes and chosen 50 tons for his 100-tons-and-below class and 1,200 for 1,000 tons and up. Days operated and cords of wood per-ton per-day are from James Mak and Gary Walton. As the text notes, these data assume that all fuel was wood and therefore overstate use.[1]

Table A2.4 contains steamboat fuel use from the 1880 census. By this time, coal dominates, but wood hangs on in rural regions, presumably where it is relatively cheap.[2]

Table A2.1. Coal and Its Competitors: Major Sources of American Energy, 1800–1940

Year	Horses & mules (M)	Wood cords (M)	Anthracite (M tons)	Bituminous (M tons)	All coal (M tons)	Oil (M bbl.)	Nat'l gas (Mmft.³)	Wind (Mm hph)	Water (Mm kwh)
1800	1,342	23,000	0	118	118	—	—	—	—
1840	4,337	77,000	1,129	1,345	2,474	—	—	—	—
1850	4,898	102,000	4,172	3,518	7,690	—	—	1,414	1,033
1860	7,399	126,000	10,984	9,057	20,041	500	—	2,120	1,436
1870	1,245	139,000	19,958	20,471	40,429	5,261	—	1,130	1,830
1880	12,781	136,000	28,650	50,757	79,407	26,286	7,073	1,140	2,025
1890	18,054	120,000	46,469	111,302	157,771	45,824	239,000	1,080	2,211
1900	20,995	100,000	57,368	212,316	269,684	63,621	235,000	1,020	4,736
1910	24,211	91,000	84,485	417,111	501,596	200,556	502,000	980	10,846
1920	25,742	83,000	89,598	568,667	658,265	442,959	812,338	800	21,929
1930	19,174	75,000	60,386	467,526	527,912	898,011	1,978,911	600	36,978
1940	14,478	70,000	51,485	460,772	512,257	1,353,214	2,733,819	230	50,781

Table A2.2. Coal and Its Competitors: Major Sources of American Energy, BCE, 1800–1940

				Thousands of tons				
Year	Horses & mules	Wood	All coal	Oil	Natural gas	Wind power	Water power	Total
1800	1,690	18,400	118	—	—	—	—	20,208
1840	5,465	61,600	2,346	—	—	—	—	69,411
1850	6,171	81,000	7,565	—	—	8,081	7,911	110,728
1860	9,323	100,080	19,781	111	—	10,409	9,448	149,153
1870	11,186	110,400	39,262	1,169	—	4,834	10,495	177,347
1880	16,104	108,800	68,216	5,841	283	4,372	10,406	214,022
1890	22,748	96,000	140,758	10,183	9,578	3,397	9,318	291,982
1900	26,454	76,900	248,970	14,138	9,418	2,606	16,215	394,701
1910	30,506	67,431	461,694	44,568	20,119	1,707	25,870	651,895
1920	32,435	61,503	655,577	98,435	32,556	1,424	33,499	915,429
1930	24,159	55,575	526,100	199,558	79,309	639	29,483	914,823
1940	18,242	51,870	510,712	300,714	109,563	232	34,023	1,025,125

Table A2.3. Wood Use on Antebellum Western Steamboats, 1811–1860

Year	No. of boats	Percent of boats in weight classes							Avg. wt. (tons)	Days op.	Cords/ ton/ day*	Cords/ boat/ year	Cords/ Year (M)
		50	150	250	350	450	550	1200					
1811–1836	658	22	51	15	8	2.6	1.4	—	172.4	106	1 in 14	2,284	1,503
1843	368	19	45	21	9	2.1	3.9	—	191.9	130	1 in 8	3,118	1,148
1857	531	7	33	25	14	14.0	11.0	—	288.0	141	1 in 8	5,076	2,695
1860	546	11	32	22	15	8.1	11.5	0.4	265.5	141	1 in 8	5,076	2,695

*For example, in 1843, a boat would consume eight cords of wood per day for each ton of boat weight.

Table A2.4. Coal and Wood Use by Steamboats, 1880

Region	States	Tons coal	Cords wood	Pct. coal*
1	New England	375,080	8,568	98.21
2	Northern Lakes	488,610	255,629	70.49
3	Upper Miss. R.	525,331	93,318	87.56
4	Ohio River	676,347	16,474	98.09
5	Middle States	1,272,857	12,642	99.21
6	Lower Miss. R.	184,951	112,254	67.32
7	Gulf of Mexico	41,993	62,556	45.63
8	S. Atlantic Coast	64,995	66,835	54.87
9	Pacific Coast	146,407	103,446	63.89
10	Upper Missouri R.	6,281	53,063	12.89
	US Total	3,782,852	784,785	85.77

Source: Tenth Census of the United States, 1880, *Report on Agencies of Transportation* (Washington, 1883), chap. 3, table 2, 44.
Note: US waters only; excludes canals and intrastate waters.
*1.25 cords of wood = one ton of coal.

Table A2.5. Energy Efficiency in Pig-Iron Production, 1880–1910

Census Year	1880	1890	1900	1910
Anthracite (tons)	3,175,182	2,012,476	995,192	306,368
Bituminous (tons)	1,051,753	551,008	932,103	1,166,135
Coke (tons)	2,128,267	9,125,935	16,461,533	31,649,865
Charcoal (bushels)	53,909,828	67,672,156	31,421,585	38,032,618
Oil (bbl.)	—	—	—	19,446
Natural Gas (Mft.³)	—	—	—	949,622
Total BCE (tons)	7,825,441	16,821363	26,881959	49,308,399
Output (tons)	3,781,021	9,906,607	16,186,502	28,730,014
BCE/Output	2.07	1.70	1.66	1.72
Bituminous share*	0.54	0.85	0.95	0.99

Source: Twelfth Census of the United States, "Iron and Steel," *Bulletin* 246 (Washington, 1902), table 8; and Thirteenth Census of the United States, 1909, Manufactures, *Reports for Principal Industries*, vol. 9 (Washington, 1913).
*Bit share = (Bituminous coal + 1.5 × Coke)/BCE.

Industrial Fuel and Power

Chapter 2 describes the efforts of pig-iron makers to improve energy efficiency and reports on one producer whose efforts to reduce the limestone in his mix improved fuel economy. Almost certainly, this resulted from a shift away from siliceous iron ores that required large amounts of limestone to flux. The following equation derives from that company's biennial observations from the second half of 1855 through 1875. The figures in parentheses are t-ratios:

$$\frac{Pounds\ of\ Coal}{Pounds\ of\ Pig} = 1.11 + \frac{.838}{(5.25)}\left(\frac{Pounds\ of\ Limestone}{Pounds\ of\ Pig}\right) - \frac{.020}{(5.77)}Trend\ \ R^2 = .52,\ N = 41$$

Since limestone use over the period fell about .448 lbs. per pound of pig, while coal use declined by .683 pounds per pound of pig, limestone accounted for .448 × .838 = .375 pounds of the total decline, or about 55 percent.[3]

Table A2.5 presents census data for pig-iron production. Similar calculations on steel making and iron and steel products are not possible owing to problems of double counting. Literary evidence presented in the text, however, suggests that the rise in fuel efficiency and shift to bituminous coal depicted in table A2.5 were typical of the industry as a whole. In computing BCE and the share of bituminous coal in BCE, I have given coke a weight of 1.5 because it then took roughly 1.5 tons of coal to produce a ton of coke. The census of 1910 lists 25,651,798 tons of pig-iron production that year, but according to the American Iron and Steel Institute, these are long (2,240 pound) tons, so I have adjusted them.[4] Iron production also consumed a small amount of wood for fuel that I have ignored. Finally, the apparent worsening of fuel economy between 1900 and 1910 probably reflects the aftereffects of the panic of 1907. Industrial production had peaked in August that year and did not entirely recover until October 1909.[5]

Table A2.6 contains estimates of fuel employed for steam power in manufacturing. Although some of the fuel was certainly wood, most was coal, so that is what I term it. The horsepower data are from the USGS. Horsepower can be measured in several ways, and I believe these are indicated horsepower (Ihp).[6]

The estimates of average coal use per horsepower-hour derive from Dewhurst, whose figures appear to be brake horsepower. For conversion to indicated horsepower, I have reduced them by

Table A2.6. Coal Used for Power in Manufacturing, 1850–1920

Census	Steam hp (M)	Coal per hph	Hours/ Day	Days operated	Horsepower-hours (M)	Coal use (tons)
1850	450	9.72	11.50	300	1,552,500	7,541,657
1860	700	8.35	11.00	300	2,310,000	9,640,785
1870	1,216	7.44	10.50	300	3,830,400	14,244,300
1880	2,185	6.52	10.30	300	6,751,650	22,008,691
1890	4,586	5.35	10.00	294	13,482,840	36,043,002
1900	8,190	5.25	9.89	289	23,408,740	61,396,443
1910	14,229	3.04	9.49	289	39,024,598	59,375,925
1920	17,038	1.94	9.09	281	43,519,993	42,214,393

Table A2.7. Coal Used for Power in Mining, 1850–1920

Census	Steam hp (M)	Coal per hph	Hours/ day	Days operated	Horsepower-hours (M)	Coal use (tons)
1850	38	9.72	11.0	216	90,288	438,597
1860	120	8.35	11.0	216	285,120	1,189,948
1870	275	7.44	9.5	216	564,300	2,098,491
1880	525	6.52	9.5	216	1,077,300	3,511,729
1890	1050	5.35	9.5	216	2,154,600	5,759,784
1900	2433	5.25	8.8	212	4,539,005	11,904,902
1910	3787	3.04	8.6	220	7,165,004	10,901,554
1920	3712	1.94	8.0	230	6,830,080	6,625,178

15 percent based on a contemporary estimate.[7] Daily hours in manufacturing and days operated are mostly from *Historical Statistics*. Horsepower-hours are simply the product of horsepower × hours per day × days worked, while coal use converted to tons multiplies that product by coal per hour. These data contain two downward biases; some historians assume that, first, the work year in the mid-nineteenth century was 309 days and, second, that coal use per day should be increased by 12 percent based on the need to start fires early or bank them at night. Collectively, these adjustments might increase estimated coal consumption by 15 percent. The data, however, ignore down time from strikes, weather, and equipment failure, which was surely considerable.[8]

The calculations for mining are similar to those for manufacturing. In table A2.7, daily hours before 1870 are estimates; for 1870 to 1890, they are from my *Safety First*; for 1900 on, from Willard Hotchkiss et al. Days operated before 1890 are also estimates; for 1890 to 1910, they are from USBM *Bulletin* 115. Days operated for 1920 are from MR.[9]

Manufactured Gas, 1890–1920

The decadal censuses include sufficient usable data to chart the evolution of the manufactured-gas industry from 1890 onward; the results are in table A2.8. I have computed the "real" price of gas using the David Solar consumer price index.[10]

Domestic and Retail Fuel and Light

The estimates of domestic and retail fuel and light for 1800 in text table 1.2 are relatively straightforward. My treatment of fuel wood again follows the census of 1880 in estimating 95 percent of it as used for domestic consumption, while I arbitrarily took domestic consumption of soft coal to be

Table A2.8. Manufactured Gas Production, 1890–1920

	Census year			
	1890	**1900**	**1910**	**1920**
Gas sales (Mft.³)	36,519,512	68,265,496	150,835,793	282,000,000
Retail customers	699,323	4,200,000	5,425,000	8,484,000
Price/Mft.³ ($)	1.42	1.04	0.92	0.92
Real price (1900 = 100)	$1.32	1.04	0.82	0.39
Fuel use				
Anthracite (tons)	313,170	—	—	1,307,383
Bituminous (tons)	1,997,385	2,487,287	4,940,598	6,193,527
Coke (tons)	76,909	217,354	591,919	1,355,322
Oil (bbl.)	867,046	4,639,359	13,769,272	20,992,223
Gasoline (bbl.)	1,235,930	—	32,088	3,002
Natural gas (Mft.³)	—	—	—	13,104,056
Other gas (Mft.³)	153,992	2,696,571	16,769,705	23,995,064
BCE (tons)	2,842,547	3,825,954	9,226,907	14,900,102
Bituminous/BCE	0.73	0.71	0.60	0.51
Fuel efficiency (Mft.³/ton BCE)	12.85	17.84	16.35	18.93

Table A2.9. Domestic Energy Use, 1860

	Total	Power	Iron	Domestic	BCE (tons)*
Wood (cords)	126,000,000	—	—	119,700,000	95,760,000
Anthracite (tons)	10,983,972	6,073,424	2,180,366	2,730,182	2,648277
Bituminous (tons)	9,115,268	4,969,165	513,358	3,632,745	3,632,745
Whale oil (bbl.)	11,257	—	—	11,257	1,251
Gas (Mft.³)	2,748,352	—	—	2,191,668	(273,958)
Kerosene (bbl.)	171,429	—	—	171,429	(189,993)
Total					102,042,273
Per capita					3.24

*Domestic only; total and per capita exclude gas and kerosene figures in parentheses, which are included in bituminous coal.

half the total. I omit what must have been minor sources of energy use such as camphene and candles for light.

The materials in table A2.9 are the basis for the 1860 estimates in text table 1.2. The estimates for coal use treat domestic and retail use as a remainder, and their accuracy therefore depends on how well other uses are measured. For total coal used to power steam engines in manufacturing and mining, see the section on stationary power in this appendix, to which I have added Fishlow's estimate of 300,000 tons of coal for railroad use. I have distributed the total in proportion to the production of hard and soft coal. Use of coal as industrial fuel for purposes other than power generation derives from calculations in the text (see chap. 1n34) for pig iron, plus the assumption that half of pig was turned into wrought iron, taking four tons of coal per ton of wrought iron.[11]

The gas estimate derives from Philadelphia's experience in 1860 extrapolated to all families with gas service that year. I have converted gas to BCE using coal consumed (8,000 ft.³/ton) rather than its likely Btu content. Use of kerosene in 1860 and my estimate of the coal it consumed are from "Coal Oil Manufacture," *Hunt's Merchants' Magazine* 42 (1860): 245. The article claimed production of 7.2 million gallons (171,429 bbl.); it also asserted that daily production of 22,750 gal. (541.7 bbl.) required 60,000 bushels (2,400 tons) of coal, or 4.43 tons per bbl. If this is accurate, the 171,429 bbl.

Table A2.10. Domestic Energy Use, 1910

	Total	Domestic	BCE (tons)*
Anthracite (tons)	84,485,236	42,768,000	41,484,960
Bituminous (tons)	417,111,142	65,561,941	65,561,941
Wood (cords)	91,000,000	84,230,000	63,172,500
Coke (tons)	44,522,150	3,273,261	3,273,261
Briquettes (tons)	179,000	179,000	179,000
Mfg. gas (Mmft.^{3})	150,836	110,230	2,314,006
Kerosene (M bbl.)	14,962	14,962	3,238,528
Gasoline (M bbl.)	11,224	11,224	2,231,412
Nat'l gas (Mmft.^{3})	509,155	169,823	6,805,987
Electricity (Mm kwh)	25,283	910	2,393,300
Total			190,654,895
Per capita			2.06

*Domestic.

of kerosene required 759,430 tons of coal, making kerosene twice as important a consumer as railroads. Since domestic bituminous coal is measured as a residual, it already includes the kerosene and gas derived from it, so they are not counted independently.[12]

There are good reasons for believing that nondomestic uses of coal have been underestimated. The coal used for iron manufacture, for example, excludes that consumed by forges, and there was an unknown quantity of coal burned in brick and glass manufacture. I ignore some coal that fired steam engines that pumped water and powered steamboats as well. It follows that if these uses are underestimated, then domestic use is overstated, which reinforces my claim in the text that per capita domestic and commercial energy consumption declined between 1800 and 1860.

Fuel used for domestic and commercial purposes in 1910 is presented in table A2.10 and is the basis for text table 2.5. Domestic anthracite excludes "steam sizes" that were rarely used at home at that time. In 1915, the USGS presented data on domestic ("and small steam trade") consumption of bituminous coal. The calculation in table A2.10 assumes that the per capita value for 1915 also held for 1910. The data on coke, from MR, are for 1915, the first year available, also converted to estimated 1910 values. Domestic wood use is again from table A1.1, while kerosene and gasoline are data for 1909 and are U.S. production minus exports. Kerosene, at least, was used almost entirely for domestic light and heat before the advent of farm tractors. Gasoline also includes naphtha and probably overstates domestic use. The totals for manufactured gas are from table A2.8. I assumed a 73 percent domestic share following Gould. I have converted kwh to BCE using the average number of pounds of coal required to produce a kwh in 1912, as 1910 data are unavailable.[13]

Appendix 3
Chapters 3 and 4

This appendix presents and explains the data underlining tables and figures of chapters 3 and 4. The topics are recovered coal wastes, energy use in manufacturing, electric utilities' fuel and power, fuel use in iron and steel making, and railroad fuel use and efficiency.

Recovered Coal Wastes

Table A3.1 underlies figure 3.1 and derives from USGS/USBM data in MR and MY. When the USGS began data collection, it took the output of washeries as synonymous with recovered culm-bank coal and reported it as such in MR. This was nearly correct, as little fresh-mined coal was then washed. But as washing of fresh-mined coal became more common, the data became less accurate, and beginning in 1923, the USGS directly queried operators on culm-bank coal.[1]

Energy Use in Manufacturing

Table A3.2 presents the data that underlie the BCE statistics in text table 3.2. The available census data on energy consumption in manufacturing from 1910 onward are for energy *use*, and these involve double counting because they include both coal and the coke and gas and internally generated electricity *derived* from it. To correct for double counting, I have omitted coke; for similar reasons, I omit, as well, mixed and manufactured gas and internally generated electricity. Since manufactured gas includes some purchases from outside the industry, this overcorrects and thus overstates conservation and coal's share of energy. Oil also includes some double counting that is

Table A3.1. Anthracite Production and Recovered Wastes, 1905–1940

Year	Production	Culm banks	Rivers	Culm & river	Pct. of production
1905	77,659,850	2,908,403	—	2,908,403	3.75
1910	84,485,236	3,625,604	102,853	3,728,457	4.41
1915	88,995,061	3,695,135	138,421	3,833,556	4.31
1920	89,598,249	6,236,192	740,453	6,976,645	7.79
1923	63,339,009	6,983,859	956,368	7,940,227	12.54
1925	61,817,149	2,034,204	1,015,708	3,049,912	4.93
1930	69,384,837	1,279,163	643,291	1,922,454	2.77
1935	52,158,783	2,723,969	590,467	3,314,436	6.35
1940	51,484,640	2,783,038	942,944	3,725,982	7.22
Average*	70,991,424	3,585,507	570,056	4,155,563	5.21

Note: All figures in tons.
*For all years, 1905–1940, not just those shown.

Table A3.2. Primary Energy Use in Manufacturing, 1910–1954

Census year	1910	1920	1930	1940	1947	1954
Bituminous coal (M tons)	151,123	188,836	196,780	138,182	210,141	176,716
Anthracite (M tons)	14,470	13,740	9,452	5,168	8,781	—
All coal (M tons)	165,593	202,576	206,232	143,350	218,922	176,716
Oil (M bbl.)	19,727	69,639	136,255	133,404	215,559	249,569
Nat'l gas (Mmft.^{3})	267,906	320,386	676,765	980,515	1,723,193	6,282,946
Mfg. gas Mmft.^{3})	—	(245,411)	(1,500,426)	(1,827,692)	(2,208,811)	—
Mixed gas (Mmft.^{3})	(308,877)	—	—	—	(1,418,879)	(2,837,443)
Electricity (Mm kwh)	—	24,859	37,394	44,849	102,822	187,027
Electricity (Mm kwh)	—	—	—	(28593)	(43,936)	(69,683)
Primary BCE (M tons)*	180,483	268,985	295,047	243,788	404,683	496,282
Coal percent of BCE	0.92	0.74	0.63	0.51	0.48	0.31
MFG index	100	150	229	225	397	501
Primary/MFG (1909 = 100)	100	101	79	70	63	56

*Primary energy excludes figures in parentheses.

Table A3.3. Central Station Sources of Power and Fuel Efficiency, 1902–1940

	Fuel generated					Fuel plus waterpower			
Year	Kwh (Mm)	BCE (M tons)	Lbs. (BCE/ kwh)	Coal (M tons)	Waterpower (Mm kwh)	Lbs. (BCE/ all kwh)	Kwh (Mm)	BCE (M tons)	Share of coal
1902	1,760	6,400	7.27	4,818	750	5.10	2,500	9,127	0.53
1907	3,440	12,350	7.18	—	2,420	4.22	5,860	21,038	—
1912	6,810	17,900	5.26	—	4,760	3.09	11,570	30,412	—
1917	14,090	24,050	3.41	21,880	11,340	1.89	25,440	43,406	0.50
1919	24,315	38,880	3.20	35,100	14,606	2.00	38,921	62,235	0.56
1920	23,600	35,791	3.03	31,640	15,800	1.82	39,400	59,753	0.53
1925	39,700	40,014	2.02	35,615	21,800	1.30	61,500	61,986	0.57
1930	59,900	47,545	1.59	40,278	31,200	1.04	91,100	72,310	0.56
1935	56,900	40,797	1.43	32,715	38,400	0.86	95,300	68,330	0.48
1940	94,500	62,942	1.33	51,494	47,300	0.89	141,800	94,446	0.55

impossible to correct. Thus, primary energy BCE includes all coal, natural gas, oil, and purchased electricity.[2]

Natural gas for 1910 and 1920 is calculated as industrial uses minus field uses but including carbon black. For 1930, it is industrial use (including carbon black) minus field and utility consumption. To convert kwh into their coal equivalent, I employed the average BCE needed to generate a kwh at utilities at the time (table A3.3, col. 3). The manufacturing index derives from Shurr et al.[3]

Electric Utilities' Fuel and Power

The electric utility fuel and power data in table A3.3 are the basis of figure 3.3; they come from several sources and do not always agree where they overlap, although the differences are small. I calculated the lbs. BCE/kwh in two ways. Column 3 shows the BCE of all fuels required to generate thermal electricity. Column 6 computes the average fuel requirement for all electricity *including* that generated by waterpower. The share of coal is its share of all BCE including this waterpower equivalent.[4]

Table A3.4. Coal Equivalent of Saved Byproducts from Coking, 1913–1940

Year	Coal for coking (M tons)		Yield (coke/coal)		Byproduct (%)	Byproducts BCE (M tons)
	Byproduct	Beehive	Byproduct	Beehive		
1913	17,095	52,144	72.4	64.6	27.46	2,600
1914	15,500	36,124	72.4	64.6	32.47	2,461
1915	19,554	42,279	72.0	65.1	33.84	3,280
1920	44,205	31,986	69.9	64.1	60.17	9,316
1925	57,110	17,423	69.9	65.2	77.85	13,786
1930	65,521	4,284	68.98	64.8	94.21	16,923
1935	49,046	1,469	69.78	62.44	97.39	12,793
1940	76,583	4,803	70.53	63.66	94.64	19,036

Table A3.5. Improving Energy Efficiency in Pig-Iron Production, 1913–1940

Year	Coal for pig (M tons)	Yield of coke	Coke for pig (M tons)	Pig iron (M tons)	Coke/ Pig	Coal/ Coke	Coal/ Pig
1913	55,594	0.669	37,192	30,877	1.20	1.49	1.80
1914	40,465	0.669	27,071	22,620	1.20	1.49	1.79
1915	49,441	0.672	33,224	30,873	1.08	1.49	1.60
1920	62,279	0.674	41,976	36,283	1.16	1.48	1.72
1925	57,038	0.688	39,242	37,405	1.05	1.45	1.52
1930	46,769	0.687	32,130	30,431	1.06	1.46	1.54
1935	29,916	0.696	20,821	21,519	0.97	1.44	1.39
1940	59,685	0.701	41,839	42,601	0.98	1.43	1.40

Fuel Use in Iron and Steel Making

The data on byproduct coking in table A3.4 derive from MR and MY. The calculations of the savings attributable to the use of byproducts cannot be checked because the USGS/USBM never explained them. The yield data for 1913 are assumed to equal those for 1914. Note that the percent byproduct is the percent of coke, which differs slightly from the percent of coal because of differences in yields.

The calculations in the text of fuel use and savings in the production of pig iron derive from table A3.5. The data on coke use in making pig iron and the yield are MR and MY; I estimated coal based on those figures.[5]

Railroad Fuel Use and Efficiency

The index of railroad fuel efficiency in text table 4.1 covers all traffic except rail motorcars but excludes stationary sources of fuel use and a small amount of gasoline. The basic data, all converted to BCE, are in table A3.6.

Coal is from Bukovsky and covers locomotive fuel for all class-I steam railroads. Fuel oil also reflects consumption by class-I railroads only; data before 1909 are from the same source; from 1909 to 1928, data are from USBM and from 1928 onward from the ICC.[6] Electricity and diesel are also from Bukovsky and the ICC. I have converted fuel oil and diesel to BCE employing 4.59 bbl./ton. To convert kwh, I have employed the average amount of fuel (excluding waterpower) that utilities used to generate a kwh from col. 3 of table A3.3.[7]

Table A3.6. Railroad Fuel Consumption, 1889–1943

Year	All coal (tons)	Fuel oil (BCE)	Diesel (BCE)	Electricity (BCE)	All fuel (BCE)	Coal pct. all fuel
1889	30,308	—	—	—	30,308	100.0
1890	34,973	—	—	—	34,973	100.0
1896	43,711	—	—	—	43,711	100.0
1900	62,480	—	—	—	62,480	100.0
1904	86,258	1,665	—	—	87,923	98.1
1909	103,900	4,855	—	—	108,755	95.5
1915	122,000	8,007	—	—	160,007	76.2
1920	139,275	11,206	—	—	150,481	92.6
1925	119,889	14,273	—	—	134,162	89.4
1930	99,539	13,743	—	536	113,819	87.5
1935	71,843	11,604	—	587	84,033	85.5
1940	79,914	14,535	301	961	95,711	83.5
1943	122,874	25,746	989	1,232	150,841	81.5

Note: All figures are M tons or percent.

Table A3.7. Railroad Use of Liquid Fuels: US Total, East, and West, 1925–1960

	US Total								
	Steam		Other Locomotives				BCE	All Loco	Percent
Year	Coal	Oil	Kwh	Oil	Diesel	Gasoline	Liquid	Fuel	Liquid
1925	119,889	2,457,827	—	—	—	—	14,273	134,699	10.6
1935	71,843	1,998,176	818,093	6,093	—	630	11,642	84,445	13.8
1945	115,293	4,413,072	1,933,725	9,218	410,230	330	27,811	144,029	19.3
1950	55,452	2,277,220	1,653,097	7,223	1,822,068	230	22,719	78,897	28.8
1960	39	98	1,049,940	89,172	3,463,620	—	18,485	18,524	99.8

East

Year	Coal	Oil	Kwh	Oil	Diesel	Gasoline	BCE Liquid	All Loco Fuel	Percent Liquid
1925	58,153	10,496	—	—	—	—	61	58,517	0.1
1935	33,889	6,348	579,792	5,268	—	309	69	34,705	0.2
1945	54,899	4,974	1,482,220	9,218	95,108	211	602	56,241	1.1
1950	26,356	12,717	1,322,540	7,223	525,578	125	2,979	29,920	10.0
1960	8	—	868,048	9,707	1,007,350	—	5,543	5,551	99.8

West

Year	Coal	Oil	Kwh	Oil	Diesel	Gasoline	BCE Liquid	All Loco Fuel	Percent Liquid
1925	34,790	2,351,814	—	—	—	—	13,657	48,558	28.1
1935	21,934	1,963,383	105,934	379	—	321	11,406	33,428	34.1
1945	33,301	4,363,494	245,772	—	226,478	119	26,515	59,931	44.2
1950	13,947	2,237,250	205,211	—	967,463	105	18,011	32,049	56.2
1960	17	98	102,969	—	1,719,089	—	8,918	8,935	99.8

Note: Data in M tons, gallons, or kwh.

Table A3.7 contains the data on regional fuel use that is the basis for figure 4.3 and derives from the ICC's *Statistics of Railways*, for various years. In converting fuels to BCE, I employed the same ratios as for table A3.6. The ICC divided the nation into three regions—East, West, and South— along with subregions, but it presented fuel data for only the major regions. Moreover, its regions only roughly coincide with state boundaries. Thus, the West is approximately that part of the United States west of the Mississippi River, while the East is the region east of the Mississippi and north of the Ohio River to Parkersburg West Virginia, and then north of a line east to southwestern Maryland and then north of the Potomac River.

Appendix 4
Chapters 5–7

Chapters 5–7 focus on the changing fuels used for cooking and heating by American households in the early twentieth century. The categories are domestic sources of energy, prices, war and the distribution of anthracite, cooking and heating, and domestic energy consumption in 1940.

Domestic Sources of Energy

The data in table A4.1, when converted to BCE, provide the basis for figure 5.2. In a few instances where data are missing, I have interpolated. The table excludes wood because it, too, was losing ground and therefore was not a source of coal's problems. Here, anthracite coal includes only domestic sizes and pea. Coke includes domestic consumption only, while all briquettes are included because they were exclusively domestic fuels. For domestic use of bituminous coal, the 1915 value includes the "small steam trade"; for 1928, it is a "rough estimate"; for 1934, on the data, include "retail deliveries." Natural gas is only that used for domestic and commercial consumption (from table A1.6). The data on fuel oil include domestic and commercial uses of fuel oil distillates (grades 1–4).[1]

Domestic consumption of kerosene and range oil are from MR and MY, and they are partly duplicative because range oil includes both Number 1 fuel oil and that portion of kerosene sold for cooking and heating. There is good reason to believe that, at least in the early years, the range oil figures are far too low.[2]

Prices

Table A4.2 contains producer prices of coal and some of its competitors. They, too, are from MR and MY, various years. For briquettes, the bureau provided two sets of prices; one for Pennsylvania

Table A4.1. Major Domestic Fuels for Heating, Cooking, and Lighting, 1906–1940

	M tons					Mmft.³		M bbl.		
Year	Hard coal	Hard coal imports	Soft coal	Briquettes	Coke	Mfg. gas	Natural gas	Kerosene	Fuel oil	Range oil
1906	36,841	—	—	—	—	—	110,406	—	—	—
1910	42,768	—	—	179	—	—	169,823	—	—	—
1915	55,041	14	71,336	222	3,637	152,890	217,201	—	—	—
1920	53,967	32	—	567	3,931	214,000	286,001	33,082	—	—
1925	41,308	383	—	839	4,479	242,000	272,146	39,969	8,830	—
1928	51,366	385	70,000	947	6,333	275,000	320,877	36,235	17,203	—
1930	48,502	675	—	1,029	8,028	301,000	376,407	34,736	28,769	—
1935	42,200	571	83,990	861	10,209	258,000	413,685	47,645	53,804	21,526
1940	39,300	135	87,700	1,028	8,344	267,000	578,290	67,662	115,533	44,692

and the other for central states. These are for Pennsylvania and are typically less than the prices in central states.

In table A4.3, coal, gas, and fuel oil prices are retail; kerosene is wholesale. Note that the rise in natural gas prices is largely a reflection of its increasingly long-distance transport.[3]

War and the Distribution of Anthracite

The data in table A4.4 exclude steam sizes, coal used on railroads and mines, and exports. As the text stresses, they reveal very large declines in the Midwest and West. For 1921, the original data are in gross (2,240 lb.) tons; I have converted them to 2,000 lb. tons.[4]

Table A4.2. Producer Prices of Coal, Coke, and Briquettes, 1900–1940 (all prices per ton)

Year	Anthracite ($)	Bituminous ($)	Anthracite/ Bituminous ($)	Briquettes ($)	Coke ($)
1900	1.49	1.04	1.43	—	—
1905	1.83	1.06	1.73	—	—
1910	1.90	1.12	1.70	—	—
1911	1.94	1.11	1.75	2.47	—
1915	2.07	1.13	1.83	2.90	—
1918	3.40	2.58	1.32	4.11	7.45
1920	4.85	3.75	1.29	5.60	8.93
1925	5.30	2.04	2.60	6.35	6.98
1930	5.11	1.70	3.01	6.22	6.03
1935	4.03	1.89	2.13	4.48	4.48
1940	3.99	1.91	2.09	4.29	4.29

Table A4.3. Retail and Wholesale Prices of Coal and Some of Its Competitors, 1913–1940

Year	Anthracite	Bituminous	Manufactured gas	Natural gas	Fuel oil	Kerosene
	$/ton		$/Mft.^{3}		$/gallon	
1913	7.79	5.48	0.95	0.27	—	0.090
1915	7.83	5.71	0.93	0.28	—	0.086
1920	12.59	8.81	1.09	0.38	—	0.171
1925	15.45	9.24	1.23	0.56	.084*	0.132
1930	14.03	8.83	1.22	0.64	.066	0.122
1935	11.38	8.29	1.05	0.69	.063	0.075
1940	11.33	8.60	0.97	0.67	.069	0.082

*Figure is for 1926.

Table A4.4. Geographic Distribution of Domestic Anthracite Consumption, 1916 and 1921

Region and states	1916 Net tons (M)	1916 Percent of total	1921 Net tons (M)	1921 Percent of total	Percent change (negative)
New England	9,949	19.5	9,402	19.3	(5.5)
Atlantic states (NY, NJ, PA, DE, MD, DC, VA)	31,454	61.6	31,744	65.3	0.9
Central states (OH, IN, IL, MI)	5,582	10.9	4,539	9.3	(18.7)
Northwest (WI, MN, ND, SD, MT)	3,209	6.3	2,565	5.3	(20.1)
Trans Mississippi (IA, MO, NE, KS)	866	1.7	366	0.8	(57.7)
Total	51,060	—	48,646	—	—

Table A4.5. Heating Characteristics by Income Levels, Four Cities, 1933

Income level	$1–499	$500–999	$1000–1499	$1,500–1,999	$2,000–2,999	$3,000–4,999	$5,000–6,999	$7,000 & up
Fargo, North Dakota								
No. of families	169	263	302	240	236	113	22	7
Use of stove (%)	24.8	18.3	7.6	2.9	0.4	0.9	0	0
Coal heat (%)	94.7	95	91.7	89.6	81.8	63.7	45.5	14.3
Gas heat (%)	0	0	0	0	0	0	0	0
Oil heat (%)	2.4	4.2	7	10	18.2	36.3	50	85.7
Trenton, New Jersey								
No. of families	729	1021	765	398	334	121	32	10
Use of stove (%)	45.5	33.7	26.4	7	3.9	3.3	0	0
Coal heat (%)	98.4	98.1	97.4	6.7	95.8	89.9	87.5	70
Gas heat (%)	0	0	0	0	0	0	0	0
Oil heat (%)	0.5	1.5	2.5	3.3	3.6	12.4	9.4	30
Cleveland, Ohio								
No. of families	8,756	10,030	7,908	5,255	4,354	2,074	511	412
Use of stove (%)	42.3	28.8	15.3	7.1	3.2	1.2	2.4	0.5
Coal heat (%)	91.2	91.6	93	93.6	63.4	87.9	74.3	50.9
Gas heat (%)	8.5	8.1	6.6	6	6	11.1	24.1	44.2
Oil heat (%)	0.1	0.1	0.2	0.2	0.3	0.7	1	4.4
Providence, Rhode Island								
No. of families	1,198	1,982	1,723	1,076	828	348	111	71
Use of stove (%)	66	60.9	42.3	26	12.1	4.9	0.9	1.4
Coal heat (%)	67.2	65.3	66.8	72.1	70.5	67.8	62.2	25.4
Gas heat (%)	0.2	0	0.1	0	0.2	0.3	0.9	7
Oil heat (%)	2.2	2.8	4.4	7.2	9.8	19.2	31.5	60.6
Kerosene heat (%)	23.5	26.1	20.3	11.8	6.4	2.6	0.9	0

Cooking and Heating

Table A4.5 omits families with no income. Accordingly, in the table, nearly all those without stoves had central heat. The fuel categories not shown include wood and "other." The type of gas—manufactured or natural—is not always specified, but of these cities, Cleveland was the only one with natural gas. These data derived from surveys of 50 cities, and although no four can be entirely representative, these seem to capture the way income and availability of oil and natural gas shaped heating choices.[5]

The data on heating equipment in table A4.6 are not entirely consistent. After 1920, the total for oil burners in use appears to include domestic and commercial conversion units and new furnaces. New orders from 1926 to 1928 also appear to be domestic and commercial. From then on, new orders include industrial burners as well. Unfortunately, there are no data on furnaces by type of

Table A4.6. Oil, Coal, and Gas Central Heating Equipment, 1920–1940

| | Conversion oil burners | | Gas central heat | | |
| | | | New orders | | |
Year	Total in use		Coal stokers	Conversion	Total
1920	7,500	—	—	—	—
1925	181,500	—	—	—	—
1926	186,275	77,700	—	—	—
1930	606,300	119,220	—	—	—
1931	712,950	83,352	11,361	—	—
1935	1,083,120	139,452	45,612	—	—
1936	1,246,400	194,016	82,704	38,000	70,000
1940	1,998,730	265,416	147,612	22,000	117,000

Table A4.7. Domestic Energy Use, 1910 and 1940

| | 1910 | | 1940 | |
Fuels	Reported	BCE (M tons)	Reported	BCE (M tons)
Bituminous coal (M tons)	65,562	65,562	87,700	87,700
Anthracite coal (M tons)	42,768	41,485	39,300	38,121
Wood (M cords)	84,230	63,173	48,500	36,375
Coke (M tons)	3,637*	3,637	9,075	9,075
Briquettes (M tons)	179	179	1,312	1,312
Fuel oil (M bbl.)	0	0	115,533	26,868
Range oil (M bbl.)	14,962	3,239	44,692	9,674
Gasoline (M bbl.)	11,224	2,231	555,340	110,406
Electricity (Mm kwh)	910	2,396	24,068	16,005
Mfg. gas (Mm therms)	606	2,314	1,172	4,473
Natural gas (Mm therms)	1,783	6,806	4,064	15,511
Mixed gas (Mm therms)	0	0	587	2,240
Total		191,019		357,760
Population (M)		92,407		132,122
Families (M)		20,183		35,153
All energy BCE per capita		2.07		2.71
All energy BCE per family		9.46		10.18
Coal tons per capita		1.17		0.96

*Data are for 1915.

fuel until after World War II. From 1933 on, stokers are class 1–3 (small domestic and commercial) stokers. Stokers for 1931 and 1932 are class 1 and 2 only.[6]

Domestic Energy Consumption in 1940

The discussion in chapter 7 of domestic energy consumption in 1940 is based on table A4.7, which, for comparative purposes, contains data for 1910 as well. For 1940, the data for solid fuels, fuel oil, and range oil also include commercial fuel. The figures for wood consumption include mill waste, as well as cordwood. Gasoline consumption includes that used for private and commercial purposes. Gas and electricity data are residential only; to convert kwh to BCE, I employed the average utility coal rate for 1940.[7]

Abbreviations

AGA	American Gas Association		ihph	indicated horsepower-hour
AGLA	American Gas Light Association		IRFA	International Railway Fuel Association
AGLJ	*American Gas Light Journal*		JEH	*Journal of Economic History*
AIEE	American Institute of Electrical Engineers		JFI	*Journal of the Franklin Institute*
			M	thousands
AIME	American Institute of Mining Engineers		ME	*Mechanical Engineering*
AISI	American Iron and Steel Institute		Mm	millions
AMC	American Mining Congress		MR	*Mineral Resources of [the] United States*
AJS	*American Journal of Science*		MY	*Minerals Yearbook*
APS	American Philosophical Society		NCGA	National Commercial Gas Association
ARA	American Railway Association		NELA	National Electric Light Association
ARRJ	*American Railroad Journal*		NGA	Natural Gas Association
ART	*American Railway Times*		NSF	*National Stockman and Farmer*
ASHVE	American Society of Heating and Ventilating Engineers		NYHT	*New York Herald Tribune*
			NYT	*New York Times*
ASME	American Society of Mechanical Engineers		PRRHML	Pennsylvania Railroad Collection, Hagley Museum and Library
ASTM	American Society for Testing Materials		psi	pounds per square inch
BCE	bituminous coal equivalent		RA	*Railway Age*
BG	*Boston Globe*		RG	*Railroad Gazette*
BHR	*Business History Review*		RI	*Report of Investigation*
CA	*Coal Age*		RME	*Railway Mechanical Engineer*
C&ME	*Chemical and Metallurgical Engineering*		SA	*Scientific American*
CT	*Chicago Tribune*		SP	Southern Pacific Railroad
E&MJ	*Engineering and Mining Journal*		T&C	*Technology and Culture*
EW	*Electrical World*		TAPPI	Technical Association of the Pulp and Paper Industry
FOJ	*Fuel Oil Journal*		TP	*Technical Paper*
ft.^{3}	cubic feet		USBLS	United States Bureau of Labor Statistics
GH	*Good Housekeeping*		USBM	United States Bureau of Mines
H&V	*Heating and Ventilating*		USDA	United States Department of Agriculture
IC	*Information Circular*			
ICC	*Interstate Commerce Commission*		USGS	United States Geological Survey
ihp	indicated horsepower			

Measurement and Conversion Factors

I have used the following definitions and estimates of the energy content of various fuels to convert them into their bituminous coal equivalents. The definitions are standard in engineering. These include long tons and tons, Btus and kilowatt-hours, candlepower and lumens, horsepower and watts. The energy-content figures come from Schurr et al., *Energy in the American Economy*, and the United States Bureau of Mines. Experts differ on these; in fact, there is often wide variation in the energy content of a given fuel, so definitions here are approximations. While these figures capture the energy *content* of various fuels, they ignore the efficiency with which the fuels are used, and where efficiencies differ, this will affect the exchange rates between fuels. Finally, throughout this book, in estimating the coal equivalent of a kilowatt-hour, I employ the then-current estimate of coal required to generate it:

Anthracite coal: 25,400,000 Btu per ton = .97 tons bituminous.

Bituminous coal: 26,200,000 Btu per ton; 25 bushels = one ton.

Barrel (one): 42 U.S. gallons.

Btu: 0.00029307107017 kwh.

Candlepower (one): 12.57 lumens.

Charcoal: 100 bushels = one ton; one ton = .93 tons bituminous.

Coke: 45 lb. per bushel; 44 bushels to the ton; one ton = one ton bituminous.

Cordwood: 1800–1899, 20,960,000 Btu/chord; one chord = .8 tons bituminous. For 1900, 20,154,000 Btu/cord; one cord = .77 tons bituminous. For 1901 onward, 19,407,000 Btu/cord; one cord = .75 bituminous.

Crude oil: 5,800,000 Btu/barrel = 4.5 bbl. = one ton bituminous.

Fuel oil (#2) & Diesel oil: 5,712,000 Btu/bbl. or 4.59 bbl. = one ton bituminous.

Gasoline: 5,208,000 Btu/bbl.; 5.03 bbl. = one ton bituminous.

Horsepower (one): 745.7 watts = .7457kw; 1 kwh = 1.341 hph.

Kerosene: 5,670,000 Btu/bbl.; 4.62 bbl. = one ton bituminous.

Kwh (one): 3,412.142 Btu; 7,678.462 kwh = one ton bituminous.

Long or gross ton (one): 2,240 lbs. = 1.12 ton.

Manufactured gas: 550 Btu/ft.^{3}; 47,636 ft.^{3} = one ton bituminous (46,181 anthracite).

Natural gas: 1,050 Btu/ft.^{3}; 24,952 ft.^{3} = one ton bituminous (24,190 = one ton anthracite).

Residual oil: 6,384,000 Btu/bbl.; 4.10 bbl. = one ton bituminous.

Therm (one) = 100,000 Btu.

Notes

Introduction

1. Henry Adams, *The Education of Henry Adams* (New York: Modern Library, 1931), 490.

2. Berton Braley, "King Coal," CA 4 (Sept. 27, 1913): 456.

3. Ise, *United States Oil Policy*. See also Hays, *Conservation*, 82–87, where Hays points out that these lands were later made available but only at higher prices.

4. Origins of the conference and attendees are from Blanchard et al., *Proceedings*. See also Dorsey, *Theodore Roosevelt*, chap. 4; "A Century of Waste," *The Independent*, Oct. 4, 1900, 2400–2401; and "The Empty Coal Cellar," AGLJ 72 (May 21, 1900): 812. Neither White nor Holmes had private-sector experience. Gifford Pinchot worked in the private sector as a forester at Vanderbilt's Biltmore Estate. The chemist Charles B. Dudley, who spoke briefly, worked for the Pennsylvania Railroad.

5. Theodore Roosevelt, "Opening Address by the President," in Blanchard et al., *Proceedings*, 3–12; "wealth" is on 3; "nowhere," "reckless," and "wise use" are on 6; "these resources" is on 7; "right of the nation" is on 10; "property rights" is on 11; "national efficiency" is on 12. Tyrrell, in his *Crisis of the Wasteful Nation*, chap. 5, provides an excellent discussion of these matters and is one of the few modern writers to stress the connection Progressives saw between energy and national power.

6. Theodore Roosevelt, "The National Editorial Association at Jamestown, VA., June 10, 1907," in *Presidential Addresses and State Papers*, vol. 6 (New York: Review of Reviews, 1910), 1308–1323; "the reckless" and "to get our people" are on 1311; "under private control" is on 1314.

7. Israel C. White, "The Waste of Our Coal Resources," in Blanchard et al., *Proceedings*, 26–37, 26, 33, 36. This resource pessimism of Roosevelt, White, and others has a lineage stretching back to Malthus at least. In modern times the pessimism

has largely shifted to environmental pollution and, most recently, climate change. For a critique, see Simon, *The Ultimate Resource*. For a broader discussion of the evolution of official thinking about the relationship between coal and national power, see Peter Schulman, *Coal and Empire* (Baltimore: Johns Hopkins University Press, 2015).

8. "Memorial of Joseph Austin Holmes," Geological Society of America *Bulletin* 27 (1916): 22–34; Parker, "Report on Operations"; "National Waste," *New York Herald Tribune* (NYHT), Oct. 6, 1907.

9. Holmes, "Mineral Resources Summary," in Gannett, *Report*, 1:95–111, 109.

10. Joseph Holmes, "How Conservation of Mineral Resources Can Be Accomplished," in Blanchard et al., *Proceedings*, 439–445, 441 (emphasis added); Joseph A. Holmes, "A Rational Basis for the Conservation of Natural Resources," AIME *Transactions* 49 (May 1909): 469–476. See also the preface by Holmes in Charles L. Parsons, "Notes on Mineral Wastes," USBM *Bulletin* 47 (Washington, 1912); Holmes, "Carbon Wastes," *Journal of Industrial and Engineering Chemistry* 4 (March 1912): 160–162; and Marius Campbell and Edward Parker, "Coal Fields of the United States," in Gannett, *Report*, 3:426–442. The bureau's purposes are from US Congress, *Bureau of Mines*. The best discussion of its founding is Graebner, *Coal Mining Safety*, chap. 1.

11. Gifford Pinchot, "The Conservation of Natural Resources," *Outlook*, Oct. 12, 1907, 291–294, 291. See also Gifford Pinchot, "The ABC of Conservation," *Outlook*, Dec. 4, 1909, 770–772, 770. In the commission's summary (which he signed), Pinchot qualified his claim that the country would run out of anthracite. The *"available and easily accessible* supplies of coal . . . [will] at the present increasing rate of production . . . be so depleted as to approach exhaustion before the middle of the next

century." See Gifford Pinchot et al., "Report," in Gannett, *Report*, 1:15 (emphasis added).

12. Van Hise, *The Conservation of Natural Resources*, 2, 19, 25, 27, 35 (fatalities on 370). Van Hise did emphasize the "smoke nuisance" and claimed that its effect on public health was "even more important" than conservation (30). He also noted that the rise in CO_2 resulting from coal burning would raise global temperatures, which he seemed to think a good thing because it would reduce the demand for coal. Van Hise's importance is from Williamson, "Prophecies of Scarcity," 106.

13. In modern times, of course, the emphasis of conservation policy has shifted away from fuels and minerals, with market value such as coal and oil and copper, toward clean air and water and other amenities. See Krutilla, "Conservation Reconsidered."

14. The emphasis on costs also helps explain the concern with national rather than international supplies of resources, for fuel purchased on international markets would be at world-market prices, erasing America's cost advantage, on which, all assumed, the country's high-wage manufacturing sector depended.

15. Pinchot et al., "Report," in Gannett, *Report*, 1:13–26, 16; Joseph A. Holmes, "Mineral Resources Summary" 101; David Day, "The Petroleum Resources of the United States," in Gannett, *Report*, 3:446–464. Pinchot later retreated from this utilitarian position; see Miller, "The Greening of Gifford Pinchot." For continuing concerns over coal, petroleum, and gas reserves, see Williamson, "Prophesies of Scarcity." Hays catches this market skepticism; see his *Conservation*, chap. 13.

16. "The salvation" is from "Waste; Waste—Nothing but Waste," *The Independent*, June 4, 1908, 1301–1302; "The great clock" is from "Conserving National Resources," *New York Observer and Chronicle* 86 (June 18, 1908): 791–792.

17. Economists contributed comparatively little to these discussions. The understanding of how efficient markets would allocate nonrenewable resources over time was then in its infancy. See, e.g., Gray, "The Economic Possibilities of Conservation"; and Hotelling, "The Economics of Exhaustible Resources." Institutionalists such as Ely, *Foundations of National Prosperity*, although more historically oriented, tended to be statists and were skeptical of markets. Accordingly, they stressed the ways that markets failed but not how they worked. There is no evidence that any of these ideas had any influence on policy makers' thinking about exhaustible resources.

See Crabbé, "The Contribution of L. C. Gray"; and Smith, "Natural Resource Theory."

18. Douglas's biography is from the James Douglas Collection. "It is better," "sin," and "every thinking man" are from James Douglas, "Conservation of Natural Resources," AIME *Transactions* 40 (1909): 419–431, 419, 424, 430. Other quotations are from James Douglas, "Introductory Address," AIME *Bulletin* 29 (May 1909), appendix, n.p.; Hays, *Conservation*, 263; and Henry Petroski, *To Engineer Is Human: The Role of Failure in Successful Design* (New York: St. Martin's, 1985).

19. For similar views, see John Birkinbine, "The American Institute of Mining Engineers and the Conservation of Resources," AIME *Transactions* 40 (1909): 412–418. The American Chemical Society, although it praised the conference, also pointed out all that chemists had done and were doing for conservation. "Report of the Committee of the American Chemical Society Appointed to Co-operate with the National Conservation Commission," *Journal of Industrial and Engineering Chemistry* 1 (Feb. 1909): 115–117. Even before the conference met, the American Society of Mechanical Engineers felt the need to point out all that engineers were doing for conservation; see *Transactions* 30 (1908): 13–25. Marius Campbell and Edward Parker, "Coal Fields of the United States," in Gannett, *Report*, 3:426–442, 430 (emphasis added), 432.

20. Cohn, in her "Utilities as Conservationists?," is one of the few modern writers to emphasize Progressives' coal worries, as well as business interests. Another is Tyrrell, *Crisis*, chap. 5. Samuel Hays, *Conservation*, emphasizes the role of technical people in the conservation movement—as does the present work—but discusses coal only in terms of federal policy toward coal lands. Desrochers discusses modern writers' skepticism of the role of markets in "How Did the Invisible Hand?" Gottlieb, in *Forcing the Spring*, and Johnson, in *Escaping the Dark, Gray City*, say little about coal or market-induced conservation. Other writers emphasize aesthetic and ethical motivations. Stoll, in *Inherit the Holy Mountain* and elsewhere, stresses its religious roots.

21. Schumpeter, *Capitalism, Socialism and Democracy*, 7, 67, 98.

22. For a good modern exposition of Schumpeter's ideas, see Rosenberg, *Schumpeter and the Endogeneity of Technology*, which contains the quotation from Schumpeter on page 10.

23. Baumol, *Free Market Innovation*; Arthur Mellen Wellington, *The Economic Theory of the*

Location of Railways, 2d ed. (New York: John Wiley, 1887), 1.

24. Rosenberg, "Direction of Technological Change"; Rosenberg, "Technological Change in the Machine Tool Industry."

25. David and Wright, "Increasing Returns," stresses the social construction of abundance; as the authors put it, abundance was "endogenous." "Invisible college" is from Derek Price, *Little Science, Big Science* (New York: Columbia University Press, 1963).

26. For the broader political and legal context within which markets operate, see Sheshinski et al., *Entrepreneurship*. I have taken the idea of market-augmenting activities from Olson, *Power and Prosperity*.

27. Jeffery Zallen's *American Lucifers: The Dark History of Artificial Light, 1750–1865* (Chapel Hill: University of North Carolina Press, 2019) emphasizes the various costs of energy.

28. The argument in the text assumes that all external benefits of conservation are captured by market prices. Also, waste has many other possible definitions: my neighbor is wasting energy by leaving his lights on all night.

29. Jevons, *The Coal Question*, 123 (emphasis his). Jevons was referring to industrial, not domestic, uses; and writing in 1865, he did not foresee the rise of oil and gas. With such substitutes available, conservation of coal would be more likely to expand its market, as Jevons claimed. For a modern assessment, see Sorrell, "Jevons' Paradox Revisited."

30. Critics such as Holmes wanted to mine thin coal seams, which was uneconomic; the process would save coal for the future but would have reduced the standard of living at the time. This "conservation ethic" is not wrong in some logical sense—see Page, *Conservation and Economic Efficiency*, chap. 9—but consider the implications. Progressives would have reduced the standard of living for people a century ago to save coal for Americans who are now far richer and no longer want to burn the coal.

31. For the likely impact of modern policies on living standards, see Greg Ip, "Who Will Pay for Green Transition?," *Wall Street Journal*, Nov. 30, 2023. One way to combine the top-down pressures for an energy transition away from carbon fuels along with the use of markets to smoke out the cheapest ways to achieve that transition is to implement a carbon tax. See N. Gregory Mankiw, "A Carbon Tax That America Could Live With," *New York Times*, Sept. 1, 2013, BU4.

Chapter 1. The Dawning of the Coal Age, 1800–1860

1. Primary energy is energy that is a natural resource; it is differentiated from secondary sources of energy that may be derived from it such as electricity or waste gases.

2. The data on waterpower in figure 1 were originally expressed as horsepower-hours. Following others' work, I have converted them into coal equivalents. For details, see appendix 1. Because data for coal and wood reflect energy content, not work output, for consistency, I have estimated the food energy it took to maintain horses and mules, not their output of horsepower.

3. "Prodigal Waste of Natural Resources," NYT, Nov. 19, 1907; Jones, *Routes of Power*, 1. For similar findings for the post-1880 years based on mineral fuels and waterpower only, see Schurr et al., *Energy in the American Economy*, chaps. 1 and 4; and Schurr, "Energy Efficiency."

4. In "Measurement and Conversion Factors," I follow conventional definitions of the average energy content of various fuels. Thus, bituminous coal is assumed to contain 26.2 Mm Btu/ton, and a ton of anthracite is equivalent to 0.97 tons of soft coal, a cord of hardwood 0.74 tons of bituminous.

5. This paragraph is based largely on Eavenson, *First Century*, chap. 3. For tariffs, see Adams, "Promotion, Competition, Captivity."

6. Eavenson, *First Century*, 157, 245–246, 300, 165, 171.

7. Pollution levels are from Davidson, "Air Pollution in Pittsburgh." For the effect of smoke on urban mortality in Britain at this time, see Beach and Hanlon, "Coal Smoke and Mortality." "Statistics of Pittsburgh, PA.," *Country Gentleman* 3 (June 29, 1854): 410–411, 411.

8. "This whole country" is quoted in Eavenson, *First Century*, 164; Davidson, "Air Pollution in Pittsburgh."

9. James Fenimore Cooper, *The Pioneers* (Philadelphia: Lea & Blanchard, 1841), 141. A ton of coal takes about 40 ft.^{3} of space, whereas a cord of wood requires 128 ft.^{3} Since a ton of coal is the energy equivalent of about 1.25 cords of hardwood, the volume of wood equivalent to a ton of coal is 160 ft.^{3}

10. Silvio Bedini, *Thinkers and Tinkers: Early American Men of Science* (New York: Scribner, 1975). By "thinkers and tinkers," Bedini meant scientists who were also technologists. I am using the term differently to refer to two groups—those with scientific training and those without, both of whom worked on energy-related subjects. Literature

on the interrelationships between knowledge, science, and the Industrial Revolution is too vast to summarize here. For a broad and elegant assessment, see Mokyr, *The Gifts of Athena*. Nathan Rosenberg has argued that America's resource abundance and its rapid exploitation encouraged the development of distinctive American technologies in wood- and metalworking; see Rosenberg, *Exploring the Black Box*, chap. 6.

11. Primary-school enrollment is from Easterlin, "Why Isn't the Whole World Developed?," table 1.

12. On consulting, see Lucier, *Scientists and Swindlers*. On the increasing importance of science, see Mowery and Rosenberg, *Technology and the Pursuit*, chaps. 2–3.

13. Hindle, *America's Wooden Age*.

14. The French traveler is quoted in Billington, *Land of Savagery*, 202. See also Nash, *Wilderness and the American Mind*. The 95-percent figure is from Schurr et al., *Energy in the American Economy*, 53. "Bulk of the fuel" is from Williams, *Americans and Their Forests*, 133. See also "Steamboats," *Western Monthly Magazine and Literary Journal* 3 (June 1834): 316.

15. The original data for figure 1.1 present figures on low, medium, and high prices for each year; where medium prices were not available, I averaged those for low and high. Where the data were for face cords (4' × 4' × 1') rather than full cords (4' × 4' × 8'), I multiplied them by eight. The data prior to 1777 are undeflated because the price index begins in 1774. For a review of fuel-wood use, see Cole, "Mystery of Fuel Wood Marketing."

16. William Jordan, "Geology and the Industrial-Transportation Revolution in Early to Mid Nineteenth Century Pennsylvania," in *Two Hundred Years of Geology in America: Proceedings of the New Hampshire Bicentennial Conference on the History of Geology*, ed. Cecil Schneer (Hanover, NH: University Press of New England, 1979), 91–103. Jordan reverses the causation and argues the importance of industrial development for geology. Most other works focus largely on the scientific consequences of the surveys; see, e.g., Millbrooke, "State Geological Surveys." Eavenson, *First Century*, table 20, reveals the beginnings of mining on a state-by-state basis. David and Wright, "Increasing Returns," emphasizes the surveys' importance in resource exploitation but focuses on metal mining.

17. Discussions of steam-powered systems of drainage and ventilation are in Daddow, *Coal, Iron, and Oil*, chaps. 22–24. See also "Mine Power and Electrification," CA 41 (Oct. 1936): 442–452; Yearly, *Enterprise and Anthracite*, 64, 122; and James

Douglas, "Conservation of Natural Resources," AIME *Transactions* 40 (1909): 419–431, 424.

18. For the early anthracite industry, see Healey, *The Pennsylvania Anthracite Coal Industry*; and Powell, *Philadelphia's First Fuel Crisis*. The pamphlet is reprinted in Jacob Cist, "Lehigh Coal," AJS 4 (1822): 8–16. The quotation is from Eavenson, *First Century*, 148.

19. For the anthracite canals, see Jones, *Economic History of the Anthracite-Tidewater Canals*. For excellent modern studies, see Jones, "A Landscape of Energy Abundance"; and Jones, *Routes of Power*.

20. For overviews of anthracite, see Adams, "Promotion, Competition, Captivity"; and Adams, *Home Fires*, chap. 3, which discusses early marketing. Binder's *Coal Age Empire* contains the material on the Franklin Institute. "Dealers of wood and coal" is from Cole, "Mystery of Firewood Marketing," 348.

21. Bogen, *The Anthracite Railroads*; Philadelphia and Reading Railroad, *Report of the President and Managers to the Stockholders* (Philadelphia, 1854). Alas, the Stourbridge Lion was a flop, being too heavy for the flimsy track; see Delaware and Hudson Company, *A Century of Progress* (Albany, 1925).

22. "Rapidly increasing" is from "Anthracite Coal Trade of the United States," *Hazard's Register* 8 (July 16, 1831): 47–48; "Anthracite Coal vs. Wood," *Hazard's Register* 13 (Nov. 15, 1834): 312–323; Massachusetts Bureau of Labor, *Sixteenth Annual Report* (Boston, 1885), 20; "Fuel Imported into Massachusetts," *Hunt's Merchants' Magazine*, May 1842, 475. On the share of wood in family fuel bills, see US Bureau of Labor Statistics (USBLS), "Cost of Living," 301.

23. Marcus Bull, "Experiments to Determine the Comparative Quantities of Heat Evolved in the Combustion of the Principal Varieties of Wood and Coal Used in the United States, for Fuel; and also to Determine the Comparative Quantities of Heat Lost by the Ordinary Apparatus Made Use of for Their Combustion," APS *Transactions* 3 (May 1826): 1–63. "Scientific disposition" is from Sinclair, *Philadelphia's Philosophical Mechanics*. Benjamin Silliman, "Appendix, May 11, 1826," AJS 11 (Oct. 1826): 98; see also "Economy of Fuel," *Genesee Farmer and Gardner's Journal* 4 (Feb. 22, 1834): 57–58.

24. Henry Darwin Rogers, *Fifth Annual Report on the Geological Exploration of the Commonwealth of Pennsylvania* (Harrisburg, 1841); Johnson, *Notes on the Use of Anthracite*; Johnson, *A Report to the Navy Department*. Biographical data on Johnson are from Samuel Parr, "A Pioneer Investigator," *Industrial and Engineering Chemistry* 18 (Jan. 1926): 94–98.

The interest of the US Navy, and more broadly of the federal government, in coal research is discussed in Peter Schulman, *Coal and Empire* (Baltimore: Johns Hopkins University Press, 2015), chap. 15. "Coal of Pennsylvania and Other States," *Hunt's Merchants' Magazine*, July 1845, 67–72; H. M. Chance, "The Relative Value of Coals to the Consumer," AIME *Transactions* 4 (1886): 19–33.

25. Brown, *Count Rumford*. The author, whose house sports a Rumford-style fireplace, can testify to its efficiency.

26. Franklin's claim is reprinted in Benjamin Franklin, "Franklin's Fireplace," *Useful Cabinet* 1 (May 1, 1808): 97–105. "The fire place" is from "Warming Houses," *The Cultivator* 4 (March 1847): 75–76; Henry David Thoreau, *Walden* (New York: E. P. Dutton, 1904), 203.

27. Production data are from Jeremiah Dwyer, "Stoves and Heating Apparatus," in DePew, *1795–1895*, 357–363, which also has a good history of stove evolution. For modern work, see Brewer, *From Fireplace to Cookstove*. The best source for stove making and marketing is Harris, "Inventing the US Stove Industry." Adams's *Home Fires* integrates fuels with developments in heating apparatus. The equation for 1830–1870, with year 1830 = 1, 1840 = 2, etc. is Stoves = 405,000 − 523,571 × YEAR + 171,429 × YEAR^2.

28. Nathaniel Hawthorne, "Fire Worship," in *Mosses from an Old Manse* (Boston: Houghton Mifflin, 1882), 159–169, 159–160; William Dean Howells, "The Country Printer," in *Impressions and Experiences* (New York: Harpers, 1896), 1–44, 7.

29. The estimates for 1800 are derived as follows. Domestic firewood is assumed to be 95 percent of the total (table A1.1); I have taken domestic coal use as one-half the total (table A1.4), while whale oil is from appendix 1. In both 1800 and 1860, I have no estimates for the vegetable and lard oil used for lighting. Also in both years, I ignore the horsepower that would have been used for domestic purposes. It seems unlikely that inclusion of any of these would reverse the conclusion of declining per capita domestic energy use.

30. For the stove's importance in forest preservation, see Hogland, "Forest Conservation and Stove Inventors." The cost of stove-size wood is from John Foster, *Report upon the Mineral Resources of the Illinois Central Railroad* (New York, 1856), 15. "Shiver and shake" is from "Untitled," *Farmer's Cabinet and American Herd Book* 9 (Nov. 15, 1844): 134.

31. Levels of PM$_{10}$ are from Nigel Bruce et al., "Indoor Air Pollution in Developing Countries: A Major Environmental and Public Health Chal-

lenge," *Bulletin of the World Health Organization* 78 (2000):1078–1092. The relative risk of exposure is from Kirk Smith, "National Burden of Disease in India from Indoor Air Pollution," *Proceedings of the National Academy of Sciences* 97, no. 24 (Nov. 21, 2000): 13286–13293. Regina Rückerl et al., "Health Effects of Particulate Air Pollution: A Review of Epidemiological Evidence," *Inhalation Toxicology* 23, no. 10 (2011): 555–592.

32. Chandler, "Anthracite Coal." The fuel used in puddling is from "A Detailed Statement of All the Rolling Mills in Eastern Pennsylvania in the Year 1850," in Convention of Iron Masters, *Documents*. The figure for western Pennsylvania rolling mills, which used predominantly soft coal, was 4.4 tons per ton of product.

33. The best modern explanation for the slow adoption of coke as a fuel for iron making is Peter Temin, *Iron and Steel*. Paskoff, *Industrial Evolution*, provides a detailed history of Pennsylvania iron making up to the Civil War. That coke was not used in 1841 is from Johnson, *Notes on the Use of Anthracite*, 4, which also records early difficulties in using that fuel. For the tangled history of who invented what and when, see Tenth Census of the United States, 1880, vol. 2, *Report on the Manufactures of the United States* (Washington, 1882), 113–117. British experiments are reported in Thomas Clark, "On the Application of the Hot Blast in the Manufacture of Cast-Iron," JFI 20 (1837): 46–52; see also "The Blast for Iron Furnaces," SA, Jan. 24, 1863, 57.

34. Pig iron by fuel type is from Swank, *History of the Manufacture of Iron*, chap. 43. The per-ton data are from John Church, "Blast Furnace Statistics," AIME *Transactions* 4 (1876): 221–226. The Pennsylvania data are from Convention of Iron Masters, *Documents*. My calculation is (641,439 × 2.12) + (641,439/2) × 4.00 = 2,642,729.

35. Overman, *The Manufacture of Iron*; Lesley, *The Iron Manufacturer's Guide*; Pennsylvania Geological Survey, *Fifth Annual Report* (Harrisburg, 1841).

36. On the importance of iron making, see Chandler, "Anthracite Coal."

37. Temin, *Iron and Steel*, chap. 3.

38. United States Treasury, *Documents Relative to the Manufactures*, 926; United States Treasury, *Steam Engines*. The fraction built since 1835 is from Hunter, *History of Industrial Power*, 2:74.

39. Hunter, *History of Industrial Power*, 2:84–102, discusses the geographical distribution of steam power; the Cincinnati manufactures quotation is on 92. See also Atack et al., "Regional Diffusion and

Adoption of the Steam Engine," 281–308; and Halsey, "The Choice between High-Pressure and Low-Pressure Steam." For Zachariah Allen's discussion, see Allen, *The Science of Mechanics*, 89–91.

40. "Coal seems" is from Pursell, *Early Stationary Steam Engines*, 85. Stevenson, *Sketch of the Civil Engineering*, 104; Clark, *History of Manufactures*, 332. On Fall River, see Smith, *Cotton Textile Industry of Fall River*, chap. 2. That new mills were coal-fired by 1848 is from Hunter, *History of Industrial Power*, vol. 1, chap. 10.

41. Steifler, *Beginnings of a Century of Steam*; and NA, *Walworth, 1842–1942* (n.p., 1945).

42. Dewhurst et al., *America's Needs*, table E, 1111. The authors estimate that 79 percent of bituminous and 61 percent of anthracite coal were used for power generation in 1860, but the derivation of the figures is unclear.

43. Robert Thurston, *Robert Fulton: His Life and Its Results* (New York: Dodd Mead, 1891), 144; Sale, *The Fire of His Genius*.

44. Brayard, "The *Savannah*"; Chapelle, "The Pioneer Steamship *Savannah*." See also Tyler, *Steam Conquers the Atlantic*, chap. 1. The poem is from "Steam against the Wind," *Army and Navy Chronicle* 3 (Feb. 1844): 147. For general discussions of the evolution of shipping during these years, in addition to Tyler, see Zimmermann, *Zimmermann on Shipping*; and Johnson and Huebner, *Principles of Ocean Transportation*. Sailing's share of US tonnage is from US Department of Commerce, *Commerce Yearbook*, 1925 (Washington, 1926), 536.

45. The definitive work on steamboats remains Hunter, *Steamboats*. The best modern treatment of Evans's work is Ferguson, *Oliver Evans*.

46. Jean Baptiste Marestier, *Memoir on Steamboats of the United States of America* (Paris: Royal Press, 1824); "Consumption of Wood," *Niles' National Register* 74 (Dec. 20, 1848): 394. That passengers helped to load wood is from "Western Steamboats," *Saturday Evening Post*, August 15, 1829, 3.

47. Hunter, in *History of Industrial Power*, vol. 2, discusses the origins and spread of expansive working in chapter 3. See also Lardner, *A Rudimentary Treatise*. Mone's *Treatise on American Engineering* contains the 40-percent claim on page 4. Renwick, *Essay on the Steamboats*. Mak and Walton, "Steamboats and the Great Productivity Surge," table 1. The authors attribute the rise in energy efficiency largely to better boats and less time spent in port.

48. William Hodge, *Papers Concerning Early Navigation on the Great Lakes* (Buffalo, NY: Bigelow Brothers, 1883), 36; "Consumption of Wood by Steam Boats," *Niles' Weekly Register* 34 (August 1828): 362. For details of the calculation, see table A2.3. This surely overstates wood use because it assumes that no boats burned coal and because Mak and Walton largely ignore increases in steam-engine efficiency.

49. United States Treasury, *Steam Engines*, 11. William W. Mather, *Second Annual Report of the Geologic Survey of the State of Ohio* (Columbus, 1838), 11. Mather's calculations were optimistic; a less-enthusiastic observer estimated that coal would take about a third the space of wood. See "Relative Value of Stone Coal and Cord Wood," *Western Journal of Agriculture, Manufactures, Mechanic Arts, Internal Improvements, Commerce, and General Literature* 2 (March 1849): 175, which also contains the captain's claim. "It will lead . . ." is from "Steamboat *North America*," *American Masonic Register and Literary Companion* 1 (May 9, 1840): 283. Hunter, *Steamboats*, 269.

50. The best descriptions of efforts to burn hard coal are Warner, "The Anthracite Burning Locomotive"; and Binder, *Coal Age Empire*, chap. 6. On the *Pennsylvania*, see also "Anthracite Coal," *Atkinson's Saturday Evening Post* 10 (July 2, 1831): 3. Alan Nevins, ed., *Diary of Philip Hone*, vol. 1 (New York: Dodd, Mead, 1927), 215. Labor saving is from "Steamboats, Anthracite Coal," *Niles' Weekly Register* 48 (April 11, 1835): 104. The 150,000 is from "Anthracite Coal," *Niles' National Register* 66 (June 29, 1844): 281. "Scarcely anything" is from Hunter, *Steamboats*, 267.

51. The Franklin Institute study is reprinted and evaluated in Sinclair, *Early Research at the Franklin Institute*.

52. The best discussion of nonfarm-work animals is Hunter and Bryant, *History of Industrial Power*, vol. 3, chap. 1.

53. "Steamboat and Railroad," SA, August 30, 1851, 397.

54. On saving space and stops, see "Coal for Locomotives," *Mining Magazine*, April 1856, 343–348. On problems of burning anthracite, see the citations in note 50 above. Savings on the Reading are from W. R. Johnson, "Use of Anthracite Coal on the Railroads," JFI 44 (August 1847): 110–113. George Whistler Jr., "Report on the Use of Anthracite Coal in Locomotive Engines on the Reading Railroad Made to the President of the Reading Railroad Company by George W. Whistler, April 20, 1849," JFI 48 (July–Sept 1849): 6–12, 78–87, 176–181. The $64 is derived from the table on page 9, where the Baltimore is shown to save about $8 per 100 tons transported compared to a wood burner, while trips

average 700 to 900 tons. Hence, ($8.00/100) × 800 = $64. Whistler estimated the extra cost or repair on pp. 81–82. Keuchel also discusses Whistler but appears to confuse gross and net costs. See also John White, "James Millholland and Early Railroad Engineering," *United States National Museum Bulletin* 252 (Washington: Smithsonian Institution Press, 1967).

55. Data on the New York Central are from Holley, *American and European Railway Practice*; "wood is constantly" is on 72.

56. Conversion costs are from Holley, *American and European Railway Practice*, 74. "Full one half" is from "The Griggs Coal Burner '*Washington*,'" ART 10 (May 22, 1858): 2. Locomotive numbers are from "The Use of Coal on Locomotives," ART 11 (Feb. 19, 1859): 2. Baltimore & Ohio Railroad, *Nineteenth Annual Report* (Baltimore, 1845), 11; Illinois Central Railroad, *Sixteenth Annual Report 1856* (Chicago, 1856), n.p. Cutting the bill by three-fourths is from "Coal Burning Locomotives," *Hunt's Merchants' Magazine*, July–Dec. 1858, 250. George Vose, *Handbook of Railroad Construction* (Boston: J. Munroe, 1857); Chicago, Burlington & Quincy Railroad *Annual Report* (Chicago, 1859), 45.

57. Holley, *American and European Railway Practice*, "general rush" is on 72. For the brick arch, see "Coal Burning Locomotives," ART 14 (Jan. 18, 1862): 22; and "Griggs' Fire-Brick Arch for Locomotives," ART 19 (Nov. 16, 1867): 366. The Boston & Providence is from Angus Sinclair, *Development of the Locomotive Engine* (New York: Sinclair Publishing, 1907), 397. The Pennsylvania's tests are from "Coal Burning Locomotives," *Hunt's Merchants Magazine*, Nov. 1859): 622; and "Coal Burning Locomotives," *DeBow's Review* 29 (1860): 118–119. Pennsylvania Railroad, *Seventeenth Annual Report* (Philadelphia, 1864), 33, 52.

58. Fishlow, *American Railroads*, 124–129.

59. Longer trains and larger cars raise energy efficiency because both increase the ratio of train payload to deadweight. Zerah Colburn, "The Economy of Railroads," ARRJ 27 (April 1, 1854): 193–198; "Coal Transportation by Railroads," *Mining Magazine*, July 1854, 99–101, 100. Energy productivity is calculated from Fishlow, "Productivity and Technological Change," table 9.

60. In England, cannel coal was the source of gas. It is a very high volatile coal, and Americans imported it for manufacturing gas. By the 1850s, domestic cannel and other bituminous coals were being employed for gas making as well. Initially, some gasworks used rosin and whale oil as a feedstock, but by the 1850s, these were on the way out.

61. Jeremy Zalen has written of the dangers resulting from artificial light of all forms; see Zalen, *American Lucifers: The Dark History of Artificial Light* (Chapel Hill: University of North Carolina Press, 2019). Of course, sunlight is also dangerous—it causes skin cancer among other things—as are most other forms of human activity.

62. This brief overview of the origins of gas light leaves out much. For more detail, see Arthur Elton, "Gas for Light and Heat," in *A History of Technology*, ed. Charles Singer et al. (Oxford: Clarendon, 1958), 4:258–276; Castaneda, *Invisible Fuel*; and Tomory, *Progressive Enlightenment*; see also Tarr, "Manufactured Gas." "Greatest improvements" is from "On the Utility of Coal Gas Light," *Select Reviews and Spirit of the Foreign Magazines* 5 (Jan. 1811): 66.

63. Melville's efforts and the quotation from the visitor are reported in "Gas Lights," *Niles' Weekly Register* 6 (May 21, 1814): 198. A paper by Murdoch was reported in detail in "Great Britain," *Christian Observer* 7 (Sept. 1808): 612. See also "Gas Lights," *Boston Weekly Magazine*, Feb. 1, 1817, 66, which focused on Philadelphia. For an early comparison of gas and candles, see "Observation on the Application of Coal Gas to the Purposes of Illumination," *Journal of Science and the Arts* 1 (Jan. 1, 1817): 71.

64. The candlepower of a gas was rated as the candle equivalent of gas, which was burned at the rate of 5 ft.^{3} per hour under controlled conditions. That is, if a gas burned at that rate was equivalent in brightness to 14 candles, it was rated as 14 candlepower gas. "European Gas Light Journals," AGLJ 1 (Sept. 1, 1859): 1; Franz Von Gerstner, "Civil Engineering," JFI 30 (Oct. 1840): 217–224, 218.

65. Tomory, *Progressive Enlightenment*, emphasizes the scientific origins of gas lighting. Consolidated Gas, *American Gas Centenary*, 249. An advertisement containing coal properties is "F. Butts & Co," AGLJ 1 (Dec. 1, 1859): 98. Jackson and Silliman are from Lucier, *Scientists and Swindlers*. "Common labor" is from "Answers to Correspondents," AGLJ 1 (Sept. 1, 1859): 9. "Our Contributors," AGLJ 1 (August 1, 1859): 1; "Scientific Staff of the American Gas Light Journal," AGLJ 2 (Feb. 1, 1860): 161; "Practical Management of Gas Works," AGLJ 2 (March 1, 1860): 176.

66. On sulfur in gas, see "On the Manufacture of Gas," JFI 14 (Nov. 1, 1847): 352–356. "Clay Retorts," AGLJ 1 (July 1, 1859): 274; see also "Practical Management of Gas Works." The weight of coke is from "Pennsylvania Coal for Gas," SA, July 29, 1854, 362. For New York, see "Gas," *United States Economist* 9 (Oct. 11, 1856): 401. Simon Garfield, *Mauve: How One Man Invented a Color That*

Changed the World (New York: Norton, 2001).
Chandler, *On the Gas Nuisance*.

67. "Nuisances Arising from Gas Works,"
Medical Times 15 (July 1857): 124; Chandler, *On the Gas Nuisance*; Marston Bogert, *Biographical Memoir of Charles Frederick Chandler, 1836–1925* (Washington: National Academy of Sciences, 1931). For a modern assessment of gas companies' environmental record, see Tarr, "Toxic Legacy."

68. The Baltimore Gas Company had no private customers, and its early distribution methods are from Consolidated Gas, *American Gas Centenary*, 249–251. Urban areas and populations are from US Department of Commerce, *Historical Statistics to 1970*, series A-43 and A-57. For antebellum per capita GDP and wage data, see Carter et al., *Historical Statistics*, series C11 and Ba 4253–4267.

69. A candlepower is about 12.57 lumens, and a 40-watt incandescent bulb yields about 460 lumens. Note that the calculations in the text are subject to rounding error.

70. For yields, see "American Gas Coals," AGLJ 1 (Feb. 1, 1860): 145; "Annual Reports of Gas Companies," AGLJ 1 (April 2, 1860): 203; "Reports of Gas Light Companies," AGLJ 2 (August 1, 1860): 42; and "Niagara-Falls, N.Y., Gas-Light Company," AGLJ 2 (March 15, 1861): 286.

71. Gesner, *Practical Treatise*. For a modern discussion, see Beaton, "Dr. Gesner's Kerosene"; and Lucier, *Scientists and Swindlers*, which contains Breckenridge on 148. The one million families is from Lebergott, *The Americans*, 323.

Chapter 2. The Age of Bituminous Coal, 1860–World War I

1. GDP data are from Angus Maddison, *World Economy Historical Statistics* (Paris: OECD, 2003). Steel production is from Peter Temin, "The Relative Decline of the British Steel Industry, 1880–1913," in *Industrialization in Two Systems*, ed. Henry Rosovsky (New York: John Wiley, 1966), 140–155. Economic historians would be leery of a claim that coal was *indispensable* to economic development during these years, for market economies are very good at innovating substitutes. Still, it is difficult to imagine what other sources of fuel might have been available to yield similar results.

2. For the origins and meaning of technological literacy, see Stevens, *Grammar of the Machine*. On scientific consulting, see Lucier, *Scientists and Swindlers*, chap. 4; see also his "Commercial Interests and Scientific Disinterestedness," *Isis* 86 (June 1995): 245–267. For engineering education, see Nienkamp, "American Land-Grant Colleges";

and Marcus, *Engineering in a Land Grant Context*. See also Sinclair, *A Centennial History*; and Mowery and Rosenberg, *Technology and the Pursuit*, chaps. 2–3. "The Engineering Schools of the United States," *Engineering News* 27 (March 19, 26, 1892): 277–278, 284–296; Ira Baker, "Engineering Education in the United States at the End of the Century," Society for the Promotion of Engineering Education, *Proceedings* 8 (1900): 11–27; Samuel Christy, "The Growth of American Mining Schools and Their Relation to the Mining Industry," AIME *Transactions* 23 (1894): 444–465.

3. Between 1855 and 1920, life expectancy at birth for Massachusetts females rose from 40.9 to 56.6 years; for males the figures are 38.7 and 54.1. US Department of Commerce, *Historical Statistics to 1970*, series B, 126, 127.

4. The verse is from "The Cost of Coal," *Littell's Living Age* (Feb. 22, 1868): 510–511. For an overview of anthracite during these years, see Healey, *Pennsylvania Anthracite Industry*. On early mine safety, see Mark Aldrich, *Safety First* (Baltimore: Johns Hopkins University Press, 1997); fatality rates are from appendix table A2.1 (therein). A comparative treatment of the issues discussed in this section is Javier Silvestre, "Productivity, Mortality, and Technology in European and US Coal Mining," in *Standard of Living: Essays on Economics, History, and Religion in Honor of John Murray*, ed. Patrick Gray et al. (Cham, CH: Springer, 2022): 345–371. H. H. Stoek, "The International Correspondence Schools, Scranton, Pa., with Special Reference to the Courses in Mining," AIME *Transactions* 28 (1898): 746–758; Watkinson, "'Education for Success.'"

5. Dates for the arrival of mechanization are from "Transportation," CA 41 (Oct. 1936): 417–424. Labor productivity is calculated as average daily output per worker and is from "Coal, Part A," MR, 1918, part 2.

6. For English experiments, see "'Dust Fuel,'" *American Artisan* 11 (Sept. 28, 1870): 194; and "Utilization of Coal Dust," E&MJ 13 (March 5, 1872): 152. An early briquetting effort is E. F. Loiseau, "Artificial Fuel," JFI 97 (1874): 111–136. Brief reviews of European efforts are "Briquettes," AGLJ 54 (May 4, 1891): 636–637; and "The Velna Process for the Manufacture of Briquettes from Waste Coal and Mineral Tar," AGLJ 65 (Nov. 23, 1896): 815–816. For tests, see "The Loiseau Artificial Fuel," *Manufacturer and Builder* 14 (Dec. 1, 1882): 283–284.

7. Platt, *A Special Report*, 31. Pumping dust back into the mine is from Pennsylvania, *Report of the Bureau of Mines* (Harrisburg, 1897), xxxv–li. Chance,

Report on the Mining Methods; Commonwealth of Pennsylvania, *Report of the Commission*; Charles Ashburner, "Brief Description of the Anthracite Coal Fields of Pennsylvania," Engineers Club of Philadelphia, *Proceedings* 4 (August 1884): 177–208, 201; R. P. Rothwell, "Remarks on the Waste in Coal Mining," AIME *Transactions* 1 (1871–1873): 55–59.

8. The classic works on the anthracite cartel are Jones, *Anthracite Coal Combination*; and Nearing, *Anthracite*. For a description of distribution about the time of World War I, see US Coal Commission, *Report*, part 2, 671–690.

9. For the strike, see Grossman, "The Coal Strike of 1902"; and Blatz, *Democratic Miners*. "Learning that" and "permanently lost" are from "Causes Alarm: Hard Coal Magnates Lose Markets," BG, July 25, 1902; "New England Faces a Hard Coal Famine," BG, July 23, 1902; "Soft Coal Eye," BG, Sept. 21, 1902; "thousands" is from "Big Oil Stove Demand," NYHT, Oct. 7, 1902; "Big Demand for Oil Stoves," CT, Oct. 7, 1902; "fruits" is from "A Big Gain for Soft Coal," NYHT, Nov. 1, 1902.

10. Eighth Census of the United States, *Manufactures in 1860* (Washington, 1865), clxv. The data on establishments—which I interpret here to mean mines—and employment are from the various censuses, as are the employment data.

11. Tenth Census of the United States, 1880, *Report on the Mining Industries of the United States* (Washington, 1886), 674. "Cutting machines" is from "Face Preparation," CA 41 (Oct. 1936): 408–418; see also MR, 1910, part 2, 53; "labor productivity" is from "Coal, Part A," MR, 1918, part 2. Modern research finds horsepower per worker the key determinant of labor productivity in soft-coal mining; see Maddala, "Productivity and Technological Change." A good review of mechanization is Hotchkiss et al., "Mechanization, Employment, and Output."

12. For marketing, see US Coal Commission, *Report*, part 3, 1789–1802. Prices are from my table A1.5. I adjusted for inflation using series Cc2 from Carter et al., *Historical Statistics*. For the boom and its impact on wages, see MR, 1903, 351–352.

13. Coal's share in total energy use is from table A2.2. Total production is from table A1.4. Complaints about smoke in Pittsburgh as early as 1800 are recorded in Davidson, "Air Pollution in Pittsburgh." "The Smoke Nuisance," CT, Sept. 20, 1890. For a sampling of modern writings, see Grinder, "Battle for Clean Air"; and Stradling, "To Breathe Pure Air."

14. The classic discussion of the rise of the large corporation and the importance of backward integration is Chandler, *The Visible Hand*. New-

berry, *Report on the Economical Geology*. For captive mines, see the section on bituminous coal in MR, 1926, part 2.

15. Lucier, *Scientists and Swindlers*, chap. 4, discusses the role of consulting scientists in coal mining. There is no good history of the evolution of nineteenth-century mining. Dix, *What's a Coal Miner to Do* has only a brief chapter on the period. The best sources are the various reports of Pennsylvania's inspectors of coal mines that begin in 1869. See also Elmer Sperry, "Recent Observations on Electricity in Coal Mines," *Journal of the Illinois Mining Institute* 2 (May 1893): 103–107.

16. "Same wasteful practices" and "prevents the use" are from T. C. Purdy, "Report on Steam Navigation in the United States," in United States Tenth Census, 1880, vol. 4, *Report on the Agencies of Transportation in the United States* (Washington, 1885): 659–703, 696, 699. Theodore Allen, "Iron Hulls for Western River Steamboats," ASCE *Transactions* 6 (1873): 1–17.

17. For discussions of the evolution of ship building and steam propulsion, see Gracie, "Twenty Years' Progress," which is the source of coal per horsepower. See, too, Johnson and Huebner, *Principles of Ocean Transportation*; Tyler, *Steam Conquers the Atlantic*; and Hunter, *History of Industrial Power*, vol. 2, 662–666, which is the source of steam pressures. Zimmermann, *Zimmermann on Shipping*, is the source of coal consumption per ton-mile. The best modern economic analysis of the shift from wind to steam is Hartley, "The Shift from Sailing Ships to Steamships."

18. North, "Ocean Freight Rates," table 2. For exports, see US Department of Commerce, *Historical Statistics to 1970*, series U 191. Bunker fuel is from US Department of Commerce, *Statistical Abstract, 1928* (Washington, 1929), table 720.

19. "Shipping and Freights," *The Economist*, Feb. 23, 1886, 32–33, 33. The fuel burned by merchant vessels is from US Department of Commerce, *Commerce Yearbook* (Washington, various years). Bunker fuels are from US Department of Commerce, *Statistical Abstract, 1940* (Washington, 1941).

20. Track for 1860 and 1880 is from United States Census, *Preliminary Report on the Eighth Census of the United States, 1860* (Washington, 1862), table 38; and Armin Shuman, "Statistical Report of the Railroads in the United States," table V, in Tenth Census of the United States, *Report on the Agencies of Transportation*, vol. 4 (Washington, 1885), 288–290. These appear to include all track. Data for

1918 are from US Department of Commerce, *Historical Statistics to 1970*, series Q 288, and are all main-track operated.

21. Shuman, "Statistical Report." The survey is in MR, 1915, part 2, 433–514. For a study of how geography and transport costs shaped coal markets in Great Britain, see Allen, "The Transportation Revolution."

22. Coal consumption is from Bukovsky, "Use and Cost of Railway Fuel," table 3.

23. Problems of boiler encrustation may be followed in an index in American Railway Master Mechanics, *Proceedings* 33 (1900): 67–94. Testing by the Chicago, Burlington and Quincy is from "Fuel Saving on Railroads," RG 24 (Feb. 1, 1895): 66–68.

24. The efficiency of locomotives is from Bruce, *Steam Locomotive in America*. Two excellent modern books on locomotive technology are Lamb, *Perfecting the American Steam Locomotive*; and Withuhn, *American Steam Locomotives*. "The Large Car Problem," RA 21 (Feb. 29, 1896): 109. "Long Locomotive Runs," RG 29 (June 25, 1897): 464–465. On the value of heavier trains, see "Comparative Coal Consumption per Ton Hauled in 1885 and 1891," RG 25 (Jan. 27, 1893): 72. "Less fuel" is from "Efficiency of Locomotive Fuels," RG 26 (Dec. 28, 1894): 887.

25. Calculations of fuel efficiency (output/fuel consumption), based on the work of Fishlow, "Productivity and Technological Change," tables 9 and 10, show worsening fuel economy from 1849 to 1910. Sinclair's remarks are from "Difficulties Accompanying the Prevention of Dense Black Smoke and Its Relation to Cost of Fuel and Locomotive Repairs," Traveling Engineers Association, *Proceedings* 22 (1914): 59. See also Angus Sinclair, "Wasting and Saving Coal with Locomotives," RG 14 (Oct. 6, 1882): 607. "Astonishing" is from "Fuel Economy," *Railway Review* 24 (Sept. 5, 1885): 425. "Desultory" is from "Fuel Economy," RG 20 (Jan. 6, 1888): 8–9.

26. The 29-percent figure is from "A Few Facts Concerning Freight Cars: Car Mileage, Repairs and Renewals," RA 19 (July 20, 1894): 406. "Proposed Freight Car Pool," RA 20 (Aug. 23, 1895): 408. The desire to raise freight car productivity also motivated carpooling proposals.

27. To avoid double counting, coke is excluded from table A3.2. For similar reasons, internally generated electricity is not included; purchased power—which, although small, should be included—is unavailable.

28. "Natural Gas in Pittsburgh," SA, June 13, 1885, 373; and "Use of Natural Gas at Pittsburgh," SA, Feb. 27, 1886, 127–129; MR, 1907, part 2, 327.

See also Tarr and Clay, "Boom and Bust." The earliest reference I have found to the technical causes of over-rapid depletion is John Carrl, *The Geology of the Oil Regions of Warren Venango Clarion and Butler Counties*, Pennsylvania Second Geological Survey (Harrisburg, 1880), chap. 5. For a fuller discussion, see "Underground Wastes in Oil and Gas Fields and Methods of Prevention," USBM TP 130 (Washington, 1916); and works cited. This volume also provides an early discussion of unit operation as a solution. See also Henry Doherty, "Suggestions for Conservation of Petroleum by Control of Production," in AIME, *Production of Petroleum in 1924* (New York, 1925), 7–19. On the rule of capture, see Williamson and Daun, *American Petroleum Industry*, vol. 1, appendix E; and Liebcap, *Contracting for Property Rights*, chap. 6.

29. Gushers were not simply the result of the American system of private ownership; they characterized Russian production as well. See A. Beebe Thompson, *The Oil Fields of Russia and the Russian Petroleum Industry* (New York: D. Van Nostrand, 1904). For comparisons with Russian practice, see also David Day, "Petroleum Resources of the United States," in Gannett, *Report*, 3:446–464.

30. For natural gas production, see table A1.6. For competition with coal, see, e.g., "Natural Gas vs. Coal," *American Artisan* 8 (March 1, 1885): 23; and "Coal vs. Natural Gas," AGLJ 51 (Sept. 23, 1889): 414.

31. The original data are in "Anthracite Iron Statistics," E&MJ 17 (June 27, 1874): 404–405. For discussion, see also "American Blast Furnace Examples," E&MJ 18 (July 1, 1874): 20–22. Benjamin West Frazier, "Economy of Fuel in Our Anthracite Blast Furnaces," AIME *Transactions* 3 (1874–1875): 157–172. Improved fuel economy is from John Church, "Blast Furnace Statistics," AIME *Transactions* 4 (1875–1876): 221–226. See also the section on industrial fuel and power in appendix 2. For a broader discussion of productivity in iron making, see Allen, "Peculiar Productivity History."

32. The classic work on the location of the industry in these years is Isard, "Some Locational Factors."

33. Osborn, *Metallurgy of Iron and Steel*, 500, 568.

34. Octave Chanute, "The Chemist as a Factor in Foundry Practice," AGLJ 59 (August 14, 1893): 222–224; Andrew Carnegie, *Autobiography* (Boston: Houghton Mifflin, 1920), 182. See also Temin, *Iron and Steel*, 156. "Forty Years of Progress in the Pig Iron Industry," *Iron Age* 57 (Jan. 2, 1896): 21–24, "appreciation" is on 22. Lowthian Bell, "Scientific

Education in America," *The Iron and Steel Institute in America* (London, 1890): 65–67, 67; see also Rosenberg, *Studies on Science*, chap. 2.

35. "The Greatest Steel Plant in the World, I," *Iron Age* 83 (Jan. 7, 1908): 1–11, "crowning feature" is on 2. "The Development of the Large Gas Engine in America," *Cassier's Magazine*, Nov. 1907, 41–54.

36. John Fritz, "The Progress in the Manufacture of Iron and Steel in America and the Relations of the Engineer to It," ASME *Transactions* 18 (1896): 39–64, 42; William Phillips, "Phosphate Slag," AIME *Transactions* 17 (1888): 84–88; Rosenberg, *Exploring the Black Box*, chap. 9.

37. There is an enormous literature—both primary and secondary—on the triumph of the steam engine and the competition among steam, waterpower, and electricity. The best of this is Hunter, *History of Industrial Power*, vol. 2, which cites much of the relevant primary literature. For a sampling of the economic discussions, see Temin, "Steam and Waterpower"; and Atack et al., "Regional Diffusion."

38. The 61 million tons should be a modest overestimate because, as noted in the text, by about 1900, some industries were generating power as a byproduct of process heat.

39. Energy used for power in 1900 includes the BCE of horses and mules, waterpower and wind power, railroads, power in manufacturing and mining, and electricity generation. These data are taken from various tables in the text and appendices. They accounted for 48 percent of BCE in 1900.

40. Hunter, *History of Industrial Power*, 2:143, dates the expansive use of steam in stationary power from the 1840s; "the impact" is on 448; the rise in steam pressures is on 675–676. On Corliss, see chap. 5 in the same volume; "gifted mechanic" is on 254, and the savings of fuel are my calculations based on information in that chapter. See also Rosenberg, *Studies on Science*, chap. 6. Horsepower data are from Daugherty et al., "Power Capacity." Fuel efficiency data are experts' estimates from Dewhurst et al., *America's Needs*, 1106.

41. "Statistics Regarding the Gas Companies of America," AGLJ 45 (Dec. 16, 1886): 360–363; "Report of Committee on Differential Rates, National Commercial Gas Association," AGLJ 102 (Jan. 11, 1915): 18–24. Regulators also sometimes helped push rates down by threatening competition; see Rose, *Cities of Light and Heat*, chap. 1.

42. Awareness that additional cooking load would lower costs is from "On the Advantages of Gas for Cooking and Heating," AGLJ 16 (Jan. 2, 1874): 9. That gas might be cheaper is from "The

Economy of Gas as a Fuel for Cooking Purposes," AGLA *Proceedings* 4 (1879): 174–181. Some companies did advertise. See "The Extension of the Uses of Gas for Purposes Other Than Illumination," AGLA *Proceedings* 9 (1890): 298–305.

43. Seventy cents would buy 550,000 Btu of manufactured gas, while $0.26 would buy 1,050,000 Btu of natural gas. Thus $1.00 bought 786,000 Btu of manufactured gas and 4.1 million Btu of natural gas. "Use of Gas for Culinary Purposes," AGLA *Proceedings* 2 (1877): 199–202; "Economy of Gas as a Fuel for Cooking Purposes," AGLA *Proceedings* 4 (1879): 174–181.

44. On early research, see Mowery and Rosenberg, *Technology and the Pursuit*, chaps. 2–3. Charles F. Chandler, "The Chemistry of Gas Lighting," *American Chemist* 6 (Jan. 1, 1876): 242–253, and (Feb. 1, 1876): 285–294. Charles Munroe, "Byproducts in Gas Manufacture," JFI 174 (July 1912): 1–34. Efforts to capture ammonia and naphthalene are briefly noted in "Our Report of Progress," AGLJ 22 (Jan. 16, 1875): 28–29, which also contains the quotation (italics in original). For one of many examples of the borrowing of British expertise, see G. E. Brown, "The Chemistry of Gas Making," AGLJ 68 (May 2, 1898): 696–697. That thfe American Gas Light Association began as a social organization is from "The Boston Meeting of the American Gas Light Association," AGLJ 35 (Dec. 16, 1991): 265. "Advance Circular from Committee on Research, American Gas Light Association," AGLJ 64 (Feb. 24, 1896): 283. Fellowships are from "President's Address," Annual Meeting of the Michigan Gas Association, *Proceedings* 9 (1902): 9–14.

45. The value of clay retorts is from "Effects of Various Temperatures in Carbonizing," AGLJ 10 (July 16, 1868): 18. A brief history is "On Water Gas," AGLA *Proceedings* 3 (1878): 196–205.

46. "The 'Water Gas' Plague," *Boston Journal of Chemistry* 5 (July 1870): 7; "Burning Water as Fuel," SA, Feb. 15, 1873, 96.

47. Silliman and Wurtz's report is in "American Water Gas Experiments," AGLJ 20 (Jan. 16, 1874): 21–22; (Feb. 16, 1874): 57–58; (March 2, 1874): 75–76; and (March 16, 1874): 93–94. "The Massachusetts Board of Gas and Electric Light Commissioners on the Subject of Water Gas," AGLJ 52 (Feb. 24, 1890): 245–247.

48. "Some Experiences with the Welsbach Incandescent Gas Lamp," AGLJ 61 (Nov. 19, 1894): 722–724. There are 12.57 lumen-hours per candlepower-hour.

49. A Welsbach mantle yielding 20.3 candlepower per ft.[3] yields 255.2 lumens. This is roughly

equivalent to a 40-watt incandescent bulb that yields 260 lumens. Note that these calculations and those in the text may vary owing to rounding.

50. Data on domestic consumption of soft coal are scarce and inconsistent; see the sections on bituminous coal in MR, 1917, part 2; and MY (1944). USBLS, "Cost of Living," table E, 391, found that urban homeowners with a median income of $1,050 consumed 4.2 tons of coal a year compared to 4.9 tons for those with a median income of $2,300. Thus, at this time, a 120-percent higher income resulted in only a 17-percent increase in coal consumption.

51. For brief histories of district heating, see Bushnell and Orr, *District Heating*, chap. 1; and National District Heating Association, *Handbook*, chap. 1.

52. MR, 1915, part 2, 464.

53. "Hot Air Furnaces," *The Cultivator* 8 (June 30, 1840): 338–339; "Hot Air Furnace," SA, August 12, 1848, 372. The early days of steam heat are from Stifler, *Beginnings of a Century of Steam and Water*; and John Allen Murphy, *Walworth, 1842–1942* (n.p.: n.p., 1945).

54. For some modern evidence on the comparative safety of central heating, see H. J. Moriske et al., "Indoor Air Pollution by Different Heating Systems: Coal Burning, Open Fireplace and Central Heating," *Toxicology Letters* 88 (1996): 349–354.

55. "A Threatened Coal Famine," RA, Nov. 10, 1916, 840; "Coal Supply Conditions Are Very Acute in Ohio," EW 70 (Oct. 20, 1917): 780–781. The number of mines is from MR, 1925, part 2, 466. "Today It Is Everybody's Business to Save Coal," EW 70 (Oct. 20, 1917): 481, contains the Columbus Railway episode. "America Faces Coal Shortage," CT, May 14, 1917; "New York Subway Operations Stopped by Coal Shortage," EW 70 (Sept. 1, 1917): 442. "Skipped stops" and the 80-million-ton deficit are from "Coal Deficit Threatens to Cause Famine," EW 71 (June 15, 1918): 1282. "Shortage of 50,000,000 Tons in Nation's Coal Supply," EW 70 (Nov. 17, 1917): 971. "Shortage of Coal Looms a Highly Important Factor in Gas Works Economy," *American Gas Engineering Journal* 113 (Sept. 18, 1920): 217–220.

56. On energy policies during World War I, see Johnson, *Politics of Soft Coal*; and Clark, *Energy and the Federal Government*.

57. On the taxation of petroleum drilling, see McDonald, *Federal Tax Treatment* and sources therein.

58. Requa, *Petroleum Resources*, 18. For equally alarmist views, see Gilbert and Pogue, "Petroleum: A Resource Interpretation."

59. MR, 1917, part 2, 904.

Chapter 3. Soft Coal in Industry, 1900–1940

1. Bituminous Coal Trade Association, *Report*; MR, 1923, part 2, 519–520.

2. Oshima, "Growth of US Factor Productivity," stresses the importance of immigration restrictions in inducing mechanization that saved unskilled labor. See also Jerome, *Mechanization in Industry*.

3. On federal policies, see Johnson, *A New Deal for Soft Coal*; Johnson, *Politics of Soft Coal*; and Clark, *Energy and the Federal Government*. See also James and Fisher, *Minimum Price Fixing*. For a modern critique of the oil industry and policy, see MacDonald, *Petroleum*. The best brief exposition of the impact of various oil policies is Kahn, "Combined Effects."

4. Engineers and chemists data are from Lebergott, *The Americans*, 349. Mowery, "Emergence and Growth," table 1; Baumol, *Free-Market Innovation Machine*, chap. 4; see also Sheshinski et al., *Entrepreneurship, Innovation*; and Mowery and Rosenberg, *Paths of Innovation*, chap. 2.

5. For origins of the fuel investigations and early work of the USGS, see US Secretary of the Interior, *Fuels and Structural Materials*. A list of topics and research institutions is in Fieldner and Emery, "Research Activities in the Mineral Industries of the United States." A good review of the bureau's work is in Rice et al., "Conservation of Coal Resources." Forty laboratories are reported in US National Resources Committee, *Energy Resources*, 103. The bureau's efforts to create liquid fuel from coal and oil shale are discussed in Stranges, "US Bureau of Mines' Synthetic Fuel Programme."

6. For an example that shares the details of pushing the technological envelope, see E. M. Gilbert, "The Susquehanna Station," ME 48 (April 1926): 362–368.

7. "Solving the Problems of the Portland Cement Association," *Engineering News-Record* 88 (June 15, 1922): 996–998. For a brief history of TAPPI and its functions, see "President's Address," TAPPI *Papers* 1 (1918): 5–6. TAPPI *Papers* 4 (1921) contains a membership directory.

8. "Ideas from Practical Men," *Power* 67 (Jan. 10, 1928): 72; "The Plant Notebook," C&ME 33 (July 1926): 433–434. The increasing number of highly specialized journals is an example of Adam Smith's dictum that "the division of labor is limited [only] by the extent of the market."

9. "Art" is from Cecil Herrington, "Pulverized Coal—Its Utilization in the Industries," *Combustion* 7 (Dec. 1922): 345–350, 345.

10. George Otis Smith and C. E. Lesher, "The Cost of Coal," AMC *Proceedings* 19 (1916): 452–464;

George Otis Smith, "Thrift in Coal," AISI *Yearbook* (1920), 472–487; "Coal is the shortest" is on 471. Smith, "Industry's Need for Oil," AISI *Yearbook* (1920), 25–34; "the decade" is on 26; "use of fuel oil" and "the use of gasoline" are on 31. For a discussion of the idea of a hierarchy of fuel uses, see Zimmermann, *Conservation*, chap. 2.

11. David White, "The Oil Supply of the World," ME 44 (Sept. 1922): 567–569. For updates, see US Congress, *Petroleum Investigation*. Experts understood that there was more oil (and gas) than in proven reserves, but they did not fully grasp that reserves are simply an inventory and largely irrelevant to the question of ultimately recoverable oil or gas. Wartime and postwar oil worries are discussed extensively in Clark, *Energy and the Federal Government*, chaps. 4, 6, and 8. For the fuels division, see "Fuel and Its Conservation," ME 43 (Jan. 1921): 22–23.

12. The recovery rates are from Rice et al., "Conservation of Coal Resources," 686–687.

13. M. S. Hachita, "Economical Use of Fuel: Remarks," ASME *Transactions* 40 (1918): 298–300.

14. Henry Kittredge, "The Utilization of Wastes and Byproducts in Manufactures," Twelfth Census of the United States, 1900, *Bulletin* 190 (Washington, 1902), 3; Platt, *A Special Report*. The commission Coxe was on was Commonwealth of Pennsylvania, *Report of the Commission*. R. W. Raymond, "Biographical Note of Eckley B. Coxe," AIME *Transactions* 25 (1896): 446–476. See also https://en .wikipedia.org/wiki/Eckley_Brinton_Coxe. Coxe's first patent was #510,565.

15. "Using Up the Culm Banks," *Coal Trade* 24 (Jan. 1, 1897): 28–31, 31; MR, 1910, part 2, 182. For a brief discussion of waste in use with an international focus, see Kershaw, *Use of Low Grade Fuels*.

16. During the interwar years, conservation motives were beginning to focus on water quality rather than coal recovery. In 1928, a joint investigation of culm and river coal by the Bureau of Mines and the State of Pennsylvania acknowledged that the old culm banks caused water pollution, which it termed a "necessary evil." See Sisler et al., "Anthracite Culm and Silt." In modern times, of course, the emphasis has entirely reversed, with stream pollution the dominant concern.

17. An early effort is reported in "Tests of an Artificial Fuel," *American Artisan* 14 (Jan. 3, 1872): 10. MR, 1907, part 2, "rock of opposition" is on 223. See also Parker, "Condition of the Coal Briquetting Industry"; and James Mills, "Binders for Coal Briquettes," USBM *Bulletin* 24 (Washington, 1911).

18. For some of the work on briquettes, see "Briquetting Tests at the United States Fuel Testing Plant, Norfolk, Virginia, 1907–1908," USBM *Bulletin* 30 (Washington, 1911).

19. For a review of government work as of 1912, see Wright, "Fuel-Briquetting" and sources cited therein. For an example of how pulverized coal processes could use otherwise wasted fine bituminous coal, see "Utilizing Coal-Mine Waste," EW 69 (June 9, 1917): 1105.

20. Shares of manufacturing are from League of Nations, *Industrialization and Foreign Trade* (New York, 1945). Productivity is from Kendrick, *Productivity Trends in the United States*, table D-II. MR, 1923, part 2, 518. For the bureau's increasingly complete discussions of efficiency and energy substitution, see the chapters on bituminous coal in *Mineral Resources* and *Minerals Yearbook* for subsequent years. See also Rice et al., "Conservation of Coal Resources"; and Tryon and Rogers, "Statistical Studies."

21. Estimates of manufacturing's share of national income for these years are from US Department of Commerce, *Historical Statistics to 1970*, series F-219.

22. J. B. Nealy, "Gas Fuel Improves the Lime Kiln Efficiency," C&ME 40 (July 1933): 356–357; "Fuel in Cement Plants," *Cement Era* 14 (Sept. 1916): 61–62, 62; Schurr, "Energy Efficiency," stresses that new sources of fuel and power raised energy efficiency. For a sampling of the large literature dealing with natural gas during these years, see Stotz and Jamison, *History of the Gas Industry*; Tussing and Barlow, *The Natural Gas Industry*; Herbert, *Clean Cheap Heat*; Castaneda, *Invisible Fuel*; and Blanchard, *Extraction State*.

23. Muller and Tarr, "McKeesport Natural Gas Boom"; William McMurray and James Lewis, "Underground Wastes in Oil and Gas Fields and Methods of Prevention," USBM TP 130 (Washington, 1916), and works cited therein. For the long, tortured history of efforts to achieve unit operations of oil and gas fields, see Hardwicke *Antitrust Laws*, chap. 3; and Liebcap, *Contracting for Property Rights*, chap. 6.

24. MR, 1921, part 2, 339–340; John Ferguson, "The Power System: Coal vs. Oil as Fuel," *Paper Industry* 4 (Oct. 1922): 921–922, 922; "Progress in Fuel Utilization in 1927," ME 50 (Jan. 1928): 22–28, 26.

25. US Federal Oil Conservation Board, *Report*, part 1, 6. On the Federal Oil Conservation Board and public worries over oil depletion, see Clark, *Energy and the Federal Government*, chaps. 6 and 8.

26. "Plenty of Fuel for Oil Engines." *Automotive Manufacturer* 68 (July 1, 1926): 21–24; "What of the

Oil Famine?," *Crockery and Glass Journal* 104 (Feb. 10, 1927): 11; George Orrok, "Fuel Utilization," ME 52 (April 1930): 334–336, 336. See also "Oil, Gas and Coal Compete as Steam Fuels," *Power* 73 (Jan. 1933): 8–9. Odell, "Facts Relating," reveals the continued assumption of diminishing natural gas supplies.

27. For a good contemporary review of the role of geology and engineering in enhancing supplies of oil and gas, see US National Resources Committee, *Energy Resources*, 143–162 and 346–369, The changing role of science is on 148.

28. Gilmer, "History of Natural Gas Pipelines."

29. Edgar Hill, "Long Distance Transportation of Natural Gas," AGA *Proceedings* 12 (1930): 288–306, provides a technical discussion of gas transmission. "Republic's New Mill Electric Welds 10 Miles of Pipe in 24 Hours," *Iron Age* 126 (August 14, 1930): 422–424; "entirely feasible" is from H. R. Moorhouse, "Distribution of Natural Gas in the United States by Long Distance Pipe Lines," ME 53 (Sept. 1931): 657–663, 662.

30. McMurray and Lewis, "Underground Wastes"; E. L. Rawlings and L. D. Wosk, "Leakage from High-Pressure Natural Gas Transmission Lines," USBM *Bulletin* 265 (Washington, 1926).

31. Rawlings and Wosk, "Leakage"; the leak discussion is on 62. James Martin, "Leak Detection on High-Pressure Transmission Lines," AGA *Proceedings* 12 (1930): 273–282.

32. The 900-mile pipeline is from Moorhouse, "Distribution," 661.

33. Gilmer, "History of Natural Gas Pipelines." In modern times, climate change is sharply raising the payoff to leak detection; see "Frackers, Shippers Eye Natural Gas Leaks as Climate Change Concerns Mount," *Wall Street Journal*, August 12, 2021.

34. The wage gain is from USBLS, "The Changing Status of Bituminous Coal Miners, 1937–1946," *Monthly Labor Review* 63 (August 1946): 165–174. Figure 3.2 is useful to spot trends; however, because it ignores location, it can't be employed to conclude, for example, that in the 1930s, gas was always cheaper at the point of consumption than was coal.

35. Yaworski et al., "Fuel Efficiency," table A-5. The figures for iron and steel and for chemicals are for census industry groups and derive from the census of manufacturers for 1929 and 1939. On the development of petrochemicals, see Haynes, *American Chemical Industry*, vol. 5, chap. 15.

36. "Report of the Committee on Central Development Laboratory," AGA *Proceedings* 2 (1920): 96–97; for a review of the work of other

associations, see "Report of Committee on Testing Laboratory," AGA *Proceedings* 6 (1924): 70–71; "Five Years of Laboratory Progress," AGA *Proceedings* 12 (1930): 21–23.

37. These developments may be followed in "Minutes of the Industrial Gas Section," AGA *Proceedings* (1923–1927); J. B. Neale (of the AGA), "Automatically Controlled Gas-Fired Japaning Ovens," *Iron Age* 127 (April 9, 1931): 1150–1153.

38. Gas sales are from Gould, *Output and Productivity*, table A-10.

39. "Broadening the Field for Industrial Gas Utilization," AGA *Proceedings* 16 (1934): 135–140, 136.

40. "Report of Ceramic Industries Committee," AGA *Proceedings* 18 (1936): 466–467, 466; "Application of Gas-Fired Radiant Tubes to Porcelain Enameling," *Iron Age* 137 (Feb. 6, 1936): 26–29; "Gas Hot Tubes for Burning Porcelain Enamel," *Iron Age* 140 (Sept. 30, 1937): 26–28; "The Fuel for Vitreous Enamel," *Gas Age* 53 (Jan. 12, 1924): 88.

41. "Report of Non-Ferrous Metals Committee," AGA *Proceedings* 18 (1936): 539–542, 539; Michael Mawhinney, "Heating Mediums," *Metal Progress* 37 (March 1940): 284.

42. For two modern assessments that provide a broad view of the technical progress of the interwar decades, see Gordon, *Rise and Fall*; and Field, *A Great Leap Forward*.

43. The seven (of 20) industry groups that were the largest energy users in 1939 were Food & Kindred Products, Paper & Pulp, Chemicals, Products of Petroleum and Coal, Stone Clay and Glass, Iron and Steel, and Non-ferrous Metals. That year, these seven were all major users of fuel for heat, as well as power, and in 1939 they accounted for 95 percent of all natural gas used in manufacturing, 87 percent of all soft coal, 98 percent of coke, 99 percent of manufactured gas, and 67 percent of purchased electricity. Between 1919 and 1939, their share of value-added manufacturing rose from 35 to 47 percent.

44. W. S. Johnson, "Generating and Using Power in Industry," C&ME 39 (April 1932): 195–199, 198. Because differences in risk ensured that the cost of capital in manufacturing exceeded that in utilities, a good investment in the latter might flunk a market test in the former. See "Powdered Coal for Small Boilers in Industrial Plants," *Power* 67 (March 20, 1928): 501–502; Allen, "Collective Invention." Scientific management entered boiler rooms, too, in the form of incentive plans for fuel economy. See "Task Setting for Firemen and Maintaining High Efficiency in Boiler Plants," ASME *Journal* 5

(Dec. 1913): 1731–1757; and "Bonus System for Firemen," *Power* 46 (Sept. 11, 1917): 356–357.

45. Alton Adams, "Twentieth Century Energy," SA, July 14, 1900, 19.

46. "The World's Biggest Gas Engines," *Power* 82 (May 13, 1930): 736–738. For a comparison of turbines and internal combustion engines, see "Oil Engine vs. Steam Power," *Power* 56 (July 18, 1922): 80–83. See also Sixteenth Census of the United States, *Manufactures*, 1939, vol. 1 (Washington, 1942), chap. 6, table 1.

47. E. L. Moultrop, "Story of First Commercial 1,200-lb Steam Plant," *Power* 67 (April 24, 1928): 712–714, 713.

48. On the growing use of electricity, see DuBoff, *Electric Power*; Devine, "From Shafts to Wires"; Passer, *Electrical Manufacturers*; and Hughes, *Networks of Power*. On fuel costs, see US Bureau of the Census, *Census of Central Electric Light*, table 112. Hunter and Bryant, *History of Industrial Power*, vol. 3, also emphasizes the comparative importance of energy costs to managers of central stations on 319–320. The quotation is from MR, 1926, part 2, 446.

49. Chandler, *Visible Hand*, notes that firms that are first movers often have to produce services and products that are ancillary to their central function. Edison's Pearl Street Station seems to have been a pioneer utility in selling steam. See https://www.energystory.org/2023/01/14/edison-and-the-dawn-of-electric-power/. This was part of the larger district-heating movement briefly noted in chapter 2. Quaker Oats is from "Utilities That Supply Steam and Power to Industry," *Power* 71 (May 27, 1930): 810–815.

50. The ten pounds of coal per kwh is my calculation based on a survey reported in "Fuel Economy of Electric Light and Power Stations," *Engineering News* 31 (March 15, 1894): 226. "Rate Reductions and Sales Increases," EW 94 (August 31, 1929): 415–416. Total factor productivity is based on data from Kendrick, *Productivity Trends*, table H-VI. Utility costs from 1883 to 1929 are from George Orrok, "Central Stations," ME 52 (April 1930): 324–334; adjusted for inflation, they fell 85 percent over that period.

51. For a good discussion of the evolution of central station technology by a contemporary, see Orrok, "Central Stations." Modern discussions of technology are Bauer et al., *Electric Power Industry*, chaps. 2–3; Passer, *Electrical Manufacturers*, chap. 21; and Hunter and Bryant, *History of Industrial Power*, vol. 3, chaps. 5–6. On the sizes of turbines versus reciprocating steam engines, see "Economical and

Safe Limits in the Size of Central Stations," AGLJ 78 (May 18, 1903): 768–770; and "A New Industrial Situation," *Electrical Age* 25 (April 21, 1900): 123. For the growth in size and efficiency, see Morrow, *Electric Power*, chap. 5, table 11.

52. Miller, *Kilowatts at Work*, chap. 19; "Test of World's Largest Boilers," *Power* 34 (Dec. 5, 1911): 840–842; "Tests of Large Boilers at the Detroit Edison Company," ASME *Journal* 33 (Nov. 1911): 1439–1455; "15 Years," *Power* 79 (Dec. 1935): 686–691, 689.

53. For steam pressures as of 1929, see "Trends in Power Plant Development in Europe and the United States," ME 51 (Oct. 1929): 727–935; and J. W. Parker, "1,000 Degree Turbine for Delray No. 3," *Power* 69 (May 28, 1929): 909–910. Miller, *Kilowatts at Work*, chap. 19, 262, 264. For average conditions about this time, see "Progress in Fuel Utilization in 1927." "To an increasing extent" is from "Progress in Fuels Engineering," ME 47 (Dec. 1925): 1124–1126, 1124. For a brief discussion of binary systems, see "Progress in Fuel Utilization in 1927"; and "Notes on the Origin and Development of Water-Tube Boilers for Power Stations," *Combustion* 11 (August 1924): 146–152. For the 1890s, see Robert Thurston, "Promise and Potency of High Pressure Steam," ASME *Transactions* 18 (1895–1896): 160–220. For reviews of metallurgical developments, see the entire issue of *Power Plant Engineering* 44 (Jan. 1940).

54. "Post-war scarcity" is from "Fifteen Years . . . ," *Power* 79 (Dec. 1935): 686–691; "alert engineers" is from A. D. Blake and P. D. Swain, "Fifty Years of Power," *Power* 52 (April 1930): 321–323, 322. Calculations of heat balances to test new equipment or processes appear in American engineering literature in the 1890s, often as references to English or German experiments. For example, H. H. Stoek, "Blast Furnace Heat Balance Sheet," E&MJ 61 (April 11, 1896): 351–352, is Stoek's translation of a French work, and his introduction suggests that he imagined readers might not be familiar with heat balances. Their use to monitor and improve factory energy use was not common before World War I.

55. Detroit Edison is from Miller, *Kilowatts at Work*, 248–249. For Dow's life, see https://ethw.org/Alex_Dow. Hirsh, *Technology and Transformation*, chap. 5, describes the culture of innovation and risk-sharing at utilities during these years.

56. "Artificial Asbestos or Mineral Wool," *Manufacturer and Builder* 10 (Jan. 1878): 5; "The Value of Covering Steam Pipes," AGLJ 50 (June 24, 1889): 822. Rockwell is from "Savings Derived from Covering Steam Pipes," EW 70 (Oct. 6, 1917):

671–672. For tests of various materials, see "Report on Heat Insulating Materials," *Power* 45 (May 1, 1917): 593–596. J. G. Coutant, "Savings with Insulating Refractories," C&ME 44 (March 1937): 1137–1138; "Insulation," *Combustion* 6 (June 1922): 272–274; B. Townshend and. E. R. Williams, "Heat Insulation Developed for Every Purpose," C&ME 39 (April 1932): 219–222; L. B. McMillan, "Heat Insulation Practice in the Modern Steam Generating Plant," *Power* 70 (July 9, 1929): 66–68.

57. "Effect of Soot on Boiler Efficiency," *Combustion* 1 (Nov. 1919): 20–22; "Mechanical Soot Blowers," *Combustion* 2 (March 1920): 10–12; "Flue Gas Analysis and Its Relation to Steam Engineering," *Combustion* 1 (August 1919): 18–20. The utility is from "Developing Efficiency in Central Stations," *Power* 44 (July 4, 1916): 8–10. "Modern Boiler Room Instruments," *Combustion* 11 (July 1924): 40–42; David Myers, "How Much Goes Up the Chimney?," *Power* 69 (March 12, 1929): 437–440. "With no other class" is from "More Power for Your Dollar," *Power Plant Engineering* 44 (Jan. 1940): 40–43, 41. "Records Turn Meter Readings into Profits," *Power* 74 (August 25, 1931): 275–277; "Thomas A. Edison Plant Departments Pay Cash for Steam and Electricity," *Power* 74 (August 12, 1931): 188–189.

58. C. T. Hirshfeld and C. L. Karr, "Economic Operation of Steam Turbo-Electric Stations," USBM TP 204 (Washington, 1918), explains the importance of internal allocation of generating capacity. For details, see Herbert Reynolds, "How to Prepare a Schedule for Economic Boiler Loading," *Power* 70 (Oct. 8 and 15, 1929): 553–557, 593–596. "Southern California Electric Power Company," *Street Railway Journal* 9 (June 15, 1899): 401–403. Line loss of 10 percent is from Commonwealth of Pennsylvania, *Giant Power*, 29; less than 7 percent is from Bauer et al., *Electric Power*, 50. The only modern history of the grid is Cohn, *The Grid*, but she says little about transmission losses.

59. "Underfeed Stokers and Midwest Coal," *Combustion* 10 (March 1924): 206–208, 206. For efficiency gains, see "Tests of an American Stoker," *Engineering Record* 39 (Feb. 25, 1899): 289; and "Mechanical Stokers," *Engineers' Society of Western Pennsylvania* 19 (1903): 169–181. "Chronological History of Stoker Development," *Power* 56 (Dec. 5, 1922): 898–899; "Good-Bye Coal Shovel in Small Power Plants," *Power* 82 (Jan. 1938): 58–60. The furniture plant is from "Dividends from Smoke Abatement," *Power Plant Engineering* 44 (Feb. 1940): 45–46.

60. "Atlas Portland Cement Company" is from "Pulverized Coal—Its Utilization in the Industries,"

Combustion 7 (Dec. 1922): 345–350. F. R. Low, "Pulverized Coal for Steam Making," ASME *Journal* 36 (August 1914): 346–352, dates early use to England in 1831. "Pulverized Fuel," ARRJ 72 (Dec. 1, 1898): 378–380, claims there were experiments that same year in Germany. "An American device" is from Lesley, *History of Portland Cement Industry*, 123. The survey is in F. P. Coffin, "The Extent of the Use of Pulverized Fuel in the Industries and Its Possibilities in the War," *General Electric Review* 21 (May 1918): 373–380.

61. For "fly ash," see Miller, *Kilowatts at Work*, chap. 19. H. G. Barnhurst, "Pulverized Coal and Its Future," *General Electric Review* 21 (Feb. 1918): 116–119, summarizes the advantages of pulverized fuel. Henry Hull, "Utilizing Coal Mine Wastes," EW (June 9, 1917): 1106. For efforts to burn coal instead of gasoline in cars, see Rudolph Pawlikowski, "An Internal Combustion Engine Using Pulverized Coal," [Second] International Conference, *Proceedings*, 1:768–791. For a good summary as of the mid-1920s, see "The Development of Pulverized Coal as a Boiler Fuel," ME 47 (Feb. 1925): 89–93.

62. "Pulverized Coal in Milwaukee," *Power* 51 (March 2, 1920): 341–342; Henry Kreisinger et al., "Tests of a Large Boiler Fired with Powdered Coal at the Lakeside Station, Milwaukee," USBM *Bulletin* 237 (Washington, 1925), 60; "Burning of Powered Fuel in Paper Mill Power Plants," TAPPI *Papers* 4 (1921): 13–20. For a retrospective review, see "Milwaukee's Contribution to Pulverized Coal Development," ME (Oct. 1940): 723–726.

63. For other early users, see Harlow Savage, "Use of Powdered Fuel under Steam Boilers," AISI *Yearbook* (1921): 152–192. The survey is from William Harvey Young, "Sources of Coal and Types of Stokers and Burners Used by Electric Public Utility Plants," Brookings Institution *Pamphlet* 2 (Washington, 1930). For water cooling, see "A Review of Recent Applications of Powdered Coal to Steam Boilers," ME 47 (Jan. 1925): 19–23. The importance of turbulence is discussed in "Progress in Fuels Engineering," ME 47 (Dec. 1925): 1124–1126. For a fascinating study of the experimentation involved in implementing pulverized fuel, see MacDonald, *Let There Be Light*, chap. 8, esp. 211.

64. "Ocular inspection" and "dull" coal are from "The Application and Earning Power of Chemistry in the Coal Mining Industry," AIME *Bulletin* 112 (1916): 711–714.

65. The interaction of fuels, stokers, and furnaces is from "Factors in the Selection of Coal," *Combustion* 2 (May 1920): 12–15; and Young, "Sources of Coal." A. C. Fieldner, "Notes on the

Sampling and Analysis of Coal," USBM TP 76 (Washington, 1914). Discussions of utility procedures for purchasing fuel in the early 1920s are in "A Rational Basis for Coal Purchase Specifications" and "Discussion," ASTM *Proceedings* 22 (1922): 557–564 and 572, respectively; "Compare Coal Analyses of 25 Laboratories," CA 23 (March 15, 1923): 451–453; George Pope, "Sampling Coal Deliveries," USBM *Bulletin* 63 (Washington, 1913). On settling during transport, see "Notes from across the Sea," CA 42 (Oct. 1937): 74–75.

66. "Coal Analyses Made at the Power Plant," *Power* 50 (August 25, 1919): 332–335, 332. "Greatest single factor" is from Morgan Smith, "Factors Governing the Purchase of Coal," ME 49 (Oct. 1927): 1063–1066, 1063. See, too, the four articles collectively entitled "Buying Coal for Profit," *Power* 73 (April 7–28, 1931): 545–665. "Coal Washing Benefits Coke Making Operation," C&ME 40 (Sept. 1933): 470–473. For the contract, see J. H. Kerrick et al., "Use Specifications for Coal," in *Coal Preparation*, ed. David Mitchell (New York: AIME, 1943), 108–126.

67. "Hotel New Yorker Goes Back to Coal," *Power* 84 (August 1940): 72–73; George Otis Smith, "Conserving Our Petroleum Supply," *The Spur*, Jan. 15, 1925, 60. Federal Oil Conservation Board, *Report*, part 1.

68. The literature on the TVA is vast; a good summary of the issues and other sources of federal power is Twentieth Century Fund, *Electric Power*, which has the 4-percent figure in table IX-2. That Pacific Coast states used oil and gas for fuel is from US Bureau of the Census, *Census of Electrical Industries, 1932*, table 22. For opposition to the TVA, see, for example, "Coal Men Protest Power Expansion," NYT, May 21, 1934. Steam power in East South Central states is from Federal Power Commission, *Electric Power Statistics, 1920–1940* (Washington, 1941), 142.

69. That the ratio of brake to indicated horsepower was about 85 percent is from Ewing, *Steam Engine*, 302. The 50-percent loss of power in transmission is from C. H. Benjamin, "Friction Horsepower in Factories," ASME *Transactions* 18 (1897): 228–235. The 67-percent figure and Baldwin Locomotive are from George Richmond, "Operating Machine Tools by Electricity," *Engineering Magazine*, Jan. 1895, 669–686. The 93-percent power loss is from John J. Flather "The Modern Power Problem I: Electric and Compressed Air," *Cassier's Magazine*, March 1903, 636–649.

70. The best source on these matters is Hunter and Bryant, *History of Industrial Power*, vol. 3. On factory electrification, see also DuBoff, *Electric Power*; and Devine, "From Shafts to Wires." B. F. Crocker et al., "Electric Power in Factories and Mills," AIEE *Transactions* 12 (March 1895): 404–416, 415; Gordon, *The Rise and Fall*, chap. 14.

71. Westinghouse is from W. M. McFarland, "Electric Power in the Machine Shop," *Cassier's Magazine*, Nov. 1902, 61–92. The company also generated its own electricity via a turbine. D. Selby-Bigge, "Electricity in the Iron and Steel Industries," *Engineering Magazine*, Dec. 1894, 401–408. GPO is from the discussion of W. H. Tapley in "The Electrical Distribution of Power in Workshops," JFI 151 (Jan. 1901): 1–28, 18.

72. The claim in the text that with electric transmission, losses are only those of the motor would not be true if the motor required belting or gearing to run the machine at its desired speed.

73. If there are transmission losses of 65 percent from steam engine to point of production, then one Ihp yields only 0.35 horsepower-hours at the machine; to achieve one hph at the machine requires $1/0.35 = 2.86$ ihph and $3 \times 2.86 =$ about 8.6 tons BCE. At three-quarters load, efficiency is $1/0.28 \times 3 = 10.7$ lbs.

74. Jerome, *Mechanization in Industry*, appendix C. With 65-percent loss, 2.11 indicated horsepower per worker yields only 0.73 horsepower at the machine, while with 40-percent loss, 4.86 horsepower yields 2.9 horsepower at the machine and $(2.9 – 0.73)/0.73 = 2.97$—a nearly 300-percent increase.

75. Kittredge, "Utilization of Wastes," 3; Campbell and Parker, "Coal Fields of the United States," in Gannett, *Report*, 3:426–443, 432; Richard Moldenke, "The Coke Industry of the United States," USBM *Bulletin* 3 (Washington, 1910), 5. Van Hise, in *Conservation of Natural Resources*, termed coking "a reckless waste of our fuel resources" (28).

76. Berton Braley, "A Blight of Waste," CA 4 (August 16, 1913): 221.

77. One generation's environmental panacea becomes the next generation's environmental poison. For a brief overview of modern efforts to regulate byproduct coke oven emissions, see John Graham and David Holtgrave, "Coke Oven Emissions: A Case Study of Technology Based Regulation," *Risk: Health, Safety and Environment* 1 (Summer 1990): 243–272.

78. On byproduct coking, see C. J. Ramsburg and F. W. Sperr, "By-Product Coke and Coking Operations," JFI 183 (April 1917): 391–431. For a good description of the process as it was employed at Henry Ford's River Rouge plant, see C. R.

Bellamy, "Progress in the Byproduct Coke Indus-try," American Institute of Chemical Engineers *Transactions* 13, part 1 (1920): 325–340. Haynes, *American Chemical Industry*, vol. 2, chap. 12, provides an overview through World War I. Coke breeze is from MR, 1928, part 2, 743.

79. Braley, "A Blight of Waste," 221.

80. The first byproduct oven is from MR, 1897, part 1, 584. MR, 1900; production is on 481–484, quotation on 483.

81. That is, $Coal = \dfrac{Coal}{Coke} \times \dfrac{Coke}{Pig \ Iron} \times Pig \ Iron$.

For the sources of the decline in coke/(pig iron), see Herman Brassert, "Modern American Blast Furnace Practice," AISI *Yearbook* (1915), 15–69; Walter Mathesius, "Uniform Coking Coal as a Factor in Blast Furnace Economy," AISI *Yearbook* (1924), 36–63; and "Coal Washing Benefits Coke-Oven Operation," C&ME 40 (Sept. 1933): 470–473. The 69-percent figure is from "Progress in Fuel Utilization in 1927." The calculation is 45 million tons × 1.42 tons coal/ton of pig, minus 0.74 × 1.42 × 45. No estimates exist for energy savings at the time.

82. The 74-percent figure is from US Energy Information Administration, "Today in Energy," May 9, 2014, https://www.eia.gov/todayinenergy /detail.php?id=16211. Scrap iron is from "How Use of Scrap Iron is Increasing," *Iron Age* 129 (Feb. 14, 1929): 471–474; and MY (1941), 526.

83. Arno Fieldner, "Recent Developments in By-Products from Bituminous Coal," USBM RI 3079 (Washington, 1931). J. A. DeCarlo and J. A. Corgan, "Coproducts of Coke in the United States, 1917–1949," USBM IC 7504 (Washington, 1949), contains sales of coke oven gases.

84. See Haynes, *American Chemical Industry*; and Ralph Landau and Nathan Rosenberg, "America's High Tech Triumph," *Invention and Technology* 6 (Fall 1990): n.p. See also "Coke," MR, 1915, part 2.

85. For Edison, see Paul Israel, *Edison: A Life of Invention* (New York: John Wiley, 1998), 404. For Edison's importance, see Meade, *Portland Cement*, 142. Ellis Stoper, "The Rotary Kiln," ASME *Journal* 32 (Oct. 1910): 1563–1578; Lesley, *History*; "Waste Heat from Cement Kilns Operates Entire Mill," C&ME 29 (July 2, 1923): 18–19. Self-generated electricity is from MR, 1930, 423. Energy use and technology in cement making are from Yaworski, *Fuel Efficiency*, tables A-3 and A-6.

86. "The Power System; the Cost of Energy," *Paper Industry* 5 (May 1923): 323–325. 324; E. L. H. Morrison, "You Get Something for Almost Nothing

by Developing Power from Process Steam," *Power* 69 (March 12, 1929): 434–435. For the claim of 100-percent return, see William Johnson, "Power Generated from Process Steam Yields Large Profit," C&ME 33 (Feb. 1926): 95–96. The issue is *Power* 70 (August 27, 1929).

87. "Thermic flywheel" is from "Chemical Engineering Advances Made by Gas Industry," C&ME 33 (Nov. 1926): 663–701, 701. For the cement plant, see "Surplus Power from Waste Heat Sold to Central Station," *Power* 68 (August 7, 1928): 220–222. See also "Where Industrial Plants and Utilities Gain by Exchange of Power," *Power* 71 (May 27, 1930): 820–826, which contains the estimate of 539 million kwh.

88. G. R. Hopkins, "Survey of Fuel Consumption at Refineries," USBM RI 2965 (Washington, 1920), 1. Refinery energy use until 1940 may be followed in publications by the same author and with the same title in the following USBM RI: 3038, 3145, 3270, 3332, 3367, 3485, 3554, and 3607. The share of gases and sludge is from Schurr et al., *Energy in the American Economy*, table A-15.

89. "Hog Fuel and Waste Heat Generate Paper Mill Steam," *Power* 67 (April 24, 1928): 723–725. For liquid waste, see "Paper Mill Power Plant," *Power* 71 (Jan. 21, 1930): 88–91; "The Value of Fuel Economizers in Paper Mill Operation," TAPPI *Papers* 5 (1922): 92–95; "An Industrial System for Fuel Gas Handling," C&ME 44 (May 1937): 235–240; and R. S. McBride, "Again Ford Shows the Way," C&ME 46 (March 1939): 150–154.

90. Investigations of smoke problems also began to suggest that pollution involved far more than just smoke. Antismoke laws, one speaker told the American Society of Mechanical Engineers in 1926, eliminated "only a part of the air pollution evil—and that part the less grave and less menac-ing to health." "If a chimney is not emitting dense smoke," he observed, "it is considered to be a good chimney regardless of the amount of . . . sulfur oxides etc., it may be emitting." H. B. Meller, "Smoke Abatement: Its Effects and Its Limita-tions," ME 28 (Nov. 1926): 1275–1283, 1275, 1283.

91. On the goal of profitability, see "Improve-ments in Heating Systems Save Four Times Their Cost," *Power* 67 (Jan. 3, 1928): 27–28; and "Instru-ments Save Money," *Power* 68 (Sept. 25, 1928): 508–510.

92. Tarr and Zimring, "Struggle for Smoke Control in St. Louis."

Chapter 4. Railroads

1. For bus and truck transport in 1940, see Barger, *Transportation Industries,* appendix F.

2. Fuel shares are from Bukovsky, "Use and Cost of Railway," table 10; and Fifteenth Census of the United States, 1929, *Manufactures: Statistics of Industries and States* (Washington, 1932), table 1.

3. Thomas Urquhart, "On the Use of Petroleum Refuse as Fuel in Locomotive Engines," Institution of Mechanical Engineers *Proceedings* (August 1884): 272–298. The other papers are "Supplementary Paper on the Use of Petroleum Refuse as Fuel in Locomotive Engines," Institution of Mechanical Engineers *Proceedings* (Jan. 1889): 36–65; and "On the Compounding of Locomotives Burning Petroleum Refuse in Russia," Institution of Mechanical Engineers *Proceedings* (Jan. 1890): 47–73.

4. Urquhart, "Use of Petroleum Refuse," 287, and "Discussion," 320.

5. For examples of American efforts, see "Petroleum in Locomotives," ART 17 (August 12, 1872): 254; and "Petroleum as Fuel for Locomotives," ART 19 (Sept. 7, 1867): 288. "Petroleum as Fuel," RG 14 (April 14, 1882): 225. A good review of international experiments is A. Morton Bell, "Liquid Fuel for Locomotives," *Cassier's Magazine,* March 1898, 371–384.

6. Charles Cochrane, "Discussion," Institution of Mechanical Engineers *Proceedings* (Jan. 1889): 65; "Petroleum Fuel for Locomotives," RG 16 (Sept. 12, 1884): 666; "Oil Fuel in Locomotives," RA 12 (Oct. 21, 1887): 749; "Liquid Fuel," *Engineering* 41 (June 11, 1886): 563. For Dudley's career, see "Charles Benjamin Dudley," *Obituary Record of Graduates of Yale University Deceased during the Year 1909–1910* (New Haven, CT: Yale University Press, 1910?).

7. Charles B. Dudley to Theodore Ely, Dec. 1, 1886, Box 465, PRRHML.

8. A. S. Vogt to Theodore Ely, June 23, 1887, Box 465, PRRHML; Charles B. Dudley, "Fuel Oil," JFI 126 (August 1888): 81–95.

9. General Superintendent of Motive Power to R. L. O'Donnell, May 19, 1919, Box 465, PRRHML.

10. Theodore Ely to Charles Pugh, June 30, 1887, Box 465, PRRHML. Oil production grew sharply in Ohio after 1885, but Ohio was in the heart of coal country, and I have found no evidence that Ohio railroads experimented with oil fuel. Oil production over time by state may be found in the petroleum chapter of MY (1937).

11. "Coal Equivalents of Locomotive Fuels and Power," RME 115 (Oct. 1941): 421–423; "Liquid Fuel in California," RG 17 (Nov. 27, 1885): 759; J. N. Clark,

"The Development of Oil Burning Practices on Locomotives," IRFA *Proceedings* 17 (1925): 136–162.

12. There are several ways to measure the relative importance of oil firing: (1) the percentage of all locomotives that burn oil; (2) oil's share of total energy use, converting it to its coal-equivalent in Btus; and (3) the percentage of all locomotive miles accomplished by oil-fired locomotives. While these measures give slightly different results, they all show similar trends. I have typically used the second approach where available but have also used the first where necessary.

13. "Oil Fuel," RG 27 (August 23, 1895): 559; "Oil for Locomotive Fuel in Southern California," RG 27 (April 5, 1895): 217; "Oil Burning Locomotives," RA 52 (March 15, 1912): 482; "Fuel Oil Burning Apparatus on the A.T. & S.F. Ry.," *Railway Review* 66 (June 12, 1920): 1009–1011; Atchison Topeka & Santa Fe Railroad, *Annual Report* (1896): 15.

14. That is: $(4,576 \times \$3.79) - (3,642 \times \$2.30) = \$8,966$; and $(\$8,966/\$500) \times 100 = 1,793\%$. The average return, which would include the cost of oil tanks and storage, was surely less but still must have ranked as a very profitable opportunity. Conversion costs varied with the size and type of locomotive while the inflation associated with World War I also blew them up. For example, circa 1920, the Santa Fe estimated that it cost about $2,500 to convert a Pacific type (4-6-2) locomotive ("Fuel Oil Burning Apparatus").

15. For descriptions of the technology, see "Oil for Locomotive Fuel"; "Oil Burning Locomotives"; "A New Arrangement of Fuel Oil Burners for Locomotives: Southern Pacific Railroad," ARRJ 78 (July 1904): 263; and "Fuel Oil for Locomotive Use," RA 58 (May 29, 1915): 1115. A good overview is Clark, "The Development of Oil Burning Practices."

16. For the Santa Fe, see "Fuel Oil Burning Apparatus." "Oil Burning Locomotives," RA 22 (Sept. 11, 1896): 309. For storage, see "Installing Facilities for Handling Locomotive Fuel Oil," RA 69 (August 27, 1920): 346–348; "How the Santa Fe Handles Oil for Locomotive Fuel," RA 85 (July 21, 1928): 109–111; and "The Design of Fuel-Oil Stations Presents Many Problems," RA 87 (July 20, 1929): 196–200. "A New Arrangement"; "Burning Crude Oil in Locomotives," RG 33 (June 21, 1901): 431. Methods of heating are from "Oil Fuel for Locomotives," ARRJ 75 (Dec. 1901): 386–391.

17. On the importance of price uncertainty, see Gerard Kuper and Daan van Soest, "Does Oil Price Uncertainty Affect Energy Use?," *Energy Journal* 27 (2006): 55–78, and sources cited therein. Price uncertainty may have inhibited modifications in

oil-fired locomotives that would have made them more difficult to convert to coal. The quotation is from "Oil Fuel on the Southern Pacific," RG 29 (Oct. 1, 1897): 693.

18. On lock-in, see, e.g., Gregory Unruh, "Understanding Carbon Lock-In," *Energy Policy* 28 (2000): 817–830; and Varun Sivaram, "Unlocking Clean Energy," *Issues in Science and Technology* 33 (Winter 2017): 31–40. For a skeptical assessment of technical barriers, see Benjamin Sovacool and Richard Hirsh, "Energy Myth Six—The Barriers to New and Innovative Energy Technologies Are Primarily Technical: The Case of Distributed Generation (DG)," in *Energy and American Society—Thirteen Myths*, ed. Benjamin Sovacool and Marilyn Brown (Dordrecht, Netherlands: Springer, 2007), 145–169.

19. Chicago, Rock Island and Pacific Railroad, *Annual Report* (1906–1913). The Northern Pacific is from "Standard Stoker Engine Placed on Tender," RME 100 (Sept. 1926): 570–571. See also Arthur Redfield, "Railroad Fuel Oil Consumption in 1926," USBM IC 6049 (Washington, 1927).

20. Eugene McAuliffe, "Petroleum—Its Origin, Production and Use as Locomotive Fuel," IRFA *Proceedings* 3 (1911): 164–206; McAuliffe, "Economic Aspects of the Fuel Oil Situation," RA 68 (June 4, 1920): 1565–1566, 1566. McAuliffe was president of the Union Pacific Coal Company, which may have colored his views. Still, his was a common assumption; see, e.g., C. E. Beecher, "Economic Aspect of the Fuel Oil Situation," RA 75 (August 25, 1923): 352–353.

21. The SP's oil shale venture is from Alderson, *Shale Oil Industry*, 31 and 42; and Russell, *History of Western Oil Shale*, 80–83.

22. For assessments by an advocate, see John Muhlfeld, "Pulverized Fuel for Locomotives," New York Railroad Club *Proceedings* 26 (1916): 4345–4354; see also W. L. Robinson, "Powdered Coal," IRFA *Proceedings* 7 (1915): 17–27; and "Report of Standing Committee on Pulverized Fuel," IRFA *Proceedings* 11 (1919): 16–27. For railroad experiments with briquettes, see C. T. Malcolmson, "Briquetted Coal and Its Value as a Railroad Fuel," IRFA *Proceedings* 1 (1909): 112–149.

23. Oil specifications and that the SP burned crude oil are from Howard Stillman, "Locomotive Practice." Eugene McAuliffe, "Petroleum—Its Origin," 183. The longer quotation is from Clark, "Development of Oil Burning Practices," 136. On proper heating, see "Operating and Maintaining Oil-Burning Locomotives," RA 73 (Nov. 18, 1922): 949.

24. Maury Klein, "Replacement Technology: The Diesel as a Case Study," *Railroad History* 162 (Spring 1990): 109–120.

25. For the tests and the Gulf, Colorado & Santa Fe, see "Fuel Oil," Traveling Engineers Association *Proceedings* (1902): 217–232; see also "Fuel Oils in Texas and California," ARRJ 75 (August 1901): 248–249; and "The New Situation as to Fuel Oil," ARRJ 75 (August 1901): 243.

26. The spreading use of oil by western carriers may be followed in the chapter on petroleum in MR, part 2, 1907–1918; see also "Use of Texas Oil by Railroads," RG 33 (Nov. 8, 1901): 767.

27. For the history of cracking, see John Enos, *Petroleum Progress and Profits* (Cambridge, MA: MIT Press, 1962).

28. The data are from US Interstate Commerce Commission (ICC), *Annual Report on the Statistics of Railways* (1944), table 155. These are weighted by 1918 freight and passenger rates from that table. Barger, *Transportation Industries*, chap. 4, table 17, also computed an index of railroad output; its use yields results are almost identical to those shown.

29. These measures are subject to several limitations. Their exclusion of fuel burned in stationary boilers implies that the shift to direct steaming put an upward bias on the gain in fuel economy. In addition, these measures ignore quality improvement in service that resulted from faster trains; heavier, safer, steel passenger cars; and air conditioning. For example, all-steel cars rose from about a quarter of the total in 1916 to 82 percent in 1940, and passenger safety improved remarkably, with the fatality rate declining from about 22 per billion passenger miles in 1907 to fewer than two in 1939. See Aldrich, *Death Rode the Rails* (Baltimore: Johns Hopkins University Press, 2006).

30. E. D. Nelson to A. W. Gibbs, May 21, 1907, reports on tests of salt and lime to reduce smoke. The "liquid composition" is from Alfred D. Chandler to A. D. Gibbs, August 4, 1908. George Koch to E. D. Nelson, Feb. 8, 1909, contains reference to the smokeless fuel compound. All of these letters are in Box 659, PRRHML. "Fuel Economy," RG 15 (Nov. 30, 1883): 790–791, describes the Pennsylvania's rating system. Difficulties with the early coal premium are discussed in Association of Transportation Officers, "Report of the Committee on Motive Power," Jan. 13, 1913, Box 720, PRRHML. The peculation is noted in L. R. Zollinger to R. K. Reading, Oct. 9, 1912, Box 720, PRRHML.

31. General Superintendent of Motive Power to H. M. Carson, Dec. 8, 1908, Box 470, reports on the circular and test plant. The (50-page) report is

contained in "Instructions for Economical Use of Coal," Box 470. The test of the locomotive is reported in B. M. Kincaid, "Fuel Losses," June 11, 1908, Box 470. All are in PRRHML.

32. Chairman to Committee on Motive Power, Dec. 9, 1914; and Charles D. Young to I. B. Thomas, Dec. 24, 1914, Box 720, both discuss the decision to study fuel departments. For the tour, see Chairman to the Secretary, Association of Transportation Officers, Jan. 18, 1915, Box 720. The tour is reported in T. H. Watkins and E. W. Smith to C. D. Young, March 16, 1915. The rejection of a fuel department is in "Report of the Committee on Motive Power," Oct. 9, 1915, Box 720. All are in PRRHML.

33. McAuliffe was also deeply concerned with conservation of human resources. When he left the railroads to become president of Union Pacific Coal in 1923, he redoubled the company's efforts to reduce accidents; see Aldrich, *Safety First* (Baltimore: Johns Hopkins University Press, 1997), chap. 6. For biographical information on McAuliffe, see *Who Was Who in America*, vol. 3, *1951–1960* (Chicago: Marquis Who's Who, 1966), 567.

34. The rate requests are discussed in Martin, *Enterprise Denied*. Railway net income is from Jones, *Railway Wages*, table 74. Richardson is quoted in "Traveling Engineers Convention," RA 49 (August 26, 1910): 350–354.

35. J. R. Scott [St. Louis–San Francisco], J. E. Ingling [Erie], and W. L. Robertson [B&O], "Remarks," Traveling Engineers Association, *Proceedings* 21 (1913): 169–174. Urban concerns with smoke are also discussed in Rosen, "Businessmen against Pollution." See also Tarr, *Search for the Ultimate Sink*, chap. 9; Stradling and Tarr, "Environmental Activism"; and Stradling, *Smokestacks and Progressives*.

36. The Lehigh Valley is from L. C. Fritch, "Opportunities for Economy on Railways," RA 51 (Nov. 24, 1911): 1059–1061. "Thick and Thin Firing of Locomotives," RA 51 (Dec. 15, 1911): 1204–1205; "Fuel Economy on the Buffalo, Rochester and Pittsburgh," RA 53 (Oct. 11, 1912): 678–679; H. C. Woodbridge, "Fuel Economy on the Buffalo, Rochester and Pittsburgh," RA 54 (Feb. 13, 1913): 297–298; General Superintendent [of the Rock Island] to A. M. Shoyer, March 4, 1909, Box 470, PRRHML.

37. "Fuel Economy on the Alton," RA 53 (Sept. 27, 1912): 584; Chicago & Alton Railroad, Office of the President, General Circular No. 10, Sept. 12, 1912, Box 470, PRRHML; "Fuel Economy on the N.P.," RA 56 (May 1, 1914): 972; "Fuel Instruction

Car on the Northern Pacific," RA 56 (May 1, 1914): 976–982.

38. "Fuel Economy on the Great Western," RA 58 (Jan. 29, 1915): 195–197. The Great Western is discussed in T. H. Watkins and E. W. Smith to C. D. Young, March 16, 1915, Box 720, PRRHML.

39. T. H. Watkins and E. W. Smith to C. D. Young, March 16, 1915, Box 720, PRRHML; L. G. Plant, "The Fuel Department: A Constructive Criticism," RA 59 (Nov. 19, 1915): 939–941.

40. "Economies Effected [sic.] by Mallet Locomotives on the New York Central & Hudson River," RA 51 (Nov. 24, 1911): 1054–1056.

41. While many of these devices improved locomotives' technical efficiency, their main payoff was their ability to increase economic efficiency by raising hauling capacity per locomotive, per worker, and per ton of fuel.

42. The Texas & Pacific is from "Report on Locomotive Economy Devices," RA 82 (May 21, 1927): 1511–1513. For similar examples, along with the savings they generated, see *Report of the Mechanical Advisory Committee to the Federal Coordinator of Transportation* (Dec. 1935), appendix A. For modern discussions of locomotive evolution during these years, see Bruce, *Steam Locomotive*; Lamb, *Perfecting the American Steam Locomotive*; and Withuhn, *American Steam Locomotives*. For the career of Goss, see *Encyclopedia of American Biography*, n.s., vol. 21 (New York: American Historical Society, 1949), 46–60.

43. "The New Haven Saves a Million Dollars in Fuel," RA 64 (Jan. 25, 1918): 207–208; "Better Air-Brake Mechanism Will Save Fuel," RA 65 (August 9, 1918): 244.

44. L. G. Plant, "Fuel Department Organization," IRFA *Proceedings* 13 (1921): 203–227.

45. "Organization and Work of Fuel Conservation Section," RA 65 (Nov. 19, 1918): 959–960. For a discussion of the 1918 meeting, see McAuliffe, *Railway Fuel*, 253–256. The Pennsylvania's committee is noted in General Superintendent of Motive Power to E. B. Whitman, Chair, Fuel Conservation Committee, July 16, 1919, Box 470, PRRHML. See also "Fuel Conservation Should Be Continued," RA 68 (Jan. 2, 1920): 85–86.

46. For labor and other costs, see Jones, *Railway Wages*, tables 72 and 73. The survey is reported in Plant, "Fuel Department Organization."

47. For lists of these advantages, see Urquhart, "Use of Petroleum Refuse"; Dudley, "Fuel Oil"; and McAuliffe, "Petroleum—Its Origin." For the MKT see "Installing Facilities for Handling Locomotive Fuel Oil," RA 69 (August 27, 1920): 346–348; and

"M-K-T Fuel Bill Takes Big Drop," RA 93 (Nov. 5, 1932): 642. "Milwaukee Buys Steam Locomotives for Fast Schedules," RA 98 (May 11, 1935): 719–721. "Northwestern Inaugurates '400' High Speed Train Service," RA 98 (Feb. 2, 1935): 188–192.

48. Atchison, Topeka & Santa Fe Railroad, *Annual Report* (1907); "Fuel Oil Burning Apparatus." The cost of handling is from "Operating and Maintaining Oil-Burning Locomotives," RA 73 (Nov. 18, 1922): 949. Coal-fired locomotives eventually offset some of these advantages, using netting to catch sparks and cinders and stokers to feed heavy locomotives.

49. "The Railways' Interest in Forest Fire Prevention," RA 52 (Feb. 9, 1912): 231–235. SP snow sheds are from Clark, "Development of Oil Burning Practices."

50. "Extending Locomotive Runs on the Santa Fe," *Railway Review* 73 (July 21, 1923): 83–88.

51. For discussion of the fuel efficiency advantages of heavier trains, see J. E. Davenport, "Effect of Tonnage on Fuel Consumption," RA 73 (July 8, 1922): 71–74, which also demonstrates that the greatest gain from heavier trains came from the reduction in labor costs. The representative from the Southern Pacific is T. H. Williams, "Securing the Maximum Efficiency in Train Loading," RA 65 (April 25, 1919): 1051–1052.

52. See Frank Russell, "Extension of Locomotive Runs," RME 98 (August 1924): 464–467. Miles per locomotive are total locomotive mileage divided by locomotives in service, both from ICC, *Statistics of Railways*. The experience of the Southern Pacific is described in T. H. Williams, "Fuel Economy in Long Locomotive Runs," IRFA *Proceedings* 19 (1927): 108–111. "Modernizing Locomotive Terminals on the Great Northern," RME 105 (June 1931): 286–290. Barger, *Transportation Industries*, chap. 4, stresses the loss of short-haul traffic.

53. The experience of the Central Pacific is described in Julius Kruttschnitt, "Railway Fuel: A Reducible 13 Percent of Operating Expenses," IRFA *Proceedings* 15 (1923): 9–29.

54. "Estimated Savings to Be Effected on Five Divisions of a Railway System by Replacing Manual Block System with Automatic Block System, 1926," ARA Signal Section, *Proceedings* 23 (March 1926): 495–497.

55. On the cost of stopping trains, see McAuliffe, *Railway Fuel*, 291–296. See also, e.g., Edgar Weston, "Further Testimony on Merits of Form 19 Order," RA 72 (Feb. 18, 1922): 417–418.

56. Some findings are contained in ARA, *On Fuel and Related Economies*, booklets 1–4 (New York, 1924), esp. booklet 2. For the Great Northern, see C. Herschel Koyl, "Treating Water Reduces Boiler Troubles," RA 66 (April 25, 1919): 1053–1056. The expert was C. R. Knowles, "The Fuel Saving Aspect of Boiler Water Treatment," IRFA *Proceedings* 15 (1923): 101–109; see also his "Water Softening as a Factor in Fuel Conservation," RA 71 (July 23, 1921): 285–287.

57. For a review of fuel economy in terminals, see Blakemore et al., *Fuel Economy*. For evidence of urban concerns with smoke, see the sources cited in note 35 above. On direct steaming, see "New Locomotive Economy Devices," IRFA *Proceedings* 20 (1928): 362–388; "Stationary Power Plants," IRFA *Proceedings* 21 (1929): 403–428; "Report on Coal-Fired Stationary Boiler Plants," IRFA *Proceedings* 22 (1930): 435–470; and "Grand Trunk Western Has Fireless Engine House in Chicago," RME 102 (April 1928): 221–223.

58. The New Haven is from "Memorandum for Board of Trustees," August 25, 1924, Box 20, Secretary's Records, Record Group I, New York, New Haven and Hartford Railroad Collection, University of Connecticut.

59. The 1920 tests are reported in D. W. Crawford, "Advantages of Stoker Fired Locomotives," RME 94 (March 1920): 127–130. For the Southern Pacific, see Kruttschnitt, "Railway Fuel." For the Erie, see R. E. Woodruff, "Address," IRFA *Proceedings* 22 (1930): 328–331. "How the Chicago Great Western Controls Use of Locomotive Fuel," RME 99 (March 1925): 144–146. Traveling engineers on the Frisco are reported in the discussion of C. F. Luddington, "Uniform Methods of Computing Fuel Consumption," IRFA *Proceedings* 6 (1914): 73–99. For the Santa Fe, see Case, *Report on the Handling of Fuel*.

60. For the safety campaigns, see Aldrich, *Safety First*, chap. 6. "The Delaware & Hudson is reported in "Practical Propaganda for Fuel Conservation," RA 70 (April 1, 1921): 843–845. A. P. Wells et al., "Fuel Economy on the Central of Georgia," IRFA *Proceedings* 14 (1922): 98–106; "Fuel and Locomotive Performance on the Central of Georgia," RA 72 (Feb. 25, 1922): 479–480; and W. A. Kline, "Fuel Losses at Locomotive Terminals," IRFA *Proceedings* 16 (1924): 87–93.

61. For the prize essay contest and the number of applicants, see McAuliffe, *Railway Fuel*, 256–257. For examples of cartoons and bulletins, see "Fuel Conservation Bulletins and Cartoons," IRFA *Proceedings* 20 (1928): 233–248; and 21 (1929): 318–325. See also "How the Chicago Great Western Controls Use of Locomotive Fuel," RME 99 (March 1925): 144–146. For the B&O's cooperative

plan, see David Vrooman, *Daniel Willard and Progressive Management on the B&O* (Columbus: Ohio State University Press, 1991). For the Texas & Pacific, see L. E. Dix, "Remarks," IRFA *Proceedings* 22 (1930): 270–272.

62. For details on sources and construction of figure 4.2, see table A3.7.

Chapter 5. Coal Departs the Urban Kitchen, 1900–1940

1. National Commercial Gas Association, *Nancy Gay*; "The Smile of Nancy Gay," *Lighting Journal* 2 (August 1, 1914): 183–184, 184.

2. Much of the secondary literature on anthracite focuses on the nineteenth century, whereas work covering the twentieth century emphasizes labor issues or the social consequences of the industry's decline rather than the decline itself. The best analyses of the decline of anthracite are Mead, *An Analysis*; and Michalik, *The Decline*.

3. The "smokeless" coals of West Virginia are an exception to the claim that soft coal was generally not branded. Contemporary descriptions of industry economics can be found in US Coal Commission, *Report*. See also US Federal Trade Commission, *Report*; and C. G. Duncan "Coal Distribution in America," *Black Diamond* 71 (August 2, 1923): 120–122, and subsequent articles by that title in the same volume. Prices are from MR, 1918, part 2, 105.

4. For the breakdown of retail prices, see US Coal Commission, *Report*, part 2, 836, table 51.

5. MR, 1911, part 2, 19; US Coal Commission, *Report*, part 1, 53, 55, 183–191. Jacob Hollander, who authored one of the commission's studies, dissented from the prevailing views on resource scarcity, ascribing the problem in anthracite to monopoly instead. See US Coal Commission, *Report*, part 2, 989–1010.

6. The forecast was for 100 million *long* tons (2,240 lbs.), but the chart is in net tons, so the forecast was about 10 percent too high. For productivity, see Barger and Schurr, *Mining Industries*, table A-5.

7. Figures in the text are from USBLS, "Cost of Living," table E. Those surveyed were white families of wage earners and individuals with low to medium salaries. Wood accounted for only about 4 percent of energy use for urban families about the time of World War I.

8. Note that the charts of anthracite in figures 5.1 and 5.2 differ because the former includes all production while the latter includes only domestic and pea sizes. MR, 1907, part 2, 168–169.

9. If gashouse coke was clean to burn, recall from chapter 1 that it was not clean to make. See Tarr, "Toxic Legacy." For a discussion of coke as a domestic fuel, see "Coke in a Residence Heater Designed for Coal," H&V 4 (May 1917): 19–20; Arno Fieldner, "Why and How Coke Should Be Used," USBM TP 242 (Washington, 1919); and Kudlich, "Fuels Available." For early domestic sales, see MR, 1912, part 2, 1166. Its geographic distribution is from MR, 1930, part 2, 536.

10. Fieldner, "Why and How," 5. For the advantages of coke as smokeless fuel, see C. G. Atwater, "Smokeless Fuel for Cities: It's Relation to the Modern Byproduct Coke Oven," *Cassier's Magazine*, August 1906, 313–321. As we saw in chapter 2, there is little evidence that municipal smoke ordinances had an important impact on consumers' fuel choices; anthracite use was widespread before such ordinances became common, and in any event, many regulations exempted households.

11. Briquette prices in table A4.2 are producer realizations, not retail prices; the relevant comparison is therefore with the mine price of soft coal. Centralization in the Lake States region is discussed in MR, 1928, part 2; and MY (1937), 963.

12. The source of the survey is Kryk et al., "Family Income," appendix table 30. The equation is

$$LN(Pct\ Solid\ Fuel) = \frac{-.82}{(-13.8)} LN(Income).\ R^2 = 0.98,$$

N = 98 (figures in parentheses are t-ratios). The equation also includes dummy variables to control for city size and location. Income data are mid-points of thirteen classes from \$250 to \$500 and from \$5,000 to \$10,000; the dependent variable is the percent of families in a given income class that burned coal, wood, etc. The survey was for 1935–1936; families sampled were white, native-born, married couples.

13. The states were New England, New York, New Jersey, Pennsylvania, Ohio, Illinois, Michigan, Wisconsin, and Minnesota, as well as Washington, D.C. Natural gas only reached Chicago, Indianapolis, Minneapolis, and Washington, D.C., in the early 1930s, while New York City and New England waited until after World War II. See Tussing and Barlow, *Natural Gas Industry*, chap. 3; and Herbert, *Clean, Cheap Heat*, chap. 4.

14. The Indiana study is Purdue University, "Fuels Used." I have ignored electricity because it was little used at this time; I have added manufactured gas on the assumption that it would have required the same Btu use as natural gas and sold at the average price in 1918. The coal used by the study

was from Ohio, and its Btu content was 24Mm Btu/ton. I have added anthracite to the table, assuming that it has 25.2 Mm Btu/ton, and so 24/25.2 = 0.952 pounds of anthracite would be equivalent to a pound of Ohio coal. As can be seen, cooking with anthracite used 6.2 times more energy than did cooking with gas (345,714 Btu/56,000 Btu). With coal selling at $9.57/ton, the price of gas per 1,000 ft.[3], that makes it cost the same as coal: $9.57/7.5 = $1.28.

15. The first year that sales of water heaters are available is 1925, and by then, for every coal- or wood-fired heater sold, there were eight sold that burned gas or kerosene. United States Bureau of the Census, *Biennial Census of Manufacturers, 1925* (Washington, 1928), 473.

16. In 1900, a Welsbach mantle yielded about 20 candlepower per ft.[3]/hr. A thousand hours, thus, yielded about 20,000 candlepower-hours (251,400 lumen-hours), and a thousand ft.[3] of gas cost $1.04. Gas thus provided about 245,000 lumen-hours per dollar. Lightbulbs yielded about 3.7 lumens/watt or 3,700 per kw. In early years, some companies charged a flat rate per light, making electricity "free" (that is, with a zero marginal cost). When metered, however, with electricity selling at $0.10/kwh in 1902, electricity provided about 36,000 lumen hours per dollar. For a case of the jitters, see "The Incandescent Lamp vs. the Welsbach Burner," *Electrical Age* 17 (April 6, 1895): 196.

17. In "The Welsbach Light," *Science*, n.s., 12 (Dec. 21, 1900): 951–956, its light is described as "cold, ghastly [and] harsh" (955). For an example of worries that gas would "vitiate" the air, see "Fresh Air as Medicine," GH, March 1895, 128–129. The USGS provided figures showing a decline in gas used for illumination in MR, part 2, for 1905, 1915, and 1919 but noted their unreliability because many companies simply categorized all domestic sales as illumination.

18. Mary Warner, "The Gas Stove and Its Practical Uses," GH, June 1894, 273–274, 273. E[llen] M[urdock], "From Long Experience with Gas," GH, May 1902, 398; Maria Parloa, *Home Economics* (New York: Century, 1910), chap. 8.

19. A good review of a large number of studies linking domestic coal use to adverse health effects is International Agency for Research on Cancer, *Personal Habits*, 515–538.

20. Lulu Eastman, "The Epic of the Gas Stove, Lines in Gas Meter," AGLJ 96 (July 10, 1911): 29.

21. "Our Experiment Station," GH, May 1909, 637; see also "Laundry Machinery and Methods," GH, Oct. 1913, 562–568. For estimates of the spread of various electric consumer goods, see Lebergott, *American Economy*. For modern discussions of household technology, see Strasser, *Never Done*; Cowan, *More Work*; and Day, "Capital-Labor Substitution."

22. For an example of a gas stove with an oven pilot light, see "Gas Range No. 765," *American Artisan* 31 (March 7, 1896): 23.

23. The coal industry did not lack for trade publications. *Coal Trade* (under various titles) dated from the nineteenth century, as did *Colliery Engineer*. *Coal Age*, which became the voice of the industry, began in 1911. The Retail Coal Merchants' Association also dated from the nineteenth century, as did its journal, *Retail Coalman*, but members' advertising consisted of little more than announcements. Although the Coal Mining Institute of America dated from 1900, it was largely a technical association and had nothing to do with marketing. In short, producers had no national spokesperson until the National Coal Association appeared in 1917, and its focus was on labor issues and public policy.

24. For early use of block rates, see "Papers Read before the First Annual Meeting of the Ohio Gas Light Association," AGLJ 42 (April 2, 1885), 174–177. Such rates were common in Massachusetts by 1910; see Massachusetts Board of Electric Light and Gas Commissioners, *Annual Report* 26 (Boston, 1910).

25. *Report of New Business Methods to the American Gas Institute* (Madison, WI: Cantwell Printing, 1906), 121. For a discussion of gas company advertising methods, see Rose, *Cities of Light*; and Rose, "Urban Environments."

26. *Report of New Business Methods*, 79, 145. The Grand Rapids company is from "Increasing Gas Sales to Existing Customers," AGLJ 91 (Dec. 27, 1909): 1388–1390. For a good, detailed description of the sales campaign in Denver about this time, see Rose, *Cities of Light*, chap. 3.

27. *Report of New Business Methods*, 144, 106, 84; "The Education of the Colored Cook," AGLJ 93 (August 22, 1910): 350–351; see also "Development of the New Business Division of the Gas Industry in the Last Fifty Years," AGLJ 93 (July 19, 1909): 268–276.

28. *Report of New Business Methods*, 237; "A Coke Campaign That Increased Sales Seventy-Five Percent," AGLJ 101 (Sept. 14, 1914): 166–167.

29. Muncie is from C. R. Heath, "Brief Tales of a Tenderfoot in Natural Gas and His Experience in Raising Rates from a Low Flat Rate to a Fifty Cent Rate," NGA *Proceedings* 4 (1912):71–77. For early rates in a number of Indiana cities and towns, see "Something about the Natural Gas Plants of Indiana," AGLJ 58 (April 17, 1893): 563–565.

30. That metering reduced consumption is from "Charges for Natural Gas for Domestic Consumption," AGLJ 55 (Dec. 21, 1891): 885. On marketing natural gas, see F. W. Stone, "Increasing Sales of Gas for Domestic Purposes," NGA *Proceedings* 2 (1907): 163–169. See also Lucius Bigelow, "The Commercial Side," NGA *Proceedings* 4 (1909): 451–458; Arnold and Clapp, "Wastes in Production," 12; and "The 'Gas Saving' Department," *Natural Gas Journal* 3 (Jan. 1922): 6–7.

31. J. H. Maxon, "Displacing Natural Gas with Manufactured Gas," NGA *Proceedings* 14 (1919): 344–353. The origin of the efficiency campaign is from US Department of the Interior, *Conference on Natural Gas Conservation* (Washington, 1920). See also "President's Address," and "Report of the Conservation Committee," NGA *Proceedings* 15 (1920): 20–24, 263–302. The Bureau of Standards critique is from I. V. Brumbaugh, "How Natural Gas Burners Can Be Improved," NGA *Proceedings* 16 (1921): 41–120. See also Wyer, "Waste and Correct Use"; and Cattell, "Natural Gas Manual." That the association distributed a half million copies of this document is from "President's Address," NGA *Proceedings* 18 (1923): 17–24.

32. Kanarek, "Disaster for Hard Coal," emphasizes the importance of that strike for the decline of anthracite but largely ignores the longer-term forces causing the industry's eclipse.

33. US Fuel Administration, *Distribution of Coal and Coke*, 14–15, 115–117; US Fuel Administration, *Fuel Facts*; Kreisinger, "Five Ways"; "Koppers Seaboard Coke," *Brooklyn Daily Eagle*, Dec. 1, 1919.

34. US Federal Trade Commission, *Report*, 120–123; US Coal Commission, *Report*, part 1, 51.

35. For Hollander, see US Coal Commission, *Report*, part 2, 989–1010, 996.

36. The industry allocations are from General Policies Committee of Anthracite Operators, *The Anthracite Emergency of 1922–1923 and How It Was Handled* (Philadelphia, 1923); "100,000 Ton Welch Anthracite Order," BG, Oct. 25, 1923. "States to Fight Coal Monopoly," BG, August 22, 1925, contains the municipalities' switch to coke.

37. "How Much Ash Is Found in Commercial Anthracite?," CA 25 (Feb. 21, 1924): 272.

38. "Very large number" is from "No Need to Worry over the Supply of Coal," BG, August 20, 1925. "Survey Shows Bitter Anti-anthracite Campaign in New England," CA 28 (Oct. 29, 1925): 605; "Thinks Anthracite Strike Will Make Soft Coal Universal Domestic Fuel; President Urged Not to Interfere," CA 28 (Nov. 12, 1925): 671. "War to end war" is from "States to Fight Coal Monopoly," BG,

August 22, 1925. "How to Use Soft Coal and Coke in the Homes," BG, Oct. 2, 1925; "West Virginia Smokeless Producers Plan Campaign to Capture New England," CA 28 (Nov. 12, 1925): 669.

39. Kudlich, "Fuels Available," 3; "Use of Coke in Homes Urged by Bureau of Mines," H&V 19 (Oct. 1922): 49.

40. "New Rates Assure Soft Coal Supply," BG, Dec. 2, 1925; Massachusetts Special Commission, *Report*, 1927, 104.

41. "Substitutes for Coal," CT, Oct. 24, 1925. "Good Coke Supply if Miners Strike," NYT, August 24, 1925, contains the quotation and figures. In the mid-1930s, *expenditures* on soft coal were about five times those on hard coal, and with soft coal cheaper, the quantity disparity would have been even greater. See "Family Expenditures in Selected Cities, 1935–36," USBLS *Bulletin* 648, vol. 1 (Washington, 1942), table 7.

42. MR, 1928, part 2, 478, 480. "Anthracite," NYHT, Dec. 21, 1925.

43. National magazine circulation is from NCGA *Proceedings* 10 (1914): 312; NCGA, *The Gas Equipment of the Home* (NCGA, 1914); "Gas Range Week," AGLJ 102 (April 26, 1915): 257.

44. "An Afternoon Off Every Day," GH, Sept. 1915, 97; "Buy a Detroit Jewel 'Special' Gas Range and Cook and Bake with Ease," GH, April 1917, 187; "The Great Majestic Combination Range," GH, Nov. 1919, 211.

45. For "Rochester Gas & Electric," see "Merchandising Gas," *American Gas Engineering Journal* 112 (June 5, 1920): 449; "Peoples Gas Light" is from "Advertising to Stimulate Sales of Merchandise and Appliances," AGA *Proceedings* 4 (1922): 131–141. "Gas Ranges and Gas Heating Appliances Are Profitable Lines for Hardware Merchants," *American Artisan and Hardware Record* 84 (July 8, 1922): 11. "You don't read" is from "Merchandising Gas," *American Gas Engineering Journal* 112 (May 15, 1920): 389–390 (italics in original).

46. Anna Peterson, "Relation of Home Cooking to Gas Sales," AGA *Proceedings* 4 (1922): 104–125; Ada Swann, "Home Service," AGA *Proceedings* 6 (1924): 536–550.

47. "Smoothtop," GH, Feb. 1928, 173. For Roper, see "Give Your Consumers Service," *Natural Gas Journal* 3 (Jan. 1922): n.p. "Gas Ranges Improved and Approved," GH, May 1930, 232–236; N. T. Sellman, "Five Years of Laboratory Progress," AGA *Proceedings* 12 (1930): 21–23. Roper's color stove is from "Climax of Kitchen Color Harmony," GH, March 1928, 301. "The Magic Chef Gas Range Burner," GH, March 1933, 229.

48. "Gives Credit to Benjamin F. Meacham for Invention," *Gas Age Record* 49 (June 3, 1922): 699.

49. The earliest Lorain advertisement I have found is "Spend Your Afternoon at the Red Cross," *Fort Wayne News & Sentinel*, May 10, 1918. Conservation would increase energy use is from "Mr. Hoover Says—," *Kalamazoo Gazette*, July 26, 1918. "Answer to the Servant Problem," *Ladies' Home Journal*, July 1920, 65. "Ends pot watching" is from "Accurately Measured Cooking Heat," *Ladies' Home Journal*, Sept. 1920, 111. For pilot-light requirements, see "Approved Gas Ranges," GH, Jan. 1928, 80–81, 170. For a color ad, see "Lorain," *McCall's*, July 1927, 39. Peterson, "Relation of Home," 112.

50. The 1925 survey is in Mary Sherman, "What Women Want in Their Homes," *Woman's Home Companion*, Nov. 1925, 28–30. Providence and Fargo are from Bush, *Consumer Use of Selected Goods*.

51. Boston City Planning Board, *Report on Real Property Inventory, Boston, Mass. 1934*, vol. 1 (Boston, 1935), table 3.

Chapter 6. "Cooking Shouldn't Cook the Cook"

1. Edith Hawley, "My Income from Home Canning," *Farmer's Wife* 33 (April 1922): 805; Edith Hawley, "Every Foot of My Garden Earns," *Farmer's Wife* 34 (March 1923): 362. For a brief discussion of the origins and activities of the canning clubs, see Ola Powell, *Successful Canning and Preserving: Practical Handbook for Schools, Clubs, and Home Use*, 2d ed. (Philadelphia: H. B. Lippincott, 1918), chap. 18. See also Aretas Nolan and James Green, *Vegetable Gardening and Canning: A Manual for Garden Clubs* (Chicago: Row Peterson, 1917). A modern discussion of the work of extension services in promoting home canning is Clifford Kuhn, "'It Was a Long Way from Perfect but It Was Working': The Canning and Home Production Initiatives in Greene County, Georgia, 1940–1942," *Agricultural History* 86 (Spring 2012): 68–90. See also Elizabeth Engelhardt, "Canning Tomatoes, Growing 'Better and More Perfect Women,'" *Southern Culture* 15 (Winter 2009): 78–92.

2. In this chapter, I will use kerosene and oil as synonyms, and I also include those stoves that burned gasoline. The reader should not confuse gasoline stoves that burned the liquid with gas stoves.

3. Historians have neglected the humble oil stove, eagerly jumping instead to that symbol of modern: electricity. Thus, writers interested in stove technology or household heating and cooking pass quickly over kerosene-fired equipment, as does the vast literature on farm and rural women.

4. For the 1860 estimate, see Lebergott, *The Americans*, 323; and Stephen F. Peckham, *Report on the Technology, Production and Uses of Petroleum and Its Products* (Bureau of the Census, Washington, 1885), 266. In 1882, these sources indicate that family consumption per year was 34 gallons, while total domestic consumption in 1882 was 22,000 42-gallon bbl./day. Thus, $(22,000 \times 365 \times 42)/34 = 9,919,491$ families. Prices are from *Derrick's Handbook of Petroleum* 1 (Oil City, PA: Derrick Publishing, 1898), 782. Gesner, *Practical Treatise*, 32. For biographical information on Gesner, see *Canadian Dictionary of Biography*, http://www.biographi.ca/en/bio/gesner_abraham_9E.html. Candlepower figures are from Gray, "Lighting Power."

5. Gesner, *Practical Treatise*, 96; Beaton, "Dr. Gesner's Kerosene."

6. For a brief discussion of kerosene's dangers, see Jane Brox, *Brilliant: The Evolution of Artificial Light* (New York: Houghton Mifflin, 2010), chap. 5. Early kerosene was probably no more dangerous than some of the illuminants it replaced. Early gas-distribution systems were also prone to blow up; see Castaneda, *Invisible Fuel*, chap. 3.

7. "Report of the Committee of the Franklin Institute on the Causes of Conflagrations and the Methods of Their Prevention," JFI 95 (April 1873): 261–287. The 6,000 figure is from "Losses of Life and Property by Kerosene Oil," *The Independent*, Nov. 23, 1876, 28. "Another Kerosene Explosion," NYHT, Dec. 17, 1866; "Another Kerosene Explosion," NYHT, Jan. 7, 1867. The fire marshal is quoted in "Fires in Brooklyn in 1870," NYHT, Jan. 19, 1871. "Incidents and Accidents," *Sycamore (IL) Whiteside Chronicle*, Sept. 20, 1870; "Current Paragraphs," *Sycamore (IL) True Republican*, March 16, 1870; "News of the Week," *Chicago Western Rural*, March 10, 1870; "Some New Epitaphs," *Puck* 1 (April 1877): 4.

8. Zachariah Allen, "Explosibility of Coal Oils," *Annual Report of the Board of Regents of the Smithsonian Institution* (Washington, 1862), 330–342; Peckham, *Report*, 223, claims that Hill assisted Allen.

9. Charles F. Chandler, *Report on the Quality of the Kerosene Oil Sold in the Metropolitan District* (New York, 1870); Charles F. Chandler, "Report on Petroleum Oil, Its Advantages and Disadvantages," *American Chemist* 2 (June 1872): 446–448; 3 (July 1872): 20–24; and 3 (August 1872): 41–43. Mary Gibson is from "A Dangerous Fraud," *Boston Journal of Chemistry* 5 (Feb. 1, 1871): 92. Data on Baltimore and Philadelphia are from "Report of the Committee of the Franklin Institute," 273–274. For Cleveland, see Jonathan Wlasiuk, "Refining Nature Standard Oil

and the Limits of Efficiency, 1863–1920" (PhD diss., Case Western Reserve University, 2012), 89.

10. Chandler, "Report on Petroleum Oil," 23. Peckham, *Report*, discusses the various tests and their complexities. See also New York State Board of Health, *Report on the Methods and Apparatus for Testing Inflammable Oils* (Albany, 1882). The best modern discussion of these early safety efforts is James McSwain, *Petroleum & Public Safety* (Baton Rouge: Louisiana State University Press, 2018), chap. 1.

11. "The Naphtha Fiend," NYHT, March 19, 1872; "Kerosene Murder," SA, Feb. 18, 1871, 119. "Dangerous Kerosene," NYT, Sept. 30, 1869, contains a list of sellers of kerosene who failed the city's flash test. The Maine law is from Chandler, *Report on the Quality*. The federal law is US Statutes at Large, 39th Cong., 2d Sess., 1867, chap. 169, sec. 29. *Acts and Resolves of the State of Massachusetts in the Year 1867* (Boston, 1867), chap. 286. New York State is from "Storing Combustibles," NYHT, July 15, 1878. New York City's ordinance is from "Board of Health," NYT, Jan. 26, 1869. Laws as of 1879 are from Peckham, *Report*, 236. For details on the state laws about this time, see New York State Board of Health, *Report on the Methods*. James Nesbitt, *Inspection of Petroleum Products: Digest of Statutes in the Several States and Canada* (New York, 1914).

12. For skepticism of the efficacy of the early laws, see "Report of the Committee of the Franklin Institute." Peckham, *Report*, is also dismissive of the regulations as of 1879. See also Wlasiuk, "Refining Nature," 89–91. The suit is *Wellington v. Downer*, 104, *Mass. Reports*, 64. *Report of the [Michigan] State Inspector of Illuminating Oils for the Six Months Ending December 31, 1879* (Lansing, 1880).

13. The classic work on the relation between branding and quality is Neil Borden, *The Economic Effects of Advertising* (Chicago: Richard D. Irwin, 1942), chap. 22. Chandler, *Visible Hand*, 324–325, briefly discusses Standard's concerns with quality, as do Hidy and Hidy, *History of the Standard Oil Company*, 99–100, 140, 299. On Standard's marketing of gasoline and interest in quality, see also Williamson, *American Petroleum Industry*, vol. 1, chap. 20, which also contains data on the price spread. A search of American periodicals and major newspapers from 1870 to 1940 reveals *no* articles claiming that Standard Oil adulterated its kerosene.

14. "The Oil Stove of '77: The Domestic," *Boston Journal of Chemistry* 12 (July 1, 1877): 1; "The Florence Oil Stove," *Forest and Stream*, August 16, 1877, 39; "Boston Gem Oil Stove," *Zion's Herald* 55 (June 6, 1878): 184.

15. The classic work on risk perception is Paul Slovic, "Perception of Risk," *Science* 236 (April 17, 1987): 280–285. See also Cass R. Sunstein and Richard Zeckhauser, "Overreaction to Fearsome Risks," *Environmental and Resource Economics* 48 (March 2011): 435–449, https://scholar.harvard.edu /files/rzeckhauser/files/overreaction_to_fearsome _risks.pdf.

16. Peckham, *Report*, 251.

17. The kerosene price in 1885 is from Williamson, *American Petroleum Industry*, 1:524; the price of bituminous coal is from the *Coal Trade Journal* 24 (Jan. 7, 1885): 694. Since a gallon of kerosene has 0.5 percent of the energy in a ton of coal (135,000/26,200,000), it should not sell at more than $0.02: $0.005 \times \$4.20 = \$0.021$.

18. Ohio soft coal used in the study contained 24,734,694 Btu of energy for cooking; thus, 55.5 gallons of oil (1.32 barrels of 42 gallons) equated to a ton. For kerosene and coal to be equally costly, with coal costing $6.50 a ton, $P_k \times 55.5 = \$6.50$, where P_k is the price of kerosene in gallons and is equal to $0.117. In the winter, all of coal's heat is valuable, and if it burns with 65 percent efficiency, while kerosene stoves are unvented and so 100 percent efficient, then a ton of Ohio coal yields $0.65 \times 24,734,694$ Btu = 16,077,551 Btu = 119.1 gallons of kerosene, so kerosene would have to sell at $0.055/gallon to be economic.

19. For the efficiency of coal furnaces as a heat source, see P. E. Fansler, *House Heating*, 2d ed., 6.

20. The calculations in figure 6.1 use the average Btu value for soft coal of 26,200,000 employed elsewhere in this work, a ton of which is equivalent to 58.8 bbl. kerosene for cooking. (That is, 26,200,000/24,734,694 = $1.06 \times 55.5 = 58.8$.) The Bureau of Labor Statistics average retail prices for soft coal begin in 1913; before that, only prices for individual cities are available. I have assumed that national prices would have changed the same way as those for Omaha, Nebraska, which are available back to 1907, and I have constructed national estimates for 1907–1912 accordingly.

21. USDA, Bureau of Agricultural Economics and Home Economics, *The Average Quantities and Values of Fuel and Other Household Supplies Used by Farm Families* (Washington, 1926).

22. Emma Gray, "The Kerosene Oil Stove," GH, Feb. 15, 1890, 181; Helen Russell, "Oil Stoves," GH, August 1891, 60–61; M. T. Rorer, "Cooking over All Sorts of Fuel," *Ladies' Home Journal*, July 1899, 28; Georgie Child, *The Efficient Kitchen* (New York: McBride, Nast, 1914), 94.

23. On the manufactured gas industry, see Castaneda, *Invisible Fuel*; and Tarr, "Transforming an Energy System."

24. "Eureka Oil Stove," *Ohio Farmer* 48 (July 17, 1875): 45; "Florence Oil Stove," *Cultivator and Country Gentleman* 42 (June 14, 1877): 387. "On the whole" is from Marion Harland, *Bits of Common Sense Series Cooking Hints* (New York: Home Topics, 1899), 118–119.

25. Aunt Betsy, "Oil Stoves," NSF 12 (May 10, 1888): 16; "The Florence Heater," *The Independent*, Sept. 27, 1877, 32; "Our Household," NSF 24 (Nov. 1, 1900): 750. In summer homes of the well-to-do, kerosene faced competition from gas manufactured from gasoline by portable gas machines. See Donald Linebaugh, *The Springfield Gas Machine: Illuminating Industry and Leisure, 1860s–1920s* (Knoxville: University of Tennessee Press, 2011), chap. 4. Howell Harris notes that wood or coal stoves also might be moved in the summer, but that surely must have been a substantial task. See his "Conquering Winter: U.S. Consumers and the Cast Iron Stove," *Building Research and Information* 36 (June 2008): 337–350.

26. "Burns 90% Air," *Lippincott's Monthly Magazine*, Dec. 1904, 117; "The Florence Heater," *The Independent*, Sept. 27, 1877, 32; "Heat Where You Want It," *Wallaces' Farmer*, Oct. 11, 1907, 1137; "What? Another Coal Bill?," *Cosmopolitan*, Oct. 1892, 29.

27. Day, "Capital-Labor Substitution," ignores oil-fired equipment.

28. The cooking test described in note 18 above found that about 0.82 gallons of kerosene, weighing about five pounds, would replace more than 29 pounds of coal. Kerosene delivery is briefly discussed in Hidy and Hidy, *History of the Standard Oil Company*. Mrs. Kellerman is from "Oil Stoves," *Ohio Farmer* 61 (May 13, 1882): 330.

29. "The Household," *New York Evangelist* 55 (July 17, 1884): 7; "The Kerosene Oil Stove," GH, Feb. 15, 1890, 181; Russell, "Oil Stoves," 60; US Department of Agriculture, "Domestic Needs," 10.

30. Russell, "Oil Stoves"; for Mrs. Kellerman, see "Oil Stoves." Nell Nichols, *The Farm Cook and Rule Book* (New York: Macmillan, 1923), 4, 13, 35, 206.

31. "Kerosene Stoves," *Cultivator & Country Gentleman* 40 (May 20, 1875): 318. The author can testify from personal experience that canning with a gas stove can also make the kitchen a hotbox; accordingly, we moved the operation to the porch and employed a portable Coleman stove.

32. "Kerosene Stoves," *Cultivator and Country Gentleman* 40 (May 20, 1875): 1164; "Oil Stoves," *Michigan Farmer* 14 (June 5, 1883): 7; "Real Labor

Savers," *Wallaces' Farmer*, Oct. 28, 1904, 1342; "Home Makers Exchange," *Orange Judd Farmer* 68 (June 26, 1920): 18.

33. Fanny Field, "Oil Stoves and Soap," *Ohio Farmer* 63 (June 30, 1883): 474; Sarah Wilcox, "Seasonable and Sensible Suggestions," NSF 20 (July 30, 1896): 377; Katherine Johnson, "Kitchen Conveniences," *Cultivator and Country Gentleman* 58 (April 27, 1893): 337.

34. Becky Sharp, "Save Yourselves," NSF 23 (July 6, 1899): 350. "If men" is from "The Saddest Are These," NSF 25 (April 11, 1901): 5; "a tired wife" is from "The Art of Keeping Cool," NSF 24 (July 19, 1900): 391. "Household Furnishing and Decoration: Summer Stoves," GH, August 3, 1899, 159. Oral Histories on Work, Box 4, Ethel Jones Hayward interview, Nov. 27, 1984; Mildred Veitch, "North Dakota's Plan," *Farmer's Wife* 17 (Oct. 1914): 139+.

35. James Evans and Rodolfo Salcedo, *Communications in Agriculture: The American Farm Press* (Ames: Iowa State University Press, 1974), has little to say about such possibilities. Evelyn Leasher, "Lois Bryan Adams and the Household Department of the 'Michigan Farmer,'" *Michigan Historical Review* 21 (Spring 1995): 100–119, yields no evidence that the journal compromised its independence.

36. "Summer Cooking Shouldn't Cook the Cook," *Ladies' Home Journal*, June 1914, 50; "use the 'Puritan'" is from "An Odd Verdict," NSF 24 (June 28, 1900): 328. "The Florence Oil Stove for Heating or Cooking," *Christian Union* 15 (March 28, 1877): 292.

37. Makers of "New Perfection" stressed that theirs was a year-round stove; see "Summer, Winter, Town, Country, Big House, Little House," *Farm Journal* 46 (July 1922): 31. "Oil Cook Stoves Are Graduating from Stage of Short Season Business," *American Artisan and Hardware Record* 81 (June 18, 1921): 23–24; "The Oil Stove with a Cabinet Top," *Ladies' Home Journal*, July 1909, 38; "A Turn of the Valve," *Maine Farmer* 68 (June 21, 1900): 5; "Florence Oil Cook Stoves," GH, May 1917, 163; and GH, June 1917, 84.

38. If the kerosene were used for cooking, 58.8 gallons or 1.4 (42-gallon) barrels would have displaced one ton of soft coal, and half of 33 million barrels would have displaced about 12 million tons of coal. Of course, kerosene could have replaced even more wood used in cooking.

39. For evaluations of oil stoves, see Snyder, "Study of Kerosene Cook Stoves"; and Snyder, "Factors Affecting." See also Monroe, "Performance Analysis"; and Demetria Taylor and Truman

Henderson, "When Your Fuel Is Oil," GH, August 1932, 90+.

40. Mildred Maddocks, "Summer Housework Made Easier," GH, August 1918, 74+. Maddocks argued that summer canning with a kerosene stove required a separate hot-water heater. In fact, many advertisements depicted heating water on the stove. "Yes You Can Have Gas," CT, Feb. 12, 1928.

41. "Admiration," *American Artisan* 39 (Feb. 3, 1900): 6; "Now Is the Time," NSF 24 (June 21, 1900): 1. In modern times, at least, white goods are often joint purchases. See Christine Bose et al., "Household Technology and the Social Construction of Housework," T&C 25 (Jan. 1984): 53–82. For fascinating similarities in modern selling to women, see Joy Parr, "Shopping for a Good Stove: A Parable about Gender, Design and the Market," in *A Diversity of Women: Ontario, 1945–1980*, ed. Joy Parr (Toronto: University of Toronto Press, 1995), 75–97. USDA "Social and Labor Needs," 21; "Florence Oil Stoves," *Ladies' Home Journal*, June 1916, 65.

42. "A Six O'Clock Start," *The Independent*, May 30, 1901, ix (italics in original); "Solve the Servant Girl Question," *Maine Farmer* (June 14, 1900): 8.

43. Funk, "What the Farm Contributes." "New Perfection Oil Cook Stoves," GH, July 1918, 136; Rapp, "Fuels Used," table 7.

44. Frederick, *New Housekeeping*, chap. 16; Frederick, *Scientific Management*, 120; Christine Frederick, "Canning Headquarters," *Hardware Dealers' Magazine*, July 1, 1924, 35–40; Christine Frederick, "Selling Cool Kitchens," *Hardware Dealers' Magazine*, May 1, 1926, 62–64. For the efficiency movement, see Samuel Haber, *Efficiency and Uplift* (Chicago: University of Chicago Press, 1964); and Daniel Nelson, *Frederick W. Taylor and the Rise of Scientific Management* (Madison: University of Wisconsin Press, 1980). On Frederick's career, see Rutherford, *Selling Mrs. Consumer*. Hawley, "Every Foot of My Garden."

45. Cowan, *More Work*, argues that new household technology led to "more" work; however, it seems more plausible that the desire to improve living standards was what motivated purchase of the new technology, rather than the other way around. On farm women's hours worked, see Vanek, "Keeping Busy," chap. 3. Among the many contemporary studies, see Ward, "The Farm Woman's Problems"; Oregon State Agricultural College Extension Station, "Use of Time"; and Arnquist and Roberts, "Present Use of Work Time."

46. Ward, "The Farm Woman's Problems"; Rankin, "Use of Time in Farm Homes." Home-produced goods is from Funk, "What the Farm Contributes." Jane Adams and Katherine Jellison emphasize the economic importance of farm women's work in Jane Adams, *Transformation of Rural Life: Southern Illinois, 1890–1910* (Chapel Hill: University of North Carolina Press, 1994); and Katherine Jellison, *Entitled to Power: Farm Women and Technology, 1913–1963* (Chapel Hill: University of North Carolina Press, 1993), respectively. In the 1930s and 1940s, the author's mother kept a roughly 50' × 100' garden; she canned and froze vegetables and jams and jellies, separated cream, fed the cow, gathered eggs, and helped slaughter the chickens. "New! Amazing New Kozy Heated Hog House," *Wallaces' Farmer*, Jan. 18, 1930, 98.

47. The many articles in *Farmer's Wife* on beekeeping suggest that it was often women's work. See, e.g., Mrs. B. J. Livingston, "Women in Rural Communities: Bees," *Farmer's Wife* 11 (Sept. 1906): 86. Consumer prices are from Carter et al., *Historical Statistics*, series Cc1-2; individual food items are from US Bureau of Labor Statistics, "Retail Prices of Food, 1890–1913," *Bulletin* 138 (Washington, 1914), 14–15. Edith Hawley, "My Income."

48. Katherine Henry, "Daughter Chooses the Farm," *Farmer's Wife* 20 (June 1917): 8; and "Mother Confesses," *Farmers Wife* 21 (March 1919): 232. For the broader context of these concerns, see Cynthia Sturgis, "'How're You Gonna Keep 'Em Down on the Farm?' Rural Women and the Urban Model in Utah," *Agricultural History* 60 (Spring 1986): 182–199. On this topic, see also John Fry, *The Farm Press, Reform, and Rural Change, 1895–1920* (New York: Routledge, 2005), chap. 7.

49. "Who Has to Get Up?," *Farmer's Wife* 26 (Feb. 1, 1924): 356; "Florence Oil Cook Stoves," *Ladies' Home Journal*, June 1917, 84; "The Oil Range for City Use," *Ladies' Home Journal*, Oct. 1915, 215; "Live Where You Want To," *Ladies' Home Journal*, June 1925, 161; "Makes Mother a Companion," *Ladies' Home Journal*, August 1925, 121. Joanne Vanek shows that urban college-educated women in the late 1920s spent much less time in food preparation and more time in family care. See table 3 in her "Household Technology and Social Status: Rising Living Standards and Status and Residence Differences in Housework," T&C 19 (July 1978): 361–375.

50. "Hold Your Man," *Ladies' Home Journal*, May 1934, 139; "See Yourself *in a* Modern Kitchen," *Better Homes and Gardens*, May 1936, 102 (italics in original). Howell Harris briefly discusses this shift to modern design in solid fuel stoves in his "The Stove Trade Needs Change Continually: Designing

the First Mass Market Consumer Durable, ca. 1810–1930," *Winterthur Portfolio* 43 (Winter 2009): 365–406.

51. "The Fuel of Today and Future," *Omaha Bee*, Sept. 23, 1888. The brick and sand are, respectively, from "Novel Way to Burn Oil," and "Burning Oil in a Cook Stove," NYHT, Oct. 8 and 11, 1902. "Substitutes for Coal in Heating and Cooking," SA, Oct. 25, 1902, 276; "Beats Coal for Heat," *Popular Mechanics*, April 1923, 188. See also "Pushes Kerosene Cook Stove Burners to Boost Oil Sales," *National Petroleum News* (August 24, 1921): 35–36.

52. For the data on stoves, see US Department of Commerce, *Biennial Census of Manufactures* (Washington, various years). Mary Neth, *Preserving the Family Farm: Women, Community, and the Foundations of Agribusiness in the Midwest, 1900–1940* (Baltimore: Johns Hopkins University Press, 1995), chap. 7, stresses the costs of electrification. Ronald Tobey, *Technology as Freedom: The New Deal and the Electrical Modernization of the American Home* (Berkeley: University of California Press, 1996), chap. 2, emphasizes inadequate wiring.

53. The 1940 Census of Housing found kerosene stoves in only about 3.3 million homes, or 9.7 percent of the total, because the census only counted the *primary* source of energy.

54. Gray, "Nebraska Rural Kitchen"; Clark and Gray, "Routine and Seasonal Work"; Rapp, "Fuels Used," table 1.

55. MY (1936), 717. Eight-cent range oil is from "Good News to Owners of Range Oil Burners," BG, Jan. 1, 1932. Monroe, "Kerosene Burners," 284. Massachusetts Special Commission, *Report* (various years), has hard-coal prices in Boston. "Range Oil Price Set at 8 and 9 Cents," BG, Nov. 11, 1938.

56. Massachusetts Department of Labor and Industries, *Annual Report: 1938* (Boston, 1938), 104. Gas consumption data are from AGA, *Historical Statistics*. The rapid growth in range oil sales also reflected the rise in sales of oil heaters.

57. It is not clear whether these figures count only the primary stove or all stoves, but I believe it is the latter, as the state totals for Indiana indicate the very high rates of possession depicted in table 6.3.

58. Urban households are from "Family Spending and Saving in Wartime," USBLS *Bulletin* 822 (Washington, 1945), table 23; rural farm and nonfarm data are in "Rural Family Spending and Saving in Wartime," USDA *Miscellaneous Publication* 520 (Washington, 1943), table 19.

59. In cooking, 1.4 bbl. would replace a ton of BCE. Half the kerosene is 22.5 million bbl., so 22.5/1.4 = 16 million tons.

60. Natural gas came slowly to rural New England. It only arrived in the author's small farm town in 1962, and the author's house, which was wired only for 50-amp service, still had a kerosene stove in 1974, when he bought it.

61. Carter et al., *Historical Statistics*, series Dh279. The author recalls use of kerosene space heaters in the 1960s. Kerosene use in residential space heating rose in the 1980s as better heaters arrived from Japan and rising fuel prices spiked the costs of central heating. Although it continues to be employed for such purposes and in cabins and mobile homes, sales have plummeted in recent years.

Chapter 7. The Battle of the Basements

1. I know of no credible estimates of the extent of domestic central heating before the 1930s. Lebergott, *American Economy*, tried to estimate its use in 1920, relying on USBLS, "Cost of Living," table D, but he misinterprets the table. It shows the "average number of rooms equipped for heating," but this tells nothing about central heating, for even if all rooms were heated, they might well all have had stoves. Moreover, Lebergott's estimate that only 1 percent of homes had central heat in 1920 implies improbably high growth rates from 1920 to 1940.

2. Sixteenth Census of the United States, 1940, *Housing* (Washington, 1943), vol. 2, part 1, table 12a. The source of the survey is Kryk, "Family Income," appendix table 30. I estimated

$$LN(Pct\ Cent\ Heat) = \frac{1.08}{(19.5)} Ln(Income)\ R^2 = 0.98;$$

N = 98. Figures in parentheses are t-ratios. The equation also includes dummy variables to control for city size and location. Income data are midpoints of 13 classes from $250 to $500 and from $5,000 to $10,000; the dependent variable is the percent of families in a given income class that had central heat. The survey was for 1935–1936; families sampled were white, native-born married couples. With incomes 76 percent higher in 1940, and with 56 percent of households then having central heat, the calculation for urban families with central heat in 1900 is 56/(1.76 × 1.08) = 29.6.

3. "Gas Heating," AGLJ (Nov. 9, 1914): 292–293; "in the next ten years" is from Irwin Moyer, "The Natural Gas Fields of Eastern United States and Their Probably Future Life," Natural Gas Association *Proceedings* 12 (1920): 44–59, 59. Odell, "Facts Relating to Substitution of Manufactured Gas for Natural Gas."

4. The space heaters are from MR, 1919, part 2, 478–480. "Combination Ranges Sell More Gas," *Gas Age* 50 (July 3, 1922): 49–50.

5. Laclede is from "Block Rates Promote Heating," *Gas Age* 45 (Feb. 10, 1920): 103–107.

6. Thomson King, *Consolidated of Baltimore* (Baltimore: Consolidated Gas, Electric Light and Power Company, 1950), chap. 14; "Manufactured Gas for Home Heating," *Gas Age* 39 (Jan. 15, 1917): 57–60; "Block Rates. Promote Heating," MR, 1928, part 2, 575–576.

7. "Block Rates Promote Heating"; "An Experiment in Residence Gas Heating at Denver Colorado," AGA *Proceedings* 5 (1923): 1083–1094; "Gas Conversions Gain Favor with Utilities," H&V 26 (Sept. 1929): 90–91. The rise of conversions is from "Gas House Heating Shows Remarkable Increase," H&V 26 (Oct. 1929): 77–78. The relative cost of gas versus anthracite is from MR, 1928, part 2, 576.

8. The quotation is from S. R. Lewis, "The Importance of a Proper Perspective," ASHVE *Transactions* 20 (1914): 249–252, 250. An early review of some of the research is F. C. Houghten, "A Study of Heat Transmission with Special Reference to Building Materials," ASHVE *Transactions* 28 (1922): 81–102. An example of its diffusion is Charles Hubbard, "Insulation in Heating Work," *Domestic Engineering* 93 (Oct. 2, 1920): 5–7. "Free" heat is from "Insulation and Gas Heating," *American Builder* 41 (April 1, 1926): 456+. For more modest claims of saving, see H. D. Bates, "Demand House Insulation," *American Builder* 36 (Nov. 1, 1923): 106–107; see also "Insulation and Fuel Cost," *American Builder* 48 (March 1930): 124–127.

9. "Better Insulated Homes on Smaller Coal Consumption," *American Builder* 26 (Jan. 1919): 104–105. Shaving a quarter off the coal bill is from "House Insulation and Fuel Economy," *National Builder* 66 (March 1923): 45. Company work is described in "In Gas Heated Homes Insulation Proves Its Economy," *American Builder* 41 (August 1, 1926): 159–162. See also "Economics of House Heating with Gas," H&V 26 (April 1929): 92–96; "Relation of House Insulation and Gas Heating," *Gas Age* 59 (May 7, 1927): 689–691; and "Insulation Booms House Heating in Denver," *Gas Age* 67 (Jan. 17, 1931): 85.

10. "Number of Gas Companies Promoting House Heating Shows Significant Growth," H&V 26 (Feb. 1929): 95–98; the 76,000 figure is from "Gas House Heating"; housing values are from Fifteenth Census of the United States, 1930, *Population*, vol. 6, table 23, and are for nonfarm homes. "Gas Heat in New England," H&V 26 (August 1929): 67–70. The limits of the market are from "The Ever-Changing Picture of Automatic Heating for Residences," H&V 30 (July 1933): 17–18. Statistics for 1940 are from

U.S. Department of Commerce, *Survey*, 1942 supplement, 112. The 2.1 million tons calculation is based on evidence that in 1920, Baltimore households required 32,000 ft.^{3} of manufactured gas to replace one ton of hard coal. See "Secondary Rates Make House Heating with Gas a Success in Baltimore," *Gas Age* 46 (Feb. 16, 1920): 108–112.

11. "Grimly it demands" is from "The Furnace," *The Advance* 68 (Feb. 1912): 148. Ella Lyle, "The Ash-less Isle," GH, Feb. 15, 1890, 174. Data in Kryk, "Family Income," appendix table 30, yield the following estimate of the impact on rising incomes on use of solid fuels for heating:

$$LN(Pct\ Solid\ Fuel) = \frac{-.81}{(13.8)} Ln(Income)\ R^2 = 0.97;$$

N = 98. Figures in parentheses are t-ratios; other variables and income are as described in note 2 above.

12. On the early use of fuel oil, see Pratt, "The Ascent of Oil." The Copley Plaza is from "Where We Saved by Changing to Oil," FOJ 2 (Dec. 1923): 20–24. "Fuel Oil Burning in New York City," FOJ 1 (March 1923): 9–12. The quotation is from the *Brooklyn Daily Eagle*, Dec. 10, 1923.

13. "Fuel Oil Waits for the Right Burner to Make It a Home Fuel," *National Petroleum News* 14 (July 26, 1922): 43–47, 43.

14. "Replacement Market," FOJ 18 (July 1939): 16; Massachusetts Special Commission, *Report* 1940, appendix 2, table 11. The text calculation equates 4.45 bbl. of oil to one ton of hard coal.

15. Senner, "The Domestic Oil Burner"; "The Development of Domestic Oil Burners from Industrial Types," H&V 23 (April 1926): 68–71.

16. Coal and oil efficiency might depend on proper firing (coal) and installation and maintenance (oil). Pratt, "The Ascent of Oil," 18, argues that economics largely determined the rise of oil, by which he seems to mean relative prices and costs, but his focus is on industrial, not domestic, uses of fuel oil.

17. Fansler, *House Heating*, 2d ed., 7. The figure of 520 is from "The Oil Burner Boom—What It Means," *Forbes*, Sept. 1, 1926, 27–30. The survey is in Fansler, *House Heating*, 3d ed., 10.

18. "History of the Williams Oil-O-Matic," scrapbook, Williams Oil-O-Matic Collection, McLean County Historical Society, McLean, IL. On Petro, see https://www.waltergrutchfield.net/petroleum.htm and sources cited therein. "Timkin Axel Co. Gets [Socony] Arrow Burner," *Brooklyn Eagle*, August 20, 1925.

19. "American Association of Oil Burner Manufacturers Formed," FOJ 2 (Oct. 1923): 17–19.

"Oil Burner Manufacturers Will Cooperate with Oil Men," "Burner Manufacturers and Petroleum Institute Have Joint Meeting," and "What the Burner and Oil Industries Can Do for Each Other," all in FOJ 2 (Jan. 1924): 17–19, 23–24, 25–26; "Does Loose Sales Talk Pay"? H&V 23 (April 1926): 57–61. "Eliminate the Fly-by-Night Burner Firms," FOJ 2 (Feb. 1924): 19–21. The code is from Fansler, *House Heating*, 3d ed., 6. Formation of the institute and association are from "Oil Burner Convention," *American Builder* 44 (Feb. 1, 1928): 186. On the work of the Oil Burner Association, see Bishop, *Retail Marketing*.

20. U.S. Federal Oil Conservation Board, *Hearings*, May 27, 1926, 27–31, 27, 28.

21. Fansler, *House Heating*, 3d ed., 29, which also contains a discussion of fuel-oil technical characteristics. "Underwriters Want Uniform Fuel Oil," FOJ 2 (Feb. 1924): 17–18; see also "Uniform Grades of Oil Assured by Commercial Standards," FOJ 8 (Sept. 1929): 33–35. U.S. National Bureau of Standards, *Standards Yearbook, 1930* (Washington, 1930), 188.

22. Fansler, *House Heating*, 3d ed., 12–13; Senner, "The Domestic Oil Burner," 19.

23. The best survey of dealers and retailing is Bishop, *Retail Marketing*. For a sampling of the many articles on dealers and service, see "Training Oil Burner Salesmen," FOJ 5 (Jan. 1927): 19–21; and "How We Licked the Service Problem," FOJ 5 (May 1927): 13–14. "Rise of a New Industry," *Printers' Ink* 169 (Oct. 4, 1934): 49–54.

24. "How to Avoid Fuel Uncertainties," *Springfield (MA) Sunday Republican*, Nov. 15, 1925.

25. National spending is from Fansler, *House Heating*, 3d ed., 10. Williams data are from R. V. Hopkins to W. W. Williams, Sept. 7, 1939, folder 18, Williams Oil-O-Matic Collection. For comparisons with other burner manufacturers, see Crowell Company, *National Markets and National Advertising* (New York, 1927). Examples of the publications include the *Boston Globe, Brooklyn Eagle, Building Age, House Beautiful*, and *Vogue*. The Oil Heating Institute ad is in FOJ 7 (Sept. 1928): 38–39. Market research is from "The Oil Burner Market Map," FOJ 7 (Jan. 1929): 40–41. See, too, "Oil-Burner Distribution Data Show Startling Sales Possibilities," H&V 26 (August 1929): 34–35; "Domestic Engineering Oil Burner Survey," *Domestic Engineering* 135 (Oct. 5, 1929): 66–67; and "55 Key Oil Burner Markets Analyzed," FOJ 11 (Oct. 1932): 19+. For Williams's advertising leverage, see "A Decade of Burner Advertising," FOJ 11 (July 1932): 20–21, 90–94.

26. "Automatic," *American Builder* 48 (Oct. 1, 1929): 141.

27. "Progress of Chicago Oil Burner Campaign," H&V 23 (Nov. 1926): 98. The price premium is from "Where the Oil Burner Stands," H&V 23 (Sept. 1926): 83–85. The $20,000 figure is from "Progress in the Oil Burner Industry," *Domestic Engineering* 128 (Sept. 7, 1929): 82, 139+. For housing values, see Fifteenth Census of the United States, 1930, *Population*, vol. 6 (Washington, 1933), table 23.

28. Men were almost certainly the source of these advertisements. See Roland Marchand, *Advertising the American Dream* (Berkeley: University of California Press, 1985). I infer the sexual division of labor in furnace tending from advertising copy. Neither households nor advertisers seemed aware that the shift from coal to oil heat would reduce indoor air pollution. Companies did stress the health benefits of oil, but these, they claimed, resulted because with automatic heat, the house would be less drafty, and children would get fewer colds. See Jan Sundell, "On the History of Indoor Air Quality and Health," *Indoor Air* 14, supplement 7 (2004): 51–58.

29. "The Urge behind Installment Selling," FOJ 7 (August 1928): 19–20, 124; "Without Buying the Burner You Can Now Have Nokol's Clean Automatic Heat Service," BG, April 24, 1928; "quiet as a drifting cloud" is from NYHT, August 22, 1929; "Quiet May," NYHT, May 8, 1928.

30. "Freedom of the Shes," BG, Feb. 5, 1935; "No Coal Shovel," *Better Homes and Gardens*, April 1926. The verse is from "A Ballad of Oil Burning," FOJ 4 (Oct. 1925): 78. "Don't Leave Them," *Literary Digest* (Oct. 9, 1926); May advertisements, NYHT, April 9 and 18, 1929; "Men May Buy," *Better Homes and Gardens*, June 1926; "The Shadow," *Literary Digest* 91 (Nov. 9, 1926).

31. "Equipped with modern" is from NYHT, Dec. 4, 1929. For analysis of advertising content, see Marchand, *Advertising the American Dream*; Rose, *Cities of Light*; and Strasser, *Never Done*. "Does she remember" is from NYHT, Sept. 25, 1929.

32. "Still Living in the Coal Age," NYHT, August 14, 1928; "Every Modern Convenience," *House Beautiful*, May 1926. Williams listed six architectural publications and 18 major national magazines in which it advertised. Fifth International Convention of the Williams Oil-O-Matic Heating Corporation, June 3d–4th, 1929," folder 2, Williams Oil-O-Matic Collection. "A Decade of Burner Advertising," FOJ 11 (July 1932): 20–21, 90+.

33. See, e.g., "What an Oil Burner Offers You," *Popular Science*, July 1926, 15+; "Home Owners Report Remarkable Oil Burner Results," *Popular Science*, Jan. 1927, 30+; P. E. Fansler, "Selecting an

Oil Burner for Your Home," GH, Sept. 1925, 86–87; P. E. Fansler, "The Truth about Oil Burners," *Garden and Home Builder*, Sept. 1926, 88+. "Pointers" is from C. S. Kauffman, "Oil Burners Are Practical for Modern Homes," *Better Homes and Gardens*, Oct. 1925, 16–18, 16.

34. "Oil Heating Increases the Size of the Small House," *American Builder* 44 (Jan. 1, 1928): 100; "Oil Burners Help Sales," *American Builder* 48 (Oct. 1, 1929): 91–94; "Oil Fuel for Domestic Purposes," *Building Age* 46 (Feb. 1, 1924): 91–92; "Facts about Fuel Oil—'A Burning Question,'" *American Builder* 33 (Sept. 1, 1922): 114–115.

35. "Dealers Increase 1929 Sales by 41%," FOJ 8 (Jan. 1930): 39–40, 84+. Clifford [illegible] to W. W. Williams, Sept 7, 1939, folder 18; and "Sales Plans and Methods Employed by Williams Oil-O-Matic," both in scrapbook, Williams Oil-O-Matic Collection. On the role of dealers in selling oil and oil burners, see the citations in note 23 above. The bond is from "How We Licked the Service Problem," FOJ 5 (May 1927): 13–14. The New Jersey story is from "Selling Oil Heat in a Metropolitan Area," FOJ 6 (June 1928): 42–44, 72–73.

36. "Insulation for Comfort," *American Builder* 52 (Oct. 1931): 48–50; "Tests Show Oil Heat Economy," *American Builder* 49 (July 1, 1930): 118+. The 55 bbl. calculation assumes 4.45 bbl. of oil is the energy equivalent of a ton of hard coal and that combustion efficiency is 70 percent versus 65 percent for coal.

37. "Solving Hard Coal's Commercial Problem," CA 22 (Oct. 26, 1932): 659; S. D. Warriner, "The Case for Anthracite Coal," *Mining Congress Journal* 10 (Jan. 1924): 32–34. "Order takers" is from "Competitive Pressure Grows in Hard Coal Markets," CA 27 (April 2, 1925): 495–498, 496. "Dead level" is from "Anthracite Operators Realize Necessity for New Sales Program," CA 33 (August 1928): 548–550, 548.

38. "New England Dealers See Menace to Anthracite in Wider Use of Soft Coal," CA 27 (April 2, 1925): 512. Parker is from "Economic Problems of the Coal Industry," CA 30 (Dec. 17, 1926): 816. Complaints about gloomy oil forecasts are in "Editorial," FOJ 1 (June 1923): 25; and "Editorial," FOJ 2 (July 1923): 27. For Dad Joiner, see https://www.tshaonline.org/handbook/entries /joiner-columbus-marion-dad. Federal Oil Conservation Board, *Report*, part 5, 7. "Annual output" is from MR, 1921, part 2, 340. Williamson, "Prophesies of Scarcity," reviews ideas about resource depletion during these years.

39. "How Much Ash Is Found in Commercial Anthracite?," CA 25 (Feb. 21, 1924): 272; "Anthracite

Trade Near Accord on Uniform Standards," CA 27 (March 19, 1925): 436, contains the quotation; "Let Fine Sizes Be Fine in Quality," CA 27 (Jan. 1, 1925): 2. Better quality is from "First Annual Anthracite Conference Staged at Lehigh University," CA 43 (June 1938): 72–74. "National Coal Association Analyzes Merchandizing Problems of the Industry," CA 34 (Nov. 1929): 677–680; "Dustproofing Coal Brings Wide Consumer Acceptance," CA 41 (June 1936): 229–230.

40. "Anthracite Industry Rests Success for Future on Engineering and Merchandising," CA 37 (Jan. 27, 1927): 107–108; "New England Dealers See Menace"; "Dealer Customer Service Plan Is Inaugurated," CA 31 (March 3, 1927): 326–327; "Anthracite as a Domestic Fuel," H&V 26 (Oct. 1929): 94–97. "How Anthracite Attacks Sales Problems," CA 33 (August 1928): 467–469. The survey is from "Anthracite Producers Awake to the Importance of Aggressive Merchandising," CA 34 (Jan. 1929): 17–18. "Anthracite Battling to Regain Lost Markets Takes the Dealer into Partnership," CA 38 (March 1933): 82–84; MR, 1926, part 2, 553; MR, 1927, part 2, 481; MR, 1929, part 2, 829–831.

41. For the Mt. Carmel Conference, see MR, 1927, part 2, 481. For the Anthracite Institute, see the chapters on anthracite in the 1927 and 1929 issues of MR, part 2.

42. "The time to meet the competition" is from "Machines and the New Competition," CA 36 (Jan. 1931): 1. "Role of Research in Winning Markets for Coal," CA 42 (Nov. 1937): 67–68. "Assisting" is from MR, 1928, part 2, 576.

43. "Buckwheat Coal Demands Real Recognition in Anthracite Merchandising Problems," CA 27 (April 30, 1925): 638–641; "How Do Small Stokers Fit into the Merchandising Program of Coal for Residential Heating?," CA 41 (August 1936): 330–331. For later stoker developments, see "Coal Stokers Are Coming Strong," *Business Week*, April 2, 1938, 30–34. "Sales of Mechanical Stokers Recede Sharply," CA 44 (Feb. 1939): 101; "Bin Feeds for Stokers," H&V 33 (October 1936): 47–49.

44. An early Reading ad is "A Name Worth Knowing," *Brooklyn Eagle*, Nov. 22, 1927, 15. "Let a little" is from "The Battle Song of the Cities," CT, March 13, 1928, 19. "Refuting False Reports Concerning Oil Burner Fires," *Domestic Engineering* 121 (Nov. 5, 1927): 21, 91; "How Anthracite Attacks Sales Problems," CA 33 (August 1928): 467–469; "Announcing 'Fyrewell.' It Makes the Best Fire Better," BG, Sept. 25, 1938, 12; "Blue Coal Is Here," BG, May 14, 1929, 10; "Trademarking Coal by Automatic Paint Machines at Blackwood Mines,"

CA 36 (April 1931): 188–189. Benjamin Sovacool briefly discusses the role of established producers in subverting energy transitions in his "How Long Will It Take? Conceptualizing the Temporal Dynamics of Energy Transitions," *Energy Research & Social Science* 13 (2016): 202–215.

45. "Statistics on Fuel Used for Heating Apartment Buildings in New York City," H&V 36 (July 1939): 60–61. New York City Office of the Coordinator for the Retail Solid Fuel Industry, *A Survey of the Fuel Situation in New York City for the Period 1926–1937 with Particular Reference to Anthracite and Bituminous Fuels* (New York, 1937), 2. For similar analysis, see Bemis, *Evolving House*, vol. 2, chap. 2. The USBM also noted the impact of apartment living on fuel demand in MR, 1928, part 2, 755. These developments harmed bituminous coal as well, but only about 20 percent of its market was domestic.

46. On cheaper, smaller burners, see "Past Year Marked by Interest in Small Burners," FOJ 8 (Jan. 1930): 20–21. "Oil Burners in 59 Key Markets," FOJ 15 (Jan. 1937): 39–49, demonstrates the eastward shift. On the integration of oil and burner dealers, see "Oil Companies, Burner Dealers Battle for Fuel Market," *Business Week*, Nov. 2, 1932, 9; and "Oil Burner Battle," *Business Week*, Dec. 29, 1934, 9–10.

47. The Fall River price war is from "A Knock-Down Fight on Price," FOJ 14 (Oct. 1935): 11–12. Credit sales are from "Key Dealers Boost Sales 41.8%," FOJ 13 (Jan. 1935): 16, 58. "Prices in 1940," FOJ 18 (Jan. 1940): 12–14; Fansler, *House Heating*, 2d ed., 14.

48. "Radio Interference Gets KO," FOJ 6 (March 1928): 17–18; "Improved Radio-Proof Transformers," FOJ 14 (April 1936): 67; "Automatic Oil Service," FOJ 7 (August 1928): 17–18, 84+; "Degree-Day Delivery Magic," FOJ 15 (March 1937): 36–37.

49. "Petropolis," *Brooklyn Eagle*, May 6, 1930; "say goodbye" is from BG, March 23, 1930; "banish dirt" is from *Brooklyn Eagle*, April 26, 1934; "Freeing the Furnace Slave," NYHT, March 29, 1934 (emphasis in original); "Mutiny," BG, June 14, 1936.

50. "Timken," CT, Sept. 24, 1939; "I Am Buying," NYHT, April 16, 1939; "Museum," CT, May 21, 1930; "Spinning Wheel," BG, August 8, 1937; "Wives Become Sitdown Strikers," BG, June 6, 1937.

51. "The First Oil Burner," *Brooklyn Eagle*, Sept. 22, 1931; "Burner Beautiful," NYHT, April 15, 1934; "Beautiful Lines," NYHT, May 10, 1934; "Beautiful Cabinet," NYHT, May 29, 1935; "Sea Green," NYHT, Jan. 24, 1937; "Link Up for

Mass Attack on Oil Burner Market," *Business Week*, Sept. 21, 1932, 6.

52. "Space Heaters Set Pace in Chicago," FOJ 14 (Jan. 1936): 28, 84–85. See also "Space Heaters Up 71% as Range Burners Tumble," FOJ 12 (Jan. 1934): 63–64; and "Space Heaters Outstrip Range Burners," FOJ 13 (Jan. 1935): 63–64. Sales figures are from the U.S. Department of Commerce, *Biennial Census of Manufactures* (various years).

53. The AGA *Proceedings* for 1920 and 1921 discuss the introduction of the therm in Britain. For a review of early American experience, see "Thermal Gas Rates," AGA *Proceedings* 15 (1933): 61–78. Donald Henry, "Small Thermal Unit for Gas Measurement," AGA *Proceedings* 18 (1936): 44–50, claims that only 61 companies in 16 states were using therms in 1936.

54. C. A. Nash, "Capturing Today's Home Heating Market," AGA *Proceedings* 15 (1933): 494–500; Edward Franck, "House Heating Progress during 1936," AGA *Proceedings* 18 (1936): 306–310, 307.

55. "Market Possibilities for Domestic Space Heating," AGA *Proceedings* 16 (1934): 579–582 has the 50-percent figure. "Study on Relation of Rates to Domestic Gas Sales," AGA *Proceedings* 18 (1936): 36–44; "3,370 Space Heaters Sold in Worst Month of Depression," AGA *Proceedings* 16 (1934): 582–584.

56. "Minneapolis Gas Challenges Oil," FOJ 16 (May 1938): 44. Grand Rapids is from Franck, "House Heating Progress during 1936." St. Louis is from E. H. Lewis, "House Heating Progress during 1936," AGA *Proceedings* 18 (1936): 310–313.

57. "'Heat a Home for So Much,' Coal Men's Best Answer to Oil," *Business Week*, April 20, 1932, 10.

58. For these and other efforts by coal to strike back, see "Coal Men Hope Automatic Stoker Will Check Oil and Gas Sales," *Business Week*, Feb. 25, 1931, 30–31; "Plan Coal Comeback," *Business Week*, Oct. 3, 1936, 29; "Soft Coal Steps Out," *Business Week*, Sept. 11, 1937, 14–15; "Coal Stokers Are Coming," *Business Week*, April 2, 1938, 30–34; and "Trademarking Coal by Automatic Paint Machines at Blackwood Mines," CA 36 (April 1931): 188–189. "MinePakt" is from "Anthracite Region Alive to New Production Methods," CA 41 (Feb. 1936): 50–53. "Dustproofing Coal Brings Wide Consumer Acceptance in Highly Competitive Domestic Markets," CA 41 (June 1936): 229–230.

59. "Anthracite Interests Organize Corporation to Direct New Promotion Plans," CA 41 (August 1936): 345–346; "Anthracite to Advertise," *Printer's Ink* 176 (July 1936): 32; "Anthracite Wakes Up," *Printer's Ink* 187 (May 25, 1939): 74–80; "Anthracite Takes Sales Story to Consumers," CA 42

(Feb. 1937): 57–58; "Automatic Anthracite," BG, Sept. 22, 1937.

60. For the advertisements, see BG, Dec. 10, 1936; and NYHT, Oct. 30, 1938. Kansas City is from "The Story of 22 Coal Men Who Refused to Take the Count," *Sales Management* 46 (June 13, 1940): 20–22.

61. Statistics of stokers and oil burners are from U.S. Department of Commerce, *Survey*, supplement 1942, 145. Gas data are from AGA, *Historical Statistics*, table 143. The 14 percent is from "The Story of 22 Coal Men." "Tomorrow's Homes," CA 42 (June 1937): 237–238, 238.

62. "New Home Bogey Licked by Stoker Installation," CA 46 (Nov. 1941): 43–44.

63. See "Ten Year Price Record Shows Oil Cheaper," FOJ 6 (Sept. 1927): 19–23. Retail prices by city are from USBLS, "Retail Prices: Food and Coal," *Serial R 1264* (Washington, 1941), tables 8–10.

64. The Peoples campaign is described in "165 House Heating Jobs per Day," AGA *Proceedings* 15 (1933): 500–503; "installed in your home" is on 501. The survey is "Market Analysis for Sales and Rate Making," AGA *Proceedings* 19 (1937): 37–49. See also "Battle of the Fuels," *Printers' Ink* 168 (Sept. 20, 1934): 45–49; and "Battle of the Basements," *Printers' Ink* 173 (Oct. 24, 1935): 81–85. "No More Shoveling," CT, March 7, 1934; "I Couldn't Keep Warm," CT, Sept. 17, 1934; "Take a Long Rest," CT, August 9, 1936; "$50,000 Chicago Oil Campaign Starts," FOJ 14 (Nov. 1935): 22.

65. Kryk et al., "Family Expenditures," 40–42. The survey covered white families in villages and small cities in the Middle Atlantic and North Central states. This and similar detailed studies provide a glimpse of the multiplicity of household heating choices.

66. "John Q Public Answers 19 Questions on Coal," CA 44 (July 1939): 33–34; "Still Open for Intensive Effort, Survey of Household Fuel Users Indicates," CA 45 (April 1940): 78–80.

67. Morrison, "Household Energy Consumption," estimates somewhat less energy use per capita than I do (about 1.9 BCE per person versus my 2.09). She appears to exclude transport, however, so her data for 1940 cannot be compared with mine.

68. Pittsburgh provides the best example. Davidson, "Air Pollution," finds modest improvement in that city before World War II. For the postwar years, see Tarr, "Changing Fuel Use." For a broader assessment, see Alan Barreca et al., "Coal, Smoke, and Death: Bituminous Coal and American Home Heating," NBER Working Paper #19881 (2014).

69. Relying on data in USBLS, "Retail Prices: Food and Coal," I found the average price of all sizes

of low-volatile coal in St. Louis for June 15 and December 15, 1940, to be about $2.69 above that of high-volatile coal. Similar calculations for Chicago, Detroit, and other large cities with smoke problems yield similar differentials.

Chapter 8. Coal Fights Back

1. More mountainous regions also used contour mining, which essentially pealed the top cover off the side of a mountain and mined a "bench" of coal. Auger mining and mountaintop removal did not become common until after World War II.

2. On the advantages of strip mining, see Harry H. Stoek, "Strip Pit Mining of Bituminous Coal–I," CA 12 (Sept. 29, 1917): 522–527; and Cash and von Bernewitz, "Methods, Cost, and Safety." Many other writers also stressed the advantages of stripping. See, e.g., Kneeland, *Practical Coal Production*, chap. 1.

3. Few modern treatments of stripping discuss the pre–World War II years. The best discussion of opposition is Chad Montrie, "Agriculture, Christian Stewardship." That strip mining was a form of conservation to the men and women of a century ago, whereas to most moderns, it is simply evil, points to one of the many ways in which the meaning of conservation has changed. As the economist John Krutilla pointed out, not until after World War II did conservation come to have its modern meaning. See his "Conservation Reconsidered."

4. For Holmes, see Gannett, *Report*, 1:101.

5. Eli Conner, "Anthracite and Bituminous Mining," CA 1 (Oct. 21, 1911): 42–45, 44. C. M. Young, "Strip Pit Mining with Steam Shovels," CA 3 (Jan. 4, 1913): 10–11, reported mining in Kansas, "where the roof is too poor to permit underground work," on 10. Ralph Mayer, "Strip Mining Where Bad Roof Prevails," CA 13 (April 20, 1918): 735–737; Frank Kneeland, "Large Stripping Operation," CA 8 (Sept. 25, 1915): 497–501, 497. "Coal Stripping, Rush Run, Ohio," CA 9 (Jan. 2, 1916): 161–163, reported that "there was no roof for the regular method of mining" on 161. F. H. King, "Steam Shovel Has Rendered Operable Many Properties Not Hitherto Worked," CA 17 (May 6, 1920): 937–938.

6. "Hazelton, Pennsylvania," CA 10 (July 29, 1916): 204; Frank Kneeland, "Wasting Strip Pit Spoil by Means of a Cableway," CA 27 (Jan. 1, 1925): 3–5; Illinois Coal Strippers Association, *Open Cut Coal Mining Industry of Illinois* (Chicago, 1939), 15; "Stripping Completes Recovery of Deep and Already-Mined Beds," CA 34 (August 1929): 602+; R. S. Weimer, "Stripping Brings New Life to Dying Coal Fields," CA 33 (April 1928): 278–281+.

7. Harold Culver, "Preliminary Report on Coal Stripping Possibilities in Illinois," Illinois State Geological Survey, Cooperative Mining Series *Bulletin* 28 (Urbana, 1925), claims 90 percent recovery for Illinois stripping. Illinois Department of Mines and Minerals, *A Compilation of the Reports of the Mining Industry of Illinois from the Earliest Records to the Close of the Year 1930* (Springfield, 1931), claims 95 to 98 percent recovery for stripping. C. A. Allen, "Coal Losses in Illinois," Illinois Geological Survey Cooperative Mining Series *Bulletin* 30 (Urbana, 1925), claimed traditional mining losses amounted to 49.7 percent of coal in place. Louis Turnbull et al., "Use of Scrapers and Other Light Earth-Moving Equipment in Bituminous Coal Strip Mining," USBM RI 4033 (Washington, 1947), 37. For an evenhanded assessment of the history of stripping in Illinois, see Hall, "Strip Mining."

8. "Real conservation" and "the world needs" are from King, "Steam Shovel Has Rendered," 938. For cooperative efforts in Illinois, see Illinois Coal Strippers Association, *Open Cut.* For an independent assessment, see Schavilje, "Reclaiming Illinois Strip Mined Coal Land."

9. Indiana is from Toenges, "Reclamation." Walter Ludwig, "Reforestation by Coal Companies in Southwestern Pennsylvania," *Journal of Forestry* 21 (May 1923): 492–496.

10. Toenges, "Reclamation." For one of several evaluations of reclamation programs, see G. A. Limstrom et al., "Reclaiming Illinois Strip Lands by Forest Planting," University of Illinois Agricultural Experiment Station *Bulletin* 547 (Champaign-Urbana, 1951). Vietor discusses the politics of modern strip-mine regulations in his *Environmental Politics.*

11. This formula is from Thomas Kennedy, "Anthracite Stripping—II," CA 13 (Jan. 5, 1918): 13–25. The economic ratio also depended on the nature of the overburden because top cover might be more or less expensive to remove. See Toenges and Anderson, "Some Aspects of Strip Mining."

12. The 1:1 ratio is from Conner, "Anthracite and Bituminous Mining." Histories of early strip mining are from Harry H. Stoek, "Steam Shovel Mining of Bituminous Coal," AIME *Transactions* 57 (1917): 514–548. Marsh, *Steam Shovel Mining,* chap. 1. Grant Holmes, "Early History of Strip Mining Full of Heartbreak—I," CA 25 (May 29, 1924): 797–800; Hollingsworth, *History of Strip Mining Machines;* Harold Culver, "Preliminary Report." That deeper coal would be better in quality but that depth likely had diminishing returns is from Turner et al., "Mining by Stripping Methods."

13. The Model 250's origins are from Hollingsworth, *History of Development.* Marion Steam Shovel Company, *Catalog 39, Steam Shovels, Dredges and Ballast Unloaders* (Marion, OH, 1911), 9, 10. That this was the first successful shovel is from F. H. King, "Steam Shovel Has Rendered."

14. The survey is in MR, 1914, part 2, 626–627. For these developments, see Hollingsworth, *History of Development,* who dates electric shovels from 1915. Stoek, "Steam Shovel Mining," has a table of early shovels and characteristics as of 1917; however, a given model of shovel tended to become larger over time. C. M. Young, "Fifteen Feet of Cover to One Foot of Coal No Deterrent to Southwest Strippers," CA 34 (August 1929): 480–483.

15. The effects of larger shovels and longer mine life on tipples is from "World's Largest Strip Mine Has Expectancy of 30 Years," CA 34 (June 1929): 335–337.

16. O. E. Kiessling et al., "Economics of Strip Coal Mining."

17. "The dispatch with which" is from MR, 1917, part 2, 942. "New Shovel Operation and 14-Yd. Dragline for Rider Work Mark Advances at Maumee Mines in Indiana," CA 47 (March 1942): 40–42. "35-yard stripping shovels" is from "Round Table Stripping Session," *Mining Congress Journal* 26 (May 1940): 41. See also "Tecumseh Electrification Provides Flexibility and Safety," *Mining Congress Journal* 26 (Sept. 1940): 52–57. For a good discussion of the evolution of technology at one mine down to 1936, see "United Electric Coal Companies' Fidelity Mine and Washery," *Mining and Metallurgy* 17 (Sept. 1936): 416–420. Mining anthracite to 400 feet and wheel excavators are from "Trends in Strip Coal Mining," *Mining Congress Journal* 31 (Feb. 1945): 36–39.

18. Comparative downtime for steam and electric shovels is from A. M. Nielsen, "Electric Shovel and Improved Shooting Cut Strip Mining Costs," CA 33 (April 1928): 210–212. For the importance of commercial (as opposed to mine-generated) power, see E. R. Lewis, "Electric Power Essential to Modern Strip Coal Mining," EW 106 (Dec. 19, 1936): 42–44+. Operators required for shovels are from Young, "Fifteen Feet of Cover." The share of electric shovels is calculated from data in MR, 1931, part 2, 449. The importance of lighter materials is from "Modernizing the Strip Pits," *Mining Congress Journal* 23 (July 1937): 65–67. "Shovel Leveling Done Automatically by Electronic Unit," CA 47 (Nov. 1942): 56–58.

19. Cost savings from trucking are from "12-Yd Electric Dragline Strips 2 1/2- to 3-Ft. Seam at Old

Glory under 40 to 58 1/2 Ft. of Cover," CA 43 (Jan. 1938): 67–69. "Transportation at Indiana Strip Mines," CA 43 (Dec. 1938): 83–86. For the shift to diesel and its cost savings, see "Truck Haulage Moves with Times in Serving Strip Pits," CA 49 (July 1944): 81–83. "80 Ton Semi-Trailer Passes Tests with Butane Fuel at Sinclair Strip Mines," CA 44 (August 1939): 39–41.

20. Machinery and horsepower are from Sixteenth Census of the United States, 1940: *Mineral Industries*, 1939, vol. 1 (Washington, 1944), tables 1 and 24 and page 229. A modern study finds horsepower is the most important determinant of productivity change in soft coal; see Maddala, "Productivity and Technological Change." For high coverage ratios, see "Operating Practices at Indiana Strip Mines," CA 43 (Dec. 1938): 77–81. Frank Kneeland, "Triple Shifting Makes Big Stripper Profitable," CA 28 (Oct. 15, 1925): 528–531. Alphonse Brodsky, "Engineering Forethought Expedites Operation of Kansas Stripping," CA 35 (Sept. 1930): 524–526. The cost of a swing is from "Boonville Is Proving Ground for New Stripping Project," CA 33 (July 1928): 417–421. H. C. Widmer, "Time Studies Increase Efficiency at Strip Mines," CA 36 (Nov. 1931): 227–230.

21. Turnbull et al., "Use of Scrapers," provides a brief history of the use of light equipment. *Coal Age* noted a bifurcation of stripping in 1937; see "Bituminous Stripping Adopts New Methods in Active Year," CA 42 (Feb. 1937): 75–77. See also "Add to Working Facilities in a Year Marked by Increased Production," CA 45 (Feb. 1940): 68–69; "Bituminous and Lignite from Strip Mining," *Mining Congress Journal* 30 (Feb. 1944): 31; and "The Future of Strip Mining in the Northern Appalachian Field," *Mining Congress Journal* 30 (May 1944): 25–27. The mine that opened in three months is from "Box Cut Yields 2,000 Tons per Day at Pond River," CA 50 (March 1945): 74–77.

22. "New 35-Cu. Yd. Shovel Boosts Tonnage at Georgetown Stripping," CA 49 (June 1944): 72–73.

23. Frank Schraeder, "A Coal Tipple for a Stripping Operation," CA 17 (April 8, 1920): 698–701; MR, 1914, part 2, 627. Productivity data for anthracite stripping are unavailable until 1930.

24. Of course, one cannot say that absent stripping, underground coal sales would have been 37 million tons greater, for companies unable to purchase coal from strip mines might have used gas or oil instead. Scott Turner et al., "Mining by Stripping Methods," 1.

25. Frances Wold, "The Washburn Lignite Coal Company: A History of Mining at Wilton, North Dakota," *North Dakota History* 43 (Fall 1976): 4–20; "Texas Lignite Field Sets Steam Shovel to Work," CA 26 (Sept. 25, 1924): 435–436; H. E. Stevens, "Railroad Opens Large Strip Mine in Rosebud Coal Field," RA 80 (Feb. 6, 1926): 370–372; "Strip Mining Coal in the Southwest," *Mining and Metallurgy* 12 (March 1931): 147–148.

26. For a discussion of mine safety during the interwar decades, see Aldrich, *Safety First* (Baltimore: Johns Hopkins University Press, 1997), chap. 6.

27. "Coal Stripping, Rush Run"; for other early awareness of the comparative safety of stripping, see Stoek, "Steam Shovel Mining"; Kneeland, *Practical Coal Production*; and Cash and von Bernewitz, "Methods, Cost, and Safety."

28. For fatality rates in a number of industries about that time, see USBLS, "Handbook of Labor Statistics, 1936 Edition," *Bulletin* 616 (Washington, 1936).

29. Detailed figures on injuries begin in 1936. The average rate for all permanent disabilities in underground mining was twice as high as in stripping (2.13 versus 1.04 per million employee hours). Relative to coal production, underground mining was nearly seven times more productive of permanent injuries (3.76 versus 0.55 per million tons).

30. "Research, Not Rhetoric, the Hope of Anthracite," CA 32 (Nov. 1927): 288. For a review of combustion research during the interwar decades, see "A Decade of Progress in Fuel Utilization," *Power Plant Engineering* 44 (Jan. 1940): 72–78. For coal research by the Bureau of Mines, see "Bibliography of United States Bureau of Mines Investigations on Coal and Its Products, 1910–1935," USBM TP 576 (Washington, 1937); and "Bibliography of Bureau of Mines Investigations of Coal and Its Products," USBM TP 639 (Washington, 1942).

31. For an in-depth discussion of the Federal Oil Conservation Board, see Clark, *Energy and the Federal Government*, chap. 6.

32. Attitudes about use of oil as fuel are on display in US Federal Oil Conservation Board, *Public Hearings*. See also US Federal Oil Conservation Board, *Reports*, 1 and 2. Gregory Unruh, "Understanding Carbon Lock-In," *Energy Policy* 28 (Oct. 2000): 817–830, notes that government policy may contribute to lock-in.

33. "Dr. Thomas Baker, Educator, Is Dead," NYT, April 8, 1939; "Coal Research at Carnegie Institute of Technology," CA 44 (August 1939): 35–38.

34. "The Coal Conference at Pittsburgh," CA 30 (Nov. 25, 1926): 3; Thomas Baker, "Address of Welcome," [First] International Conference *Proceedings*, 1–4, 2.

35. Marius Campbell, "Our Coal Supply: Its Quantity, Quality and Distribution," [First] International Conference *Proceedings*, 5–64, 62; John Hays Hammond and Frederick G. Tryon, "National Supplies of Power," [First] International Conference *Proceedings*, 192–209, 196; Campbell, "Coal Fields of the United States," 26.

36. "Preparation Held Wide Attention in 1930," CA 36 (Feb. 1931): 66–68; "Fitting Preparation to Market Demands," CA 35 (June 1930): 362–364.

37. For some of these complexities, see "Fitting Preparation"; see also "And Now Mechanical Preparation Sweeps the Decks," CA 34 (Jan. 1929): 11–13, 11.

38. US Federal Oil Conservation Board, *Report* 4, 8.

39. Samuel Parr, "Anthracizing of Bituminous Coal," Illinois State Geological Survey *Bulletin* 4 (Urbana, 1907): 196–197; Samuel Parr, "The Modification of Illinois Coal by Low Temperature Distillation," University of Illinois Engineering Experiment Station *Bulletin* 24 (Urbana, 1908). For Parker, see "Coalite," *Gas World* 46 (June 8, 1907): 715–716. In the United States, low-temperature carbonization languished until World War I revived interest in the process. An apparently promising version of the technology produced "Carbocoal," which had US Fuel Administration financial backing but ultimately failed despite the rise in hard-coal prices. See "Government Interested in Carbocoal," RA 65 (August 30, 1918): 406; and "Artificial Anthracite," H&V 8 (Jan. 1921): 44.

40. Fieldner, "Low Temperature Carbonization." "Six gallons" is from A. C. Fieldner and R. L. Brown, "The Complete Utilization of Coal and Motor Fuel," *Blast Furnace and Steel Plant* 14 (March 1926): 138–140. "Dr. Arno C. Fieldner, Melchett Medalist 1942," *Journal of the Institute of Fuel* 16 (April 1942): 4.

41. Thomas Baker, "Address of Welcome," [Second] International Conference *Proceedings*, 1:1–8, 2.

42. "The Ability That Narrowed the Coal Market Will Expand It," CA 33 (Dec. 1928): 726–728; the quotation is from "Signposts to Stabilization," CA 33 (Dec. 1928), n.p.

43. Frederic G. Tryon and H. O. Rogers, "Analysis of the Consumption of Bituminous Coal in the United States," [Second] International Conference *Proceedings*, 1:139–170, 168–169. Arthur D. Little and R. V. Kleinschmidt, "Coal Consumption as Affected by Increased Efficiency and Other Factors," [Second] International Conference *Proceedings*, 1:110–119, 118.

44. "Institute Hard-Coal Research," CA 33 (Oct. 1928): 652; "Anthracite Experiments with Research," CA 35 (Jan. 1930): 12–13. These various research programs discovered that coal dust could be used as a filtration agent. They also investigated possible uses of ash as fertilizer and its potential as a lightweight aggregate in concrete mix.

45. See the following essays in volume 1 of the [Second] International Conference *Proceedings*: W. A. Darrah, "Economics of Low Temperature Coal Treatment," 242–268, 265; W. H. Allen, "Low Temperature Distillation by the Carbocite Process," 403–412; R. P. Soule, "The 'K.S.G.' Low Temperature Carbonization Plant at New Brunswick, N.J.," 494–507; and J. N. Vandegrift, "Commercial Aspects of Low Temperature Coal Distillation by the International Bituminoil Corporation," 546–569.

46. L. E. Young, "Proposals for Stabilization of the Bituminous Coal Industry," [Third] International Conference *Proceedings*, 1:53–81. A good statement of the conservationists' position is George Rice and A. C. Fieldner, "Conservation of Coal Resources," Third World Power Conference *Transactions*, vol. 6 (Washington, 1938), 671–721.

47. Fieldner, "Low Temperature Carbonization," 46.

48. The reference to Fanny Brice comes from the [Third] International Conference *Proceedings*, vol. 1. R. P. Soule, "Lessons from Low Temperature Carbonization," [Third] International Conference *Proceedings*, 1:272–298, 272. International research on coal was reviewed annually by Arno Fieldner in "Recent Developments in Coal Preparation and Utilization," MY (1933–1937); and Arno Fieldner, "Developments in Coal Research and Technology in 1937 and 1938," USBM TP 613 (Washington, 1940). See also Arno Fieldner and George Rice, "Research and Progress in the Production and Use of Coal," National Resources Planning Board, *Technical Paper* 4 (Washington, 1941).

49. For the bureau's work, see the references in note 30 above. "Anthracite Research Seeks New Markets for Both Fuel and Non-fuel Uses," CA 39 (Feb. 1934): 58–59. The research committee is from "Why the Coal Industry Cannot Afford to Neglect Research," CA 34 (Jan. 1929): 29. "Research in Coal Makes Fresh Progress in 1930," CA 35 (Feb. 1931): 72–75; "Coal Research at Battelle Memorial Institute," CA 44 (June 1939): 35–38. See also the articles on research in the February issues of *Coal Age* from 1931 to 1940.

50. For the Carnegie laboratory, see "More Projects Than Ever Are Listed in 1937," CA 43

(Feb. 1938): 76–79, which has the contract; and "Coal Research at Carnegie Institute of Technology," CA 44 (August 1939): 35–38. For examples of coal research, see "New Research Institutions," CA 45 (Feb. 1940): 87–94, which also contains the discussion of nylon, Lucite, and Koroseal. On the ability to derive formerly coal-based products more cheaply from oil and gas, see Spitz, *Petrochemicals*, xiii.

51. For work on oil from coal, see Fieldner, "Recent Developments"; and Fieldner, "Developments in Coal Research." For modern studies, see Beaver, "Failure to Develop Synthetic Fuels"; and Stranges, "Bureau of Mines' Synthetic Fuel Programme." For a broader perspective, see Vietor, *Energy Policy*. Grossman, *U.S. Energy Policy*, provides a scathing assessment of postwar energy policy.

52. Gustav Egloff, "Motor Fuel Economy of Europe," *Industrial and Engineering Chemistry* 30 (Oct. 1938): 1091–1104. For Japan, see Fieldner, "Developments in Coal Research."

53. Fieldner's estimate is in "Developments in Coal Research," 68. Gasoline weighs six lbs. per gallon, and consumption in 1938 is from Carter et al., *Historical Statistics*, series Df473. The cost of coal-based gasoline is from Egloff, "Motor Fuel Economy"; refinery gasoline prices are for 1939 and are from MY (1940), 997.

54. Stranges, "Bureau of Mines' Synthetic Fuel Programme."

55. Clark, *Energy and the Federal Government* terms the NRC's impact "undetectable" on 297.

56. National Resources Committee, *Energy Resources*; "oil must be regarded" is on 11; a 20-year time horizon for gas is on 143. The studies of petroleum reserves are part 1, section 2, chapter 2, and part 2, section 1, chapter 2. For a sunnier and more accurate assessment of oil availability, see American Petroleum Institute, *American Petroleum Industry* (New York, 1935), chap. 3.

57. National Resources Committee, *Energy Resources*; "free play" is on 30; planning and research are on 4; "can extend the reserves" is on 14.

58. Fieldner, "Recent Developments," 450.

59. For the career of synfuels see the citations in note 51 above.

Appendix 1. Basic Data

1. I took the figures from Sam Schurr et al., *Energy in the American Economy*, appendix 1, table 1, but they derive from Reynolds and Pierson, "Fuel Wood." For 1800, I used their data (Schurr has no figures for that date), employing Schurr's methods. The figures appear to be the results of a model that Reynolds and Pierson never presented.

For a critique of these data, see Warde, "Firewood Consumption."

2. For 1867 onward, the data on horses and mules are from US Department of Commerce, *Historical Statistics to 1970*, series K570 and K572. For 1840–1860, they are from "Horses, Mules and Motor Vehicles," USDA *Statistical Bulletin* 5 (Washington, 1925), table 1. For 1800, I took the 1840 ratio of horses to population and multiplied it by the population of 1800. The number of oxen from 1850 onward are from the decadal censuses. The numbers for 1800 and 1840 I estimated by taking the ratio of oxen to horses in 1850 and applying it to the earlier years. The source for KCalories is C. F. Langworthy, "A Digest of Recent Experiments on Horse Feeding," USDA *Experiment Station Bulletin* 125 (Washington, 1903), table 5.

3. Alan Olmstead and Paul Rhode, *Creating Abundance* (New York: Cambridge University Press, 2008), chap. 12, notes the early importance of oxen and describes breeding programs to improve horses. Brodell and Jennings, *Work Performed*, table 6, demonstrates that horses worked more hours on farms without tractors.

4. Brodell and Jennings, *Work Performed*, found 835 to be average hours worked in 1942, which is quite close to my estimate noted in the text. Frederic Dewhurst et al., *America's Needs*, 1103, employed 500 horse workhours per year; Dewhurst et al.'s work derived from C. D. Kinsman, "An Appraisal of Power Used on Farms in the United States," USDA *Bulletin* 1348 (Washington, 1925), whose data probably reflect the postwar agricultural depression.

5. Nonfarm work animals are from Dewhurst et al., *America's Needs*, 1103 and 1108, but are originally from USDA. Dewhurst assumed nonfarm work animals put in the same hours as men, but that is implausible as a horse cannot supply horsepower-hours for a full day. See Robert Thurston, *The Animal as a Machine and a Prime Mover* (New York: John Wiley, 1894). In the 1880s, horses employed by the North Chicago City Railway worked an average of two hours eighteen minutes a day. See Augustine Wright, *American Street Railways* (Chicago: Rand McNally, 1888), 194–195. For discussions of urban work animals, see Tarr and McShane "The Horse."

6. Lance Davis et al., *In Pursuit of Leviathan* (Chicago: University of Chicago Press, 1997), table 9.8; Nordhaus, *Do Real Output Measures Capture Reality?*, table 1.3.

7. Eavenson, *First Century*, tables 20 and 21. Net imports are from his table 21 to 1939 and from USGS/USBM, MR, and MY thereafter.

8. Schurr et al., *Energy in the American Economy*, appendix I, table I.

9. Schurr et al., *Energy in the American Economy*, appendix I, tables 1 and 6.

10. Waterwheels are from Dewhurst et al., *America's Needs*, table C, 1109; water turbines are from Schurr et al., *Energy in the American Economy*, 491–493.

11. Dewhurst et al., *America's Needs*, table B, 1109. Horsepower and hours are from Kinsman, "An Appraisal of Power," 8. The tonnage of sailing vessels and horsepower-hours are also from Dewhurst, et al., *America's Needs*, table B, 1109 and table L, 1116.

Appendix 2. Chapters 1 and 2

1. Hunter, *Steamboats on Western Rivers*, appendix table 29; Mak and Walton, "Steamboats and the Great Productivity Surge," appendix 1, note c.

2. T. C. Purdy, "Report of Steam Navigation in the United States," in Tenth Census of the United States, *Report on Agencies of Transportation of the United States*, vol. 4 (Washington, 1882), chap. 3, table 2, 44.

3. The equation in the text is fitted using ordinary least squares. The data are from John Church, "Blast Furnace Statistics," AIME *Transactions* 4 (1875–1876): 221–226.

4. AISI, *Annual Statistical Report*, 1918 (Washington, 1919), 10.

5. For industrial production, see Carter et al., *Historical Statistics*, series Cb28.

6. See Daugherty and Horton, "Power Capacity," table 5.

7. My adjustment from brake to indicated horsepower is based on J. A. Ewing, *The Steam Engine and Other Heat Engines* 3d ed. (Cambridge: Cambridge University Press, 1910), 302. The calculation is a bit tricky. Suppose we know that it took 10 lbs. of coal per brake hph and that brake hp was 15 percent less than indicated horsepower. Accordingly, it requires 10 lbs. of coal to produce $1/.85 = 1.18$ indicated hph, so coal per ihph = $10/1.18 = 8.5$ lbs./ihph.

8. Dewhurst et al., *America's Needs*, 1,106 presents coal use data. Daily hours in manufacturing to 1914 are from Carter et al., *Historical Statistics, Millennial ed.*, Series Ba4545 and Ba4553. For 1919, series BA4575 reports a 50-hour workweek and I assumed that this was for 5.5 days. Days in operation for 1890 on are from series Ba4552; before that I assumed a 300-day work year. The need to pad coal use by 12 percent is from Charles Emery, "The Cost of Steam," ASCE *Transactions* 12 (1883): 425–431.

9. Note that the horsepower estimates are for all mines and quarries, while the estimates for average days worked and hours per day are based on coal mining only. See Mark Aldrich, *Safety First* (Baltimore: Johns Hopkins University Press, 1997), table A2.2; and Hotchkiss et al., "Mechanization, Employment, and Output," table B1. Albert Fay, "Coal Mine Fatalities in the United States, 1870–1914," BM *Bulletin* 115 (Washington, 1916), table 2.

10. Carter et al., *Historical Statistics, Millennial ed.*, Series Cc2. For similar calculations of fuel efficiency see Jacob Gould, *Output in the Electric and Gas Utilities*, Table 24; Gould is also the source for the 1919 totals that correct for double counting.

11. Fishlow, "Productivity and Technological Change," table 9.

12. Philadelphia's 37, 580 customers consumed 389 million ft.^{3} of gas in 1859, or an average 10,362.4 ft.^{3} each. See *Annual Report of the Trustees of the Philadelphia Gas Works* (Philadelphia, 1860), 40, 44. I calculate the number of customers nationwide to be 211,502 from "Untitled," AGLJ 1 (August 1859): 1–3. If Philadelphia was typical, the 211,502 customers nationwide consumed 2,191,668,325 ft.^{3} of gas that year. In Philadelphia, total gas use equaled 1.254 times that burned by private customers; accordingly, nationwide total consumption was 2,748,352,079 ft.^{3}, and municipal use, therefore, was 556,683,755 ft.^{3}

13. My figures for coal differ from those in William Shaw, *Value of Commodity Output Since 1869* (National Bureau of Economic Research, 1947), table 2-10. For anthracite, Shaw includes pea sizes of in-domestic use, but that was uncommon in 1910. Shaw's calculation of bituminous consumption in 1910 relies on US Fuel Administration, *Distribution of Coal and Coke*, 12. I rely on MR, 1915, part 2, 471, which is substantially greater, because it includes coal burned for "domestic and small steam trade" and is more consistent with later figures for domestic consumption (see table A4.1) that include all retail deliveries. The gas data derive from Jacob Gould, *Output in the Electric and Gas Utilities*, table A 10, and my calculations; kerosene and gasoline are from United States, *Census of Manufactures, 1909*, vol. 10, "Petroleum Refining," tables 16 and 18. Natural gas is from MR, 1910, part 2, 303. Electricity use is for 1912 and is from Carter et al., *Historical Statistics*, series Db228–229.

Appendix 3. Chapters 3 and 4

1. River coal is entirely from Joseph Cogan, "Dredging Pennsylvania Anthracite," BM IC 7213 (Washington, 1949).

2. Most of the figures derive from US Census of Manufactures: 1954, vol. 1 *Summary Statistics*, chap. 8, table 1a. For 1920, the electricity estimate is from US Department of Commerce, *Historical Statistics to 1970*, series S-124, interpolated between 1917 and 1920.

3. Natural gas for 1910 and 1920 is from MR, 1910, part 2, 305; and MR, 1923, part 2, 350; for 1930, data are from MR, 1930, part 2, 464. For 1954, natural gas is the census estimate for purchased gas used for fuel power and raw materials. Manufactured gas for 1920 is the census report for all gas minus calculated natural gas. For 1940, manufactured gas includes some mixed gas, while mixed gas for 1954 is the census gas total for that year minus natural gas. Schurr et al., *Energy in the American Economy*, 79.

4. Kilowatt-hours and fuel consumption for 1920 onward are from Federal Power Commission, *Consumption of Fuel for Production of Electric Energy, 1954* (Washington, 1954). For 1902–1917, I have relied on Gould, *Output in the Electric and Gas Utilities*, tables 16 and 17. For 1919, see USGS, *Annual Report* 44 (1923), 68; and MR, 1929, part 2, 777.

5. Pig-Iron production is from Carter et al., *Historical Statistics*, series Db74.

6. Bukovsky, "Use and Cost of Railway Fuel," table 3, col. 4; Arthur Redfield, "Railroad Fuel Oil Consumption in 1928," USBM IC 6228 (Washington, 1930); ICC, *Statistics of Railways*, various years.

7. On the difficulties of making these conversions, see Bukovsky, "Use and Cost of Railway Fuel"; and "Coal Equivalents of Locomotive Fuels," RME 114 (Nov. 1940): 501–503; and RME 115 (Oct. 1941): 421–423.

Appendix 4. Chapters 5–7

1. Through 1934, anthracite is from USGS/ USBM, MR, and MY, various years. Coke and briquettes are from the same source, as is domestic use of bituminous coal; for 1915 (472), 1928 (753), and 1944 (831). From 1935 to 1940, anthracite is from A. T. Coumbe and A. F. Avery, "Fuels Consumed for Residential and Commercial Space Heating, 1935–1951," USBM IC 7657 (Washington, 1953). Domestic consumption of manufactured gas is from Gould, *Output in the Electric and Gas*

Utilities, tables 25 and 26. Fuel oil is from Bishop, *Retail Marketing*.

2. Calculations based on the likely number of oil stoves in use, along with typical figures for fuel use, suggest the range oil figures are a substantial undercount.

3. Coal prices are from USBLS, "Retail Prices, 1890–1928," *Bulletin* 465 (Washington, 1929), table 14; and USBLS, "Retail Prices of Food and Coal," *Bulletin* 707 (Washington, 1942), table 9. Coke, briquette, and natural gas prices are from USGS/USBM, MR, and MY, various years. Manufactured gas prices are from "Retail Prices of Gas in the United States," *Monthly Labor Review* 23 (August 1926): 176; and Gould, *Output in the Electric and Gas Utilities*, table 25. Fuel oil is from Oil Industry Information Committee, *Petroleum Industry Record* (New York, 1949). Kerosene prices are from Williamson, *American Petroleum Industry*, vol. 2; and American Petroleum Institute, *Petroleum Facts and Figures* (New York, various years).

4. The data in table A4.4 are from MR, 1917, part 2, 1245; and US Coal Commission, *Report*, part 2, 685.

5. Table A4.5 derives from United States Bureau of Foreign and Domestic Commerce, *Consumer Use of Selected Goods and Services by Income Classes* (Washington, 1935–1937).

6. After 1920, the total for oil burners in use is from Arthur E. Pew et al., "Peacetime Fuels from War Equipment," American Petroleum Institute, *Proceedings* (1945): 53–56. For 1920, the total in use is from "Replacement Market," FOJ 18 (July 1939): 15. New orders from 1926 to 1928 are reported in "Born as a Makeshift," *Business Week*, August 3, 1932, 6. From then on, new orders are from the *Survey of Current Business*, supplement, 1942. From 1933 on, stokers are also from that source and FOJ 19 (Jan. 1941): 40. Gas central heat equipment is from AGA, *Historical Statistics*, table 143.

7. For 1940 data, see the following. Solid fuels, fuel oil, range oil, and wood consumption all come from Coumbe and Avery, "Fuels Consumed." Gasoline is from US Department of Commerce, *Highway Statistics Summary to 1955* (Washington, 1957), table G-221; electricity is from its *Historical Statistics to 1970*, series S-121. Gas consumption is from AGA, *Historical Statistics*, tables 90–92.

Selected Bibliography

The following is a list of the most important works that I consulted. It is by no means complete, but it includes the basic sources that I have relied on for the evidence and ideas that inform this work. For individual articles from the technical press, and for various other works too numerous to cite here, the reader should consult the notes.

Archival and Unpublished Materials

James Douglas Collection, MS 1301, Arizona Historical Society. https://arizonahistoricalsociety.org/.

Mead, Richard. "An Analysis of the Decline of the Anthracite Industry since 1921." PhD diss., University of Pennsylvania, 1935.

Michalik, Benjamin. "The Decline of Anthracite, 1913–1955." PhD diss., Fordham University, 1957.

Millbrooke, Anne. "State Geological Surveys in the Nineteenth Century." PhD diss., University of Pennsylvania, 1981.

Mowery, David. "The Emergence and Growth of Industrial Research in American Manufacturing, 1899–1945." PhD diss., Stanford University, 1981.

Oral Histories on Work and Daily Life in the Brandywine Valley. Accession 1970-370. Hagley Museum and Library, Wilmington, Delaware.

Pennsylvania Railroad Collection. Hagley Museum and Library, Wilmington, Delaware.

Vanek, Joann. "Keeping Busy: Time Spent in Housework, United States, 1920–1970." PhD diss., University of Michigan, 1973.

Williams Oil-O-Matic Collection. McLean County Historical Society, McLean, Illinois.

Primary Printed
Government Documents

Arnold, Ralph, and Frederick Clapp. "Wastes in the Production and Utilization of Natural Gas and Means for Their Prevention." USBM TP 38. Washington, 1913.

Arnquist, Inez, and Evelyn Roberts. "The Present Use of Work Time of Farm Homemakers." Washington State University. Agricultural Experiment Station. *Bulletin* 234. Pullman, 1929.

Blakemore, M. N., et al. *Fuel Economy on Railroads of the United States, 1918–1937*. Works Progress Administration National Research Project. New York, 1937.

Blanchard, Newton, et al., eds. *Proceedings of a Conference of Governors in the White House*. Washington, D.C. May 13–15, 1908. Washington, 1909.

Brodell, A. F., and R. D. Jennings. *Work Performed and Feed Utilized by Horses and Mules*. FM 44. USDA. Bureau of Agricultural Economics. Washington, 1942.

Bukovsky, Alexis. "Use and Cost of Railway Fuel and Problems in Fuel Statistics." Interstate Commerce Commission, *Statement* 4428. Washington, 1944.

Bush, Ada. *Consumer Use of Selected Goods and Services by Income Class*. Market Research Series 5.1–5.11. US Bureau of Foreign and Domestic Commerce: Washington, 1937.

Campbell, Marius. "The Coal Fields of the United States." USGS *Professional Paper* 104. Washington, 1929.

Case, R. C. *Report on the Handling of Fuel, Layout of Engine Changing Stations, and Other Miscellaneous Developments in the United States of America*. N.p., 1925.

Cash, F. E., and M. W. von Bernewitz. "Methods, Cost, and Safety in Stripping and Mining Coal, Copper Ore, Iron Ore, Bauxite, and Pebble Phosphate." USBM *Bulletin* 298. Washington, 1929.

Cattell, R. A. "Natural Gas Manual for the Home." USBM TP 325. Washington, 1922.

Chance, H. Martyn. *Report on the Mining Methods and Appliances Used in the Anthracite Coal Fields*. Harrisburg: Pennsylvania Second Geological Survey, 1883.

Clark, Ruth, and Gretta Gray. "The Routine and Seasonal Work of Nebraska Farm Women." University of Nebraska Agricultural Experiment Station, *Bulletin* 238. Lincoln, 1930.

Commonwealth of Pennsylvania. *Giant Power: The Report of the Giant Power Survey Board*. Harrisburg, 1925.

Commonwealth of Pennsylvania. *Report of the Commission Appointed to Investigate the Waste of Coal Mining with a View to the Utilizing of the Waste*. Philadelphia, 1893.

Daugherty, C. R., et al. "Power Capacity and Production in the United States." USGS *Water Supply Paper* 579. Washington, 1928.

Fieldner, Arno. "Low Temperature Carbonization of Coal." USBM TP 396. Washington, 1926.

Fieldner, Arno. "Why and How Coke Should Be Used for Domestic Heating." USBM TP 242. Washington, 1920.

Fieldner, Arno, and Alden Emery. "Research Activities in the Mineral Industries of the United States." USBM. IC 6637. Washington, 1932.

Funk, W. C. "What the Farm Contributes Directly to the Farmer's Living." *Farmers Bulletin* 635. Washington, 1914.

Gannett, Henry, ed. *Report of the National Conservation Commission*. 3 vols. Washington, 1909.

Gilbert, Chester, and Joseph Pogue. "Petroleum: A Resource Interpretation." United States National Museum, *Bulletin* 102, part. 6. Washington, 1918.

Gray, Greta. "The Lighting Power of Nebraska Rural Homes by Kerosene and Gasoline Lamps." University of Nebraska Agricultural Experiment Station *Bulletin* 225. Lincoln, 1928.

Grossman, Jonathan. "The Coal Strike of 1902: Turning Point in US Policy." *Monthly Labor Review* 78 (Oct. 1978): 21–28.

Hotchkiss, Willard, et al. "Mechanization, Employment, and Output per Man in Bituminous-Coal Mining." Vol. 1. WPA National Research Project *Report* E-9. Philadelphia, 1939.

International Agency for Research on Cancer. *Personal Habits and Indoor Combustions*. Monograph 100E. Lyon, France: IARC, 2012.

Johnson, Walter. *A Report to the Navy Department of the United States on American Coals Applicable to Steam Navigation and to Other Purposes.* 28th Cong., 1st sess., S. Rept. 386. Washington, 1844.

Kiessling, O. E., et al. "The Economics of Strip Coal Mining." USBM *Economic Paper* 11. Washington, 1931.

Kreisinger, Henry. "Five Ways of Saving Fuel in Heating Houses." USBM TP 199. Washington, 1918.

Kryk, Hazel, et al. "Family Expenditures for Housing and Household Operations, Five Regions." USDA *Miscellaneous Publication* 432. Washington, 1941.

Kryk, Hazel, et al. "Family Income and Facilities, Five Regions." USDA *Miscellaneous Bulletin* 399. Washington, 1940.

Kudlich, Rudolf. "Fuels Available for Domestic Use as Substitutes for Anthracite Coal." USBM RI 2520. Washington, 1923.

Massachusetts. Special Commission on the Necessaries of Life. *Report.* Boston, 1920–1929.

Monroe, Merna. "Kerosene Burners in a Wood Cook Stove." Maine Agricultural Experiment Station *Bulletin* 433 (1945): 274–285.

Monroe, Merna. "Performance Analysis of Selected Types of Kerosene Stoves." Maine Agricultural Experiment Station *Bulletin* 394 (1939): 433–520.

Odell, William. "Facts Relating to the Production and Substitution of Manufactured Gas for Natural Gas." USBM *Bulletin* 301. Washington, 1929.

Ohio State University Department of Home Economics. *Kitchen Tests of Relative Cost of Natural Gas, Soft Coal, Oil, Gasoline and Electricity for Cooking.* Columbus, 1918.

Oregon State Agricultural College Extension Station. "Use of Time by Oregon Farm Homemakers." *Bulletin* 256. Corvallis: Oregon State Agricultural College, 1929.

Parker, Edward W. "Condition of the Coal Briquetting Industry in the United States." In *Contributions to Economic Geology*, edited by Marius Campbell, 460–485. USGS *Bulletin* 316, part 2. Washington, 1907.

Parker, Edward W. "Report on the Operations of the Coal-Testing Plant of the United States Geological Survey at the Louisiana Purchase Exposition, St. Louis Mo., 1904." USGS *Professional Paper* 48. Washington, 1906.

Perrot, George, and H. W. Clark. "Smokeless Fuel for Salt Lake City." USBM RI 2341. Washington, 1923.

Platt, Franklin. *A Special Report to the Legislature upon the Causes, Kinds and Amount of Waste in Mining Anthracite.* Harrisburg: Pennsylvania Second Geological Survey, 1881.

Rankin, J. O. "The Use of Time in Farm Homes." University of Nebraska Agricultural Experiment Station *Bulletin* 230. Lincoln, 1928.

Rapp, Miriam. "Fuels Used for Cooking Purposes in Indiana Rural Homes." Purdue University Agricultural Experiment Station *Bulletin* 339. Lafayette, IN: Purdue University Agricultural Experiment Station, 1930.

Requa, M. L. *Petroleum Resources of the United States.* 64th Cong., 2d sess., S. Doc. 363. Washington, 1916.

Reynolds, R. V., and Albert H. Pierson. "Fuel Wood Used in the United States, 1630–1930." USDA *Circular* 641. Washington, 1942.

Senner, Arthur. "The Domestic Oil Burner." USDA *Circular* 405. Washington, 1927.

Sisler, James, et al. "Anthracite Culm and Silt." Pennsylvania Geological Survey Fourth Series. *Bulletin* M-12. Harrisburg, 1928.

Snyder, Edna. "Factors Affecting the Performance of Kerosene Cook Stoves." University of Nebraska Agricultural Extension Station. *Bulletin* 64. Lincoln, 1932.

Snyder, Edna. "A Study of Kerosene Cook Stoves." University of Nebraska Agricultural Extension Station. *Bulletin* 48. Lincoln, 1930.

Toenges, Albert. "Reclamation of Stripped Coal Land." USBM RI 3440. Washington, 1939.

Toenges, Albert, and Robert Anderson. "Some Aspects of Strip Mining of Bituminous Coal in Central and South Central States." USBM IC 6959. Washington, 1937.

Turner, Scott, et al. "Mining Bituminous Coal by Stripping Methods." USBM IC 6383. Washington, 1930.

US Bureau of Labor Statistics. "Cost of Living in the United States." *Bulletin* 357. Washington, 1924.

US Bureau of Labor Statistics. "Technological Change and Productivity in the Bituminous Coal Industry, 1920–1960." *Bulletin* 1305. Washington, 1961.

US Bureau of the Census. *Census of Central Electric Light and Power Stations.* Washington, various years.

US Bureau of the Census. *Census of Manufactures Mining and Transportation.* Washington, various years.

US Coal Commission. *Report Transmitted Pursuant to the Act Approved September 22, 1922*, parts 1–3. Washington, 1925.

US Congress. *Bureau of Mines.* 61st Cong., 2d sess., S. Rept. 353 to accompany H.R. 13915. Washington, 1910.

US Congress. House Committee on Interstate and Foreign Commerce. Hearings before a Subcommittee on H.R. 441. *Petroleum Investigation*, part 2. 73d. Cong. Washington, 1934.

US Department of Agriculture. Bureau of Agricultural Economics and Home Economics. *The Average Quantities and Values of Fuel and Other Household Supplies Used by Farm Families.* Washington, 1926.

US Department of Agriculture. "Consumption of Firewood in the United States." Forest Service *Circular* 181 (Washington, 1910).

US Department of Agriculture. "Domestic Needs of Farm Women." *Report* 104. Washington, 1915.

US Department of Agriculture. Bureau of Home Economics. "Farm Housing Survey." *Miscellaneous Publication* 323. Washington, 1939.

US Department of Agriculture. "Social and Labor Needs of Farm Women." *Report* 103. Washington, 1915.

US Department of Agriculture. "Use of Wood for Fuel." USDA. *Bulletin* 753. Washington, 1919.

US Department of Commerce. *Historical Statistics of the United States, Colonial Times to 1970.* Washington, 1975.

US Department of Commerce. *Survey of Current Business.* Washington, various years.

US Department of the Interior. *Conference on Natural Gas Conservation.* Washington, 1920.

US Federal Oil Conservation Board. *Public Hearings.* Washington, 1926.

US Federal Oil Conservation Board. *Report*, parts 1–5. Washington, 1926–1932.

US Federal Trade Commission. *Report on Premium Prices for Anthracite, July 6, 1925.* Washington, 1925.

US Fuel Administration. *The Distribution of Coal and Coke*, part 1. Washington, 1919.

US Fuel Administration. *Fuel Facts.* Washington, 1918.

US Interstate Commerce Commission. *Annual Report on the Statistics of Railways in the United States.* Washington, various years.

US National Resources Committee. *Energy Resources and National Policy.* Washington, 1939.

US Secretary of the Interior. *Fuels and Structural Materials.* 59th Cong., 1st sess., S. Doc. 214, 1906.

US Treasury Department. *Documents Relative to the Manufactures of the United States.* Vol. 1. 22d Cong., 1st sess., H.D. 308. Washington, 1833.

US Treasury Department. *Steam Engines.* 25th Cong., 3d sess., H.D. 21. Washington, 1839.

Ward, Florence. "The Farm Woman's Problems." USDA *Circular* 148. Washington, 1920.

Wyer, Samuel. "Waste and Correct Use of Natural Gas in the Home." USBM TP 257. Washington, 1920.

Wright, C. L. "Fuel-Briquetting Investigations." USBM *Bulletin* 58. Washington, 1913.

Yaworski, Nicholas, et al. "Fuel Efficiency in Cement Manufacture, 1909–1935." WPA National Research Project. *Report* E-5. Philadelphia, 1938.

Scientific and Technical Proceedings and Transactions

American Gas Association; American Gas Light Association; American Gas Institute; American Institute of Mining Engineers; American Mining Congress; American Philosophical Society; American Society of Mechanical Engineers; American Society of Heating and Ventilating Engineers; American Society for Testing Materials; International Railway Fuel Association; National Commercial Gas Association; Natural Gas Association; Technical Association of the Pulp and Paper Industry [TAPPI] *Papers.*

Scientific and Technical Journals

American Artisan; *American Builder*; *American Gas Engineering Journal*; *American Gas Light Journal*; American Iron and Steel Institute *Yearbook*; *American Journal of Science*; *American Railroad Journal*; *American Railway Times*; *Black Diamond*; *Blast Furnace and Steel Plant*; *Building Age*; *Cassier's Magazine*; *Chemical and Metallurgical Engineering*; *Coal Age*; *Coal Trade*; *Colliery Engineer*; *Combustion*; *DeBow's Review*; *Domestic Engineering*; *Electrical World*; *Engineering and Mining Journal*; *Engineering Magazine*; *Engineering News*; *Fuel Oil Journal*; *Gas Age*; *General Electric Review*; *Hazard's Register*; *Heating and Ventilating*; *Hunt's Merchants' Magazine*; *Iron Age*; *Journal of the Franklin Institute*; *Journal of Industrial and Engineering Chemistry*; *Mechanical Engineering*; *Metal Progress*; *Mines and Minerals*; *Mining and Metallurgy*; *National Builder*; *National Petroleum News*; *Niles Weekly Register*; *Paper Industry*; *Power*; *Power Plant Engineering*; *Railroad Gazette*; *Railway Age*; *Railway Mechanical Engineer*; *Railway Review*; *Scientific American.*

Farm Journals

Cultivator and Country Gentleman; *Farmer's Wife*; *National Stockman and Farmer*; *Ohio Farmer*; Wallaces' Farmer.

Newspapers and Popular Magazines

Better Homes and Gardens; *Boston Globe*; *Brooklyn Eagle*; *Chicago Tribune*; Library of Congress, Chronicling America: Historic American Newspapers, https://chroniclingamerica.loc.gov/; *Good Housekeeping*; Illinois Digital

Newspaper Collections, https://idnc.library.illinois.edu/; *Ladies' Home Journal*; *New York Herald Tribune*; *New York Times*; *Outlook*; *Washington Post*.

Books, Articles, Reports

Alderson, Victor. *The Shale Oil Industry*. New York: Frederick Stokes, 1920.

Allen, Zachariah. "Explosibility of Coal Oils." Smithsonian Institution, *Annual Report*. Washington, 1862: 330–342.

Allen, Zachariah. *The Science of Mechanics*. Providence: Hutchins and Cory, 1829.

Atcheson Topeka & Santa Fe Railroad. *Annual Report*. 1896–1941.

Bemis, Albert. *The Evolving House*. Vol. 2, *The Economics of Shelter*. Cambridge, MA: MIT Press, 1934.

Bituminous Coal Trade Association. *Report on the Present and Future of the Bituminous Coal Trade*. New York, 1908.

Bogen, Julius. *The Anthracite Railroads*. New York: Ronald Press, 1927.

Bushnell, S. Morgan, and Fred Orr. *District Heating*. New York: Heating and Ventilating Magazine, 1915.

Chandler, Charles. *On the Gas Nuisance in New York*. New York: Appleton, 1870.

Chandler, Charles. *Report on the Quality of the Kerosene Oil Sold in the Metropolitan District*. New York, 1870.

Chicago, Rock Island and Pacific Railroad. *Annual Report*. 1905–1922.

Clark, Victor S. *History of Manufactures in the United States*. Vol. 1. Washington: Carnegie Institution, 1916.

Consolidated Gas, Electric Light & Power Co. of Baltimore. *American Gas Centenary, 1816–1916*. Baltimore, 1916.

Convention of Iron Masters. *Documents Relating to the Manufacture of Iron in Pennsylvania*. Philadelphia, 1850.

Daddow, Samuel. *Coal, Iron, and Oil*. Philadelphia: J. B. Lippincott, 1866.

DePew, Chauncey, ed. *1795–1895: One Hundred Years of American Commerce*. New York: D. O. Haynes, 1895.

Ely, Richard T. *The Foundations of National Prosperity*. New York: Macmillan, 1917.

Ewing, J. A. *The Steam Engine and Other Heat Engines*. Cambridge: Cambridge University Press, 1910.

Fansler, P. E. *House Heating with Fuel Oil*. 2d and 3d eds. New York: Heating and Ventilating Magazine, 1925–1927.

Frederick, Christine. *The New Housekeeping: Efficiency Studies in Home Management*. New York: Doubleday Page, 1913.

Frederick, Christine. *Scientific Management in the Home*. Chicago: American School of Home Economics, 1920.

Gesner, Abraham. *A Practical Treatise on Coal, Petroleum, and Other Distilled Oils*. New York: Baillière Brothers, 1860.

Gracie, Alexander. "Twenty Years' Progress in Marine Construction." Smithsonian Institution, *Annual Report*. Washington, 1913: 687–704.

Gray, Gretta. "The Nebraska Rural Kitchen." *Journal of Home Economics* 19 (Sept. 1927): 504–512.

Gray, L. C. "The Economic Possibilities of Conservation." *Quarterly Journal of Economics* 27 (May 1913): 497–519.

Gray, L. C. "Rent under the Assumption of Exhaustibility." *Quarterly Journal of Economics* 28 (May 1914): 466–489.

Holley, Alexander. *American and European Railway Practice in the Economic Generation of Steam*. New York: D. Van Nostrand, 1861.

Hotelling, Harold. "The Economics of Exhaustible Resources." *Journal of Political Economy* 39 (April 1931): 137–175.

Illinois Coal Strippers Association. *Open Cut Coal Mining Industry of Illinois*. Chicago: The Association, 1939.

International Conference on Bituminous Coal. First, *Proceedings*, 1926; Second, *Proceedings*, 1928; Third, *Proceedings*, 1931. Multiple vols. Pittsburgh: Carnegie Institute of Technology, 1926–1932.

Ise, John. *The United States Oil Policy*. New Haven, CT: Yale University Press, 1926.

Jevons, William Stanley. *The Coal Question*. 2d ed. London: Macmillan, 1866.

Johnson, Emery, and E. Grover Huebner. *Principles of Ocean Transportation*. New York: D. Appleton, 1918.

Johnson, Walter. *Notes on the Use of Anthracite in the Manufacture of Iron, with some Remarks on Its Evaporative Power*. Boston: Charles Little and James Brown, 1841.

Jones, Chester. *Economic History of the Anthracite-Tidewater Canals*. Philadelphia: University of Pennsylvania Press, 1908.

Jones, Eliot. *The Anthracite Coal Combination in the United States*. Cambridge, MA: Harvard University Press, 1914.

Jones, Harry. *Railway Wages and Labor Relations, 1900–1946*. New York, 1947.

Kansas City Southern Railroad. *Annual Report*. 1901–1920.

Kershaw, John. *The Use of Low Grade and Waste Fuels for Power Generation*. London: Constable, 1920.

Kneeland, Frank. *Practical Coal Production: Getting Out the Coal*. New York: McGraw Hill, 1926,

Lardner, Dionysius. *A Rudimentary Treatise on the Steam Engine*. 4th ed. London: John Weale, 1854.

Lesley, J. Peter. *The Iron Manufacturer's Guide*. New York: John Wiley, 1859.

Lesley, Robert. *History of the Portland Cement Industry in the United States*. Chicago: International Trade Press, 1924.

Marsh, Robert. *Steam Shovel Mining*. New York: McGraw Hill, 1920.

McAuliffe, Eugene. *Railway Fuel*. New York: Simmons-Boardman, 1927.

Meade, Richard. *Portland Cement*. Easton, PA: Chemical Publishing, 1906.

Mitchell, David, ed. *Coal Preparation*. New York: American Institute of Mining and Metallurgical Engineers, 1943.

Mone, Frederick. *Treatise on American Engineering*. New York Samuel Congdon, 1854.

Morrow, L. W. *Electric Power Stations*. New York: McGraw Hill, 1927.

National Commercial Gas Association. *The Story of Nancy Gay*. New York, 1914.

National District Heating Association. *Handbook*. 2d ed. Greenville, OH: The Association, 1932.

National Industrial Conference Board. *The Competitive Position of Coal in the United States*. New York: NICB, 1931.

Nearing, Scott. *Anthracite: An Instance of Natural Resource Monopoly*. Philadelphia: John Winston, 1915.

Newberry, J. S. *Report on the Economical Geology of the Route of the Ashtabula and New Lebanon Railroad*. Cleveland, 1857.

Osborn, Henry. *The Metallurgy of Iron and Steel*. Philadelphia: H. C. Baird, 1869.

Overman, Frederick. *The Manufacture of Iron*. Philadelphia: Henry C. Baird, 1850.

Renwick, James. *Essay on the Steamboats of the United States*. London: n.p., 1830.

Rice, George, et al. "Conservation of Coal Resources." Third World Power Conference. *Proceedings*. Vol. 6. Washington, 1938: 671–721.

Schavilje. J. P. "Reclaiming Illinois Strip Mined Coal Land with Trees." *Journal of Forestry* 39 (August 1941): 714–719.

Sinclair, Angus. *Development of the Locomotive Engine*. New York: Sinclair Publishing, 1907.

Southern Pacific Railroad. *Annual Report*. 1900–1941.

Stevenson, David. *Sketch of the Civil Engineering of North America*. London: John Weale, 1838.

Stocking, George. *The Oil Industry and the Competitive System: A Study in Waste*. Boston: Houghton Mifflin, 1925.

Swank, James. *History of the Manufacture of Iron in All Ages*. Philadelphia, AISI, 1892.

Tryon, Frederick G., and H. O. Rogers. "Statistical Studies of Progress in Fuel Economy." Second World Power Conference. *Transactions*. Vol. 6. Berlin, 1930: 343–365.

Van Hise, Charles. *The Conservation of Natural Resources*. New York: MacMillan, 1910.

Young, William Harvey. "Sources of Coal and Types of Stokers and Burners Used by Electric Public Utility Plants." *Brookings Institution Pamphlet* 2. Washington, 1930.

Zimmermann, Erich. *Zimmermann on Shipping*. New York: Prentice-Hall, 1923.

Secondary Works
Books

Adams, Sean. *Home Fires: How Americans Kept Warm in the Nineteenth Century*. Baltimore: Johns Hopkins University Press, 2014.

American Gas Association. *Historical Statistics of the Gas Industry*. New York: American Gas Association, 1956.

Barger, Harold. *The Transportation Industries, 1889–1946*. New York: National Bureau of Economic Research, 1951.

Barger, Harold, and Sam Schurr. *The Mining Industries, 1899–1939*. New York: NBER, 1944.

Barnett, Harald, and Chandler Morse. *Scarcity and Growth: The Economics of Natural Resource Availability*. Baltimore: Johns Hopkins University Press, 1963.

Bauer, John, et al. *The Electric Power Industry: Development Organization and Public Policies*. New York: Harper Brothers, 1939.

Baumol, William J. *The Free Market Innovation Machine*. Princeton, NJ: Princeton University Press, 2002.

Bedini, Silvio. *Thinkers and Tinkers: Early American Men of Science*. New York: Scribner's, 1975.

Billington, Ray Allen. *Land of Savagery, Land of Promise*. New York: Norton, 1981.

Binder, Frederick. *Coal Age Empire*. Harrisburg: Pennsylvania Historical and Museum Commission, 1974.

Bishop, Harvey. *Retail Marketing of Furnace Oil*. Cambridge, MA: Harvard Business School, 1946.

Blanchard, Charles. *The Extraction State: A History of Natural Gas in America*. Pittsburgh, PA: University of Pittsburgh Press, 2021.

Blatz, Perry. *Democratic Miners: Work and Labor Relations in the Anthracite Coal Industry, 1875–1925*. Albany: State University of New York Press, 1994.

Brewer, Priscilla. *From Fireplace to Cookstove*. Syracuse, NY: Syracuse University Press, 2000.

Brown, Sanborn. *Count Rumford: Physicist Extraordinary.* New York: Anchor, 1962.

Bruce, Alfred. *The Steam Locomotive in America: Its Development in the Twentieth Century.* New York: Norton, 1952.

Carter, Susan, et al., eds. *Historical Statistics of the United States, Millennial Edition.* New York: Cambridge University Press, 2006.

Castaneda, Christopher J. *Invisible Fuel: Manufactured and Natural Gas in America, 1800–2000.* New York: Twayne, 1999.

Clark, John. *Energy and the Federal Government.* Urbana: University of Illinois Press, 1987.

Cohn, Julie. *The Grid: Biography of an American Technology.* Cambridge, MA: MIT Press, 2017.

Cowan, Ruth. *More Work for Mother: The Ironies of Household Technology from the Open Hearth to the Microwave.* New York: Basic Books, 1983.

Davis, Lance, et al. *In Pursuit of Leviathan.* Chicago: University of Chicago Press, 1997.

Dewhurst, J. Frederic, et al. *America's Needs and Resources: A New Survey.* New York: Twentieth Century Fund, 1955.

Dix, Keith. *What's a Coal Miner to Do? The Mechanization of Coal Mining.* Pittsburgh, PA: University of Pittsburgh Press, 1988.

Dix, Keith. *Work Relations in the Coal Industry: The Hand-Loading Era, 1880–1930.* Morgantown: University of West Virginia Press, 1988.

Dorsey, Leroy. *Theodore Roosevelt, Conservation, and the 1908 Governors' Conference.* College Station: Texas A&M University Press, 2016.

DuBoff, Richard. *Electric Power in American Manufacturing, 1889–1958.* New York: Arno, 1979.

Eavenson, Howard. *The First Century and a Quarter of American Coal Industry.* Pittsburgh, PA: Privately printed, 1942.

Ferguson, Eugene. *Oliver Evans: Inventive Genius of the American Industrial Revolution.* Wilmington, DE: Eleutherian Mills Hagley Foundation, 1980.

Field, Alexander. *A Great Leap Forward: 1930s Depression and US Economic Growth.* New Haven, CT: Yale University Press, 2011.

Fishlow, Albert. *American Railroads and the Transformation of the Ante-bellum Economy.* Cambridge, MA: Harvard University Press, 1965.

Goodwin, Craufurd D., et al., eds. *Energy Policy in Perspective: Today's Problems, Yesterday's Solutions.* Washington: Brookings Institution, 1981.

Gordon, Robert. *The Rise and Fall of American Growth: The U.S. Standard of Living since the Civil War.* Princeton, NJ: Princeton University Press, 2016.

Gottleib, Robert. *Forcing the Spring: The Transformation of the American Environmental Movement.* Washington: Island Press, 1993.

Gould, Jacob. *Output and Productivity in the Electric and Gas Utilities, 1899–1942.* New York: National Bureau of Economic Research, 1946.

Grossman, Peter. *U.S. Energy Policy and the Pursuit of Failure.* New York: Cambridge University Press, 2013.

Hardwicke, Robert. *Antitrust Laws, et al. vs. Unit Operation of Oil or Gas Pools.* New York: American Institute of Mining and Metallurgical Engineers, 1948.

Haynes, Williams. *American Chemical Industry.* Vol. 5, *The Merger Era.* New York: D. Van Nostrand, 1948.

Hays, Samuel. *Conservation and the Gospel of Efficiency: The Progressive Conservation Movement, 1890–1920.* Cambridge, MA: Harvard University Press, 1959.

Healey, Richard. *The Pennsylvania Anthracite Coal Industry, 1860–1902.* Scranton, PA: University of Scranton Press, 2007.

Herbert, John. *Clean Cheap Heat*. Westport, CT: Praeger, 1992.

Hidy, Ralph, and Muriel Hidy. *History of the Standard Oil Company: Pioneering in Big Business, 1882–1911*. New York: Harper & Brothers, 1955.

Hindle, Brooke. *America's Wooden Age*. Tarrytown, NY: Sleepy Hollow Restorations, 1975.

Hirsh, Richard. *Technology and Transformation in the American Electric Utility Industry*. New York: Cambridge University Press, 1989.

Hollingsworth, John. *History of Development of Strip Mining Machines*. South Milwaukee, WI: Bucyrus-Erie, 1960.

Hughes, Thomas P. *Networks of Power: Electrification in Western Society, 1880–1930*. Baltimore: Johns Hopkins University Press, 1983.

Hunter, Louis. *A History of Industrial Power in the United States, 1780–1930*. Vol. 1, *Waterpower in the Century of the Steam Engine*; Vol. 2, *Steam Power*. Charlottesville: University Press of Virginia, 1979, 1985.

Hunter, Louis. *Steamboats on Western Rivers*. Cambridge, MA: Harvard University Press, 1949.

Hunter, Louis, and Lynwood Bryant. *A History of Industrial Power in the United States, 1780–1930*. Vol. 3, *Transmission of Power*. Cambridge, MA: MIT Press, 1991.

James, Charles, and Waldo Fisher. *Minimum Price Fixing in the Bituminous Coal Industry*. Princeton, NJ: Princeton University Press, 1955.

Jerome, Harry. *Mechanization in Industry*. New York: National Bureau of Economic Research, 1934.

Johnson, Arthur. *Development of American Petroleum Pipelines*. Ithaca, NY: Cornell University Press, 1956.

Johnson, Benjamin Heber. *Escaping the Dark, Gray City: Fear and Hope in Progressive Era Conservation*. New Haven, CT: Yale University Press, 2017.

Johnson, James. *A New Deal for Soft Coal*. New York: Arno, 1979.

Johnson, James. *The Politics of Soft Coal*. Urbana: University of Illinois Press, 1979.

Jones, Christopher. *Routes of Power: Energy and Modern America*. Cambridge, MA: Harvard University Press, 2014.

Kendrick, John W. *Productivity Trends in the United States*. Princeton, NJ: Princeton University Press, 1961.

Lamb, J. Parker. *Perfecting the American Steam Locomotive*. Bloomington: Indiana University Press, 2003,

Landes, David, et al., eds. *The Invention of Enterprise*. Princeton, NJ: Princeton University Press, 2010.

Lebergott, Stanley. *The American Economy: Income, Wealth and Want*. Princeton, NJ: Princeton University Press, 1976.

Lebergott, Stanley. *The Americans: An Economic Record*. New York: Norton, 1984.

Liebcap, Gary. *Contracting for Property Rights*. New York: Cambridge University Press, 1989.

Lucier, Paul. *Scientists and Swindlers: Consulting on Coal and Oil in America, 1820–1890*. Baltimore: Johns Hopkins University Press, 2008.

MacDonald, Forrest. *Let There Be Light*. Madison, WI: American History Research Center, 1957.

Marcus, Alan, ed. *Engineering in a Land Grant Context*. Lafayette, IN: Purdue University Press, 2004.

Martin, Albro. *Enterprise Denied*. New York: Columbia University Press, 1971.

McDonald, Stephen. *Federal Tax Treatment of Income from Oil and Gas*. Washington: Brookings Institution, 1963.

McDonald, Stephen. *Petroleum Conservation in the United States: An Economic Analysis*. Baltimore: Johns Hopkins University Press, 1971.

Miller, Raymond. *Kilowatts at Work*. Detroit: Wayne State University Press, 1959.

Mokyr, Joel. *The Gifts of Athena: Historical Origins of the Knowledge Economy*. Princeton, NJ: Princeton University Press, 2004.

Morrow, L. W. *Electric Power Stations*. New York: McGraw Hill, 1927.

Mowery, David, and Nathan Rosenberg. *Technology and the Pursuit of Economic Growth*. New York: Cambridge University Press, 1989.

Nash, Roderick. *Wilderness and the American Mind*. New Haven, CT: Yale University Press, 1967.

Nye, David. *Consuming Power: A Social History of American Energies*. Cambridge, MA: MIT Press, 1998.

Olson, Mancur. *Power and Prosperity*. New York: Basic Books, 2000.

Page, Talbot. *Conservation and Economic Efficiency*. Baltimore: Johns Hopkins University Press, 1977.

Paskoff, Paul. *Industrial Evolution*. Baltimore: Johns Hopkins University Press, 1983.

Passer, Harold. *The Electrical Manufacturers, 1875–1900*. Cambridge, MA: Harvard University Press, 1953.

Powell, H. Benjamin. *Philadelphia's First Fuel Crisis*. University Park: Pennsylvania State University Press, 1978.

Pursell, Carroll. *Early Stationary Steam Engines in America*. Washington: Smithsonian Institution, 1969.

Rose, Mark. *Cities of Light and Heat: Domesticating Gas and Electricity in Urban America*. University Park: Pennsylvania State University Press, 1995.

Rosenberg, Nathan. *Exploring the Black Box*. New York: Cambridge University Press, 1994.

Rosenberg, Nathan. *Schumpeter and the Endogeneity of Technology: Some American Perspectives*. New York: Routledge, 2014.

Rosenberg, Nathan. *Studies on Science and the Innovation Process: Selected Works by Nathan Rosenberg*. Singapore: World Scientific Publishing, 2010.

Russell, Paul. *History of Western Oil Shale*. East Brunswick, N.J.: Center for Professional Advancement, 1980.

Rutherford, Janice. *Selling Mrs. Consumer: Christine Frederick and the Rise of Household Efficiency*. Athens: University of Georgia Press, 2003.

Sale, Kirkpatrick. *The Fire of His Genius: Robert Fulton and the American Dream*. New York: Free Press, 2001.

Schumpeter, Joseph. *Capitalism, Socialism and Democracy*. 5th ed. London: George Allen & Unwin, 1976.

Schumpeter, Joseph. *The Theory of Economic Development*. Cambridge, MA: Harvard University Press, 1934.

Schurr, Sam, et al. *Energy in the American Economy, 1850–1975*. Baltimore: Resources for the Future, 1960.

Sheshinski, Eytan, et al. *Entrepreneurship, Innovation and the Growth Mechanism of the Free Enterprise Economies*. Princeton, NJ: Princeton University Press, 2007.

Simon, Julian. *The Ultimate Resource*. Princeton, NJ: Princeton University Press, 1996.

Simpson, R. David, et al., eds. *Scarcity and Growth Revisited: Natural Resources and the Environment in the New Millennium*. Washington: Resources for the Future, 2005.

Sinclair, Bruce. *A Centennial History of the American Society of Mechanical Engineers*. Toronto: University of Toronto Press, 1980.

Sinclair, Bruce. *Early Research at the Franklin Institute*. Philadelphia: Franklin Institute, 1966.

Sinclair, Bruce. *Philadelphia's Philosophical Mechanics*. Baltimore: Johns Hopkins University Press, 1974.

Smith, Thomas R. *The Cotton Textile Industry of Fall River, Massachusetts*. New York: King's Crown, 1944.

Soltow, Lee, and Edward Stevens. *The Rise of Literacy and the Common School in the United States: A Socioeconomic Analysis to 1870*. Chicago: University of Chicago Press, 1981.

Spitz, Peter. *Petrochemicals: The Rise of an Industry*. New York: John Wiley & Sons, 1988.

Steifler, Susan. *The Beginnings of a Century of Steam and Water Heating by the H. B. Smith Company*. Westfield, MA: H. B. Smith, 1960.

Stevens, Edward. *The Grammar of the Machine: Technical Literacy and Early Industrial Expansion in the United States*. New Haven, CT: Yale University Press, 1995.

Stoll, Mark. *Inherit the Holy Mountain: Religion and the Rise of American Environmentalism*. New York: Oxford University Press, 2015.

Stotz, Louis, and Alexander Jamison. *History of the Gas Industry*. Privately printed, 1938.

Stradling, David. *Smokestacks and Progressives*. Baltimore: Johns Hopkins University Press, 1999.

Strasser, Susan. *Never Done: A History of American Housework*. New York: Pantheon, 1982.

Tarr, Joel. *The Search for the Ultimate Sink: Urban Pollution in Historical Perspective*. Akron, OH: University of Akron Press 1996.

Tarr, Joel, and Carl Zimring. "The Struggle for Smoke Control in St. Louis." In *Common Fields: An Environmental History of St. Louis*, edited by Andrew Hurley, 199–220. St. Louis: Missouri Historical Society, 1997.

Temin, Peter. *Iron and Steel in Nineteenth-Century America*. Cambridge, MA: MIT Press, 1964.

Tomory, Leslie. *Progressive Enlightenment: The Origins of the Gaslight Industry, 1780–1820*. Cambridge, MA: MIT Press, 2012.

Tussing, Arlon, and Connie Barlow. *The Natural Gas Industry: Evolution, Structure and Economics*. Cambridge, MA: Ballinger, 1984.

Twentieth Century Fund. *Electric Power and Government Policy*. New York: Twentieth Century Fund, 1948.

Tyler, David. *Steam Conquers the Atlantic*. New York: D. Appleton-Century, 1939.

Tyrrell, Ian. *Crisis of the Wasteful Nation*. Chicago: University of Chicago Press, 2015.

Vietor, Richard. *Energy Policy in America since 1945*. New York: Cambridge University Press, 1984.

Vietor, Richard. *Environmental Politics and the Coal Coalition*. College Station: Texas A&M University Press, 1980.

Wellock, Thomas. *Preserving the Nation: The Conservation and Environmental Movements, 1870–2000*. Wheeling, IL: Harlan Davidson, 2007.

Williams, Michael. *Americans and Their Forests*. New York: Cambridge University Press, 1988.

Williamson, Harold, and Arnold Daun. *The American Petroleum Industry*. Vol. 1, *The Age of Illumination, 1859–1899*; Vol. 2, *The Age of Energy*. Evanston, IL: Northwestern University Press, 1959–1963.

Withuhn, William. *American Steam Locomotives: Design and Development*. Bloomington: Indiana University Press, 2019.

Wright, Lawrence. *Home Fires Burning: The History of Domestic Heating and Cooking*. London: Routledge & Kegan Paul, 1964.

Yearly, Clifton. *Enterprise and Anthracite*. Baltimore: Johns Hopkins University
    Press, 1961.
Zimmermann, Erich. *Conservation in the Production of Petroleum*. New Haven,
    CT: Yale University Press, 1957.

Articles and Essays
Adams, Sean. "Promotion, Competition, Captivity: The Political Economy of
    Coal." *Journal of Policy History* 18 (Jan. 2006): 74–95.
Allen, Robert. "Backward into the Future: The Shift to Coal and Implications
    for the Next Energy Transition." *Energy Policy* 50 (April 2012): 17–23.
Allen, Robert. "Collective Invention." *Journal of Economic Behavior and
    Organization* 4 (March 1983): 1–24.
Allen, Robert. "The Peculiar Productivity History of American Blast Furnaces,
    1840–1913." JEH 37 (Sept. 1977): 605–633.
Allen, Robert. "The Transportation Revolution and the English Coal Industry,
    1695–1842." JEH 83 (Dec. 2023): 1175–1220.
Atack, Jeremy, et al. "The Regional Diffusion and Adoption of the Steam Engine
    in American Manufacturing," JEH 40 (June 1980): 281–308.
Beach, Brian, and Walker Hanlon. "Coal Smoke and Mortality in an Early
    Industrial Economy." *Economic Journal* 128 (Nov. 2018): 2652–2674.
Beaton, Kendall. "Dr. Gesner's Kerosene: The Start of American Oil Refining."
    BHR 29 (March 1955): 28–53.
Beaver, William. "The U.S. Failure to Develop Synthetic Fuels in the 1920s." *The
    Historian* 53 (Winter 1991): 241–254.
Brayard, Frank. "The *Savannah*: Illustrious Failure." *Powerships* (Spring 2019):
    24–27.
Bush, Jane. "Cooking Competition: Technology on the Domestic Market in the
    1930s." *Technology and Culture* 24 (April 1983): 222–245.
Chandler, Alfred. "Anthracite Coal and the Beginnings of the Industrial
    Revolution in the United States." BHR 46 (Summer 1972): 141–181.
Chapelle, Howard. "The Pioneer Steamship *Savannah*: A Study for a Scale
    Model." *U.S. National Museum Bulletin* 228 (1963): 62–80.
Cohn, Julie. "Utilities as Conservationists? The Paradox of Electrification
    during the Progressive Era in North America." In *Green Capitalism?*, edited
    by Hartmut Berghoff and Adam Rome, 94–112. Philadelphia: University of
    Pennsylvania Press, 2017.
Cole, Arthur H. "The Mystery of Fuel Wood Marketing in the United States."
    BHR 44 (Autumn 1970): 339–359.
Crabbé, Philippe. "The Contribution of L. C. Gray to the Economic Theory of
    Exhaustible Natural Resources and Its Roots in the History of Economic
    Thought." *Journal of Environmental Economics and Management* 10 (Sept.
    1983): 195–220.
David, Paul, and Gavin Wright. "Increasing Returns and the Genesis of
    American Resource Abundance." *Industrial and Corporate Change* 6
    (March 1997): 203–234.
Davidson, Cliff. "Air Pollution in Pittsburgh: A Historical Perspective." *Journal
    of the Air Pollution Control Association* 29 (1979): 1034–1041.
Day, Tanis. "Capital-Labor Substitution in the Home." *Technology and Culture*
    33 (April 1992): 302–327.
Desrochers, Pierre. "How Did the Invisible Hand Handle Industrial Waste?
    By-Product Development before the Modern Environmental Era." *Enterprise
    and Society* 8 (June 2007): 348–374.
Devine, Warren. "From Shafts to Wires: Historical Perspectives on Electrifica-
    tion." JEH 43 (June 1983): 347–372.

Easterlin, Richard. "Why Isn't the Whole World Developed"? JEH 41 (March 1981): 1–19.

Fishlow, Albert. "Productivity and Technological Change in the Railroad Sector, 1840–1910." In *Output, Employment and Productivity in the United States after 1800*, edited by Dorothy S. Brady, 585–646. New York: NBER, 1966.

Gilmer, Robert. "The History of Natural Gas Pipelines in the Southwest." *Texas Business Review* 55 (May-June 1981): 129–135.

Graebner, William. "Great Expectations: The Search for Order in Bituminous Coal, 1890–1917." BHR 48 (Spring 1974): 49–72.

Grinder, Dale. "The Battle for Clean Air: The Smoke Problem in Post–Civil War America." In *Pollution and Reform in American Cities, 1870–1930*, edited by Martin Melosi, 83–103. Austin: University of Texas Press, 1980.

Hall, Greg. "Strip Mining and Reclamation in Fulton County Illinois: An Environmental History." *Journal of the Illinois State Historical Society* 108 (Spring 2015): 54–73.

Halsey, Harlan. "The Choice between High-Pressure and Low-Pressure Steam Power in America in the Early Nineteenth Century." JEH 41 (Dec. 1981): 723–744.

Harris, Howell. "Inventing the U.S. Stove Industry, c. 1815–1875: Making and Selling the First Universal Consumer Durable." BHR 82 (Winter 2008): 701–733.

Hartley, C. Knick. "The Shift from Sailing Ships to Steamships, 1850–1890: A Study in Technological Change and Its Diffusion." In *Essays on a Mature Economy: Britain after 1840*, edited by Donald McCloskey, 215–235. London: Methuen, 1971.

Hogland, William. "Forest Conservation and Stove Inventors, 1789–1850." *Forest History Newsletter* 5 (Winter 1962): 2–8.

Isard, Walter. "Some Locational Factors in the Iron and Steel Industry since the Early Nineteenth Century." *Journal of Political Economy* 56 (June 1948): 201–217.

Jones, Christopher. "The Carbon-Consuming Home: Residential Markets and Energy Transitions." *Enterprise and Society* 12 (Dec. 2011): 790–823.

Jones, Christopher. "A Landscape of Energy Abundance: Anthracite Coal Canals and the Roots of American Fossil Fuel Dependence, 1820–1860." *Environmental History* 15 (July 2010): 449–484.

Kahn, Alfred E. "The Combined Effects of Prorationing, the Depletion Allowance and Import Quotas on the Cost of Producing Crude Oil in the United States." *Natural Resources Journal* 10 (Jan. 1970): 53–61.

Kanarek, Harold. "Disaster for Hard Coal: The Anthracite Strike of 1925–1926." *Labor History* 14 (Jan. 1974): 44–62.

Kerker, Milton. "Science and the Steam Engine." *Technology and Culture* 2 (Autumn 1961): 381–390.

Keuchel, Edward. "Coal Burning Locomotives: A Technological Development of the 1850s." *Pennsylvania Magazine of History and Biography* 94 (Oct. 1970): 484–495.

Krutilla, John. "Conservation Reconsidered." *American Economic Review* 57 (Sept. 1967): 777–786.

Maddala, G. S. "Productivity and Technological Change in the Bituminous Coal Industry, 1919–54." *Journal of Political Economy* 73 (August 1965): 352–365.

Mak, James, and Gary Walton. "Steamboats and the Great Productivity Surge in River Transportation." JEH 32 (Sept. 1972): 619–640.

McShane, Clay, and Joel A. Tarr. "The Centrality of the Horse in the Nineteenth-Century American City." In *The Making of Urban America*, 2d ed., edited by Raymond Mohl, 105–130. Wilmington, DE: Scholarly Books, 1997.

Miller, Char. "The Greening of Gifford Pinchot." *Environmental History Review* 16 (August 1992): 1–20.

Montrie, Chad. "Agriculture, Christian Stewardship, and Aesthetics: Ohio Farmers' Opposition to Coal Surface Mining in the 1940s." *Ohio History* 111 (Winter-Spring, 2002): 44–63.

Morrison, Bonnie. "Household Energy Consumption, 1900–1980." In *Energy in Transport*, edited by George Daniels and Mark Rose, 179–200. Beverly Hills, CA: Sage, 1982.

Muller, Nicholas. and Joel Tarr. "The McKeesport Natural Gas Boom, 1919–1921." *Journal of Energy History* 4 (Sept. 2020): n.p.

Munn, Robert. "The Development of Strip Mining in Southern Appalachia." *Appalachian Journal* 3 (August 1975): 87–95.

Nienkamp, Paul. "American Land-Grant Colleges and American Engineers." *American Educational History Journal* 37 (March 2010): 313–330.

Nordhaus, William. "Do Real Output and Real Wage Measures Capture Reality? The History of Lighting Suggests Not." *Cowles Foundation Paper* 957. New Haven, CT: Yale University Press, 1998.

North, Douglass C. "Ocean Freight Rates and Economic Development, 1750–1913." JEH 18 (Dec. 1958): 537–555.

Oshima, Harry. "The Growth of U.S. Factor Productivity: The Significance of New Technologies in the Early Decades of the Twentieth Century." JEH 44 (March 1984): 161–170.

Pratt, Joseph. "The Ascent of Oil: The Transition from Coal to Oil in Early Twentieth Century America." In *Energy Transitions: Long Term Perspectives*, edited by Lewis Perelman et al., 9–32. Boulder, CO: Westview Press, 1981.

Rose, Mark. "Urban Environments and Technological Innovation: Energy Choices in Denver and Kansas City, 1900–1940." *Technology and Culture* 25 (July 1984): 503–539.

Rosen, Christine. "Businessmen against Pollution in Late Nineteenth Century Chicago." BHR 69 (Autumn 1995): 351–397.

Rosenberg, Nathan. "The Direction of Technological Change: Inducement Mechanisms and Focusing Devices." *Economic Development and Cultural Change* 18 (Oct. 1969): 1–24.

Rosenberg, Nathan. "Technological Change in the Machine Tool Industry, 1840–1910." JEH 23 (Dec. 1963): 414–446.

Rovang, Sarah. "The Grid Comes Home: Wiring and Lighting the American Farmhouse." *Building & Landscapes* 23 (Fall 2016): 65–88.

Schurr, Sam. "Energy Efficiency and Productive Efficiency: Some Thoughts Based on American Experience." *Energy Journal* 3 (July 1982): 3–14.

Smith, Gerald A. "Natural Resource Theory of the First Conservation Movement." *History of Political Economy* 14 (Winter 1982): 483–495.

Sorrell, Steve. "Jevons' Paradox Revisited: The Evidence for Backfire from Improved Energy Efficiency." *Energy Policy* 37 (2009): 1456–1469.

Stradling, David. "To Breathe Pure Air: Cincinnati's Smoke Abatement Crusade, 1904–1916." *Queen City Heritage* 55 (Spring 1997): 2–18.

Stradling, David, and Joel Tarr. "Environmental Activism, Locomotive Smoke, and the Corporate Response: The Case of the Pennsylvania Railroad and Chicago Smoke Control." BHR 73 (Winter 1999): 677–704.

Stranges, Anthony. "The US Bureau of Mines' Synthetic Fuel Programme, 1920s–1950s: German Connections and American Advances." *Annals of Science* 54 (1997): 29–68.

Tarr, Joel. "Changing Fuel Use Behavior and Energy Transitions: The Pittsburgh Smoke Control Movement, 1940–1950." *Journal of Social History* 14 (Summer 1981): 561–580.

Tarr, Joel. "Manufactured Gas, History of." In *Encyclopedia of Energy*. Vol. 3. Edited by Cutler Cleveland. Boston: Elsevier, 2004.

Tarr, Joel. "Toxic Legacy: The Environmental Impact of the Manufactured Gas Industry in the United States." *Technology and Culture* 55 (Jan. 2014): 107–147.

Tarr, Joel. "Transforming an Energy System: The Evolution of the Manufactured Gas Industry and the Transition to Natural Gas in the United States (1807–1954)." In *The Governance of Large Technical Systems*, edited by Olivier Coutard, 19–37. London: Routledge, 1999.

Tarr, Joel, and Karen Clay. "Boom and Bust in Pittsburgh Natural Gas History: Development, Policy and Environmental Effects, 1878–1920." *Pennsylvania Magazine of History and Biography* 139 (Oct. 2015): 323–342.

Tarr, Joel, and Clay McShane. "The Horse as an Urban Technology." *Journal of Urban Technology* 15 (April 2008): 5–17.

Temin, Peter. "Steam and Waterpower in the Early Nineteenth Century." JEH 26 (June 1966): 187–205.

Warde, Paul. "Firewood Consumption and Energy Transition: A Survey of Sources, Methods and Explanations in Europe and North America." *Historia Agraria* 77 (April 2019): 7–32.

Warner, Paul. "The Anthracite Burning Locomotive." *Railway and Locomotive Historical Society Bulletin* 52 (May 1940): 11–28.

Watkinson, James. "'Education for Success': The International Correspondence Schools of Scranton, Pennsylvania." *Pennsylvania Magazine of History and Biography* 120 (Oct. 1996): 343–369.

Williamson, Harold. "Prophecies of Scarcity or Exhaustion of Natural Resources in the United States," *American Economic Review* 35 (May 1945): 97–109.

Index

Adams, Henry, 1
Allen, Zachariah, 30, 156
American Association for the Advancement of Science, 18
American Gas Association, 75, 139–40, 146, 178, 235; marketing and research, 84–86, 140, 142, 147–148, 180, 198–200, 234
American Gas Light Journal, 2, 42, 47, 51, 53, 63; importance of, 40–41
American Institute of Electrical Engineers, 47
American Institute of Mining Engineers, 47, 49
American Mining Congress, 2, 76, 192
American Philosophical Society, 17, 23
American Railway Times, 37
American Society for Testing Materials, 90, 95
American Society of Civil Engineers, 18, 47
American Society of Mechanical Engineers, 2, 36, 37, 47, 76, 90–92
anthracite coal, 1, 16, 21; cartelization of, 6, 68, 131; conservation and waste, 49, 77–79, 251; for cooking, 1, 137–138; domestic consumption, 1, 23–27, 66–67, 255; geography and, 19, 21–24, 29, 67, 130, 232, 256; for heating, 1, 23; imports, 144–145; income effects, 49–50, 67, 134, 287n12; labor productivity, 48, 132; mechanization, 19–20, 48; mining methods and dangers, 21, 46; preparation and marketing, 22, 34, 131, 191–195, 201–202; prices, 21, 49–50, 132, 202, 242, 256, 130; problems with burning, 22; quality problems, 143–144, 192–193; railroad use, 36; reserve estimates and pessimism, 3–4, 131–132; statistics of production, 21, 67, 130, 241–242; steamboat use, 34–35; transportation of, 21–24, 29, 56; unions and strikes, 50, 144–145,

186; World War I and, 67, 143–144, 151. *See also* coal heating; coal research; coal stoves and furnaces; iron and steel making; strip mining
anthracite coal industry: domestic marketing, 22, 27–29, 66–67, 131, 191–193, 201–202; structure, 19, 21, 48, 49, 131
Atchison Topeka & Santa Fe Railroad, 110, 113, 119, 121

Baker, Thomas S., 222, 225, 229
Baumol, William, 8, 74
Bell, Lowthian, 59
bituminous coal, 1–5, 16; byproducts, 41, 64, 98–99; cartelization, 4, 68, 209, 226; conservation and wastes, 48, 78–79, 221, 225–226; domestic consumption, 1, 26, 66, 132–133, 255, 287n7; for heating, 1, 21, 58; income effects, 49–50, 67, 134, 184, 276n50, 287n12; industrial fuel, 21, 29; labor productivity, 51, 84, 210, 215, 217–218, 220–221; mechanization, 19–20, 51, 209, 213–19; mining methods and dangers, 3–4, 48, 51, 219–220; national power and, 1–5, 76; preparation and marketing, 51, 78, 201–202, 223–224; prices, 52, 67, 132, 202, 210, 218, 242, 256; pulverized coal, 93–95; purchasing, 40, 63, 95–96; quality, 51, 57, 68; railroad use, 36–38, 56; reserve estimates and pessimism, 3–4, 76, 221–222; statistics of production, 21, 67, 73, 241–242; transportation of, 55–56; unions and strikes, 50; wages, 52, 84, 216; World War I and, 68–69, 73–74, 76–77, 90–91. *See also* coal research; coal stoves and furnaces; iron and steel making; manufacturing; railroads; strip mining
bituminous coal, international conferences, 222–227

Bituminous Coal Association, 73
bituminous coal industry: domestic marketing, 51–52; structure, 50, 52–53, 67–68, 273n10
Boston Globe, 50, 144
Boulton & Watt Company, 32, 39
Braley, Berton, 1, 99
briquettes: domestic consumption and cost, 134, 255–266; in Europe, 48–49; statistics of production, 79
Bull, Marcus, 23–24

Campbell, Marius, 6, 222
Carnegie, Andrew, 59
Carnegie Institute of Technology, 222, 227
cement making, 52, 60; energy efficiency, 84, 98, 101–102; pulverized coal, 93; trade association, 75
central heating, 1, 67, 160, 167, 177–180; automatic heat, 182, 187, 192–194; income effects, 1, 67, 178, 257; insulation and, 179–180, 197, 199–200; statistics of, 203–204. *See also* gas heating; oil heating
Chandler, Charles F.: role in gas industry, 41; role in kerosene safety, 154, 156
charcoal: energy content, 263; in iron industry, 14, 27, 29, 59
Chicago Burlington & Quincy Railroad, 37, 121
Chicago, Rock Island & Pacific Railroad, 111, 115–116, 118, 121
Chicago Tribune, 50, 52, 68, 145, 188, 190
Clermont, 31
coal, 16–17, 46; advantages vs wood, 17, 44, 267n9, 270n49; for factory power, 31, 47. *See also* anthracite coal; bituminous coal; coal heating; coal industry; coal mining; coal research; coal stoves and furnaces

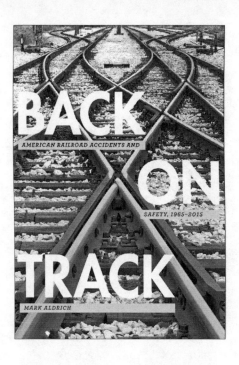
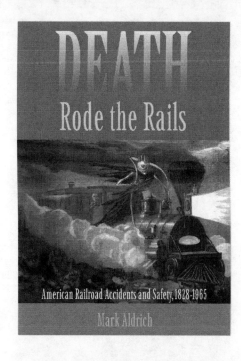

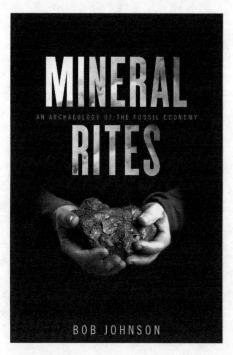